Human Biogeography

Human Biogeography

Alexander H. Harcourt

UNIVERSITY OF CALIFORNIA PRESS

Berkeley · Los Angeles · London

University of California Press, one of the most distinguished university presses in the United States, enriches
lives around the world by advancing scholarship in the
humanities, social sciences, and natural sciences. Its ac-
tivities are supported by the UC Press Foundation and
by philanthropic contributions from individuals and in-
stitutions. For more information, visit www.ucpress.edu.

University of California Press
Berkeley and Los Angeles, California

University of California Press, Ltd.
London, England

Library of Congress Cataloging-in-Publication Data

Harcourt, A. H. (Alexander H.)
  Human biogeography / Alexander H. Harcourt.
    p. cm.
  Includes bibliographical references and index.
  ISBN 978-0-520-27211-8 (cloth : alk. paper)
    1. Human geography. 2. Physical anthropology.
  3. Biogeography. I. Title.
GF41.H375 2012
304.2—dc23

                        2011052055

19 18 17 16 15 14 13 12
10 9 8 7 6 5 4 3 2 1

Cover image: Detail from map of native languages and
language families from *Handbook of North American
Indians, Vol. 17, Languages,* ed. Ives Goddard.
Smithsonian Institution (GPO), 1996.

Contents

Acknowledgments   *vii*

1. Biogeography and Humans: An Introduction   *1*
  Why Humans?   2
  Human Taxonomic Terminology   4
  Human Variation   5
  Organization of the Chapters   7
  Sources and Methods   9

**PART ONE. WHY AND HOW ARE WE WHERE WE ARE?
HISTORICAL BIOGEOGRAPHY OF HUMANS**

2. Origins and Dispersal   *15*
  Regional Variation as Evidence of Origins and Dispersal   *17*
  Origins and Global Dispersal   22
  Sex Differences in Dispersal   47
  Other Species as Evidence of Human Origins and Dispersal   48

3. Climate, and Hominin Evolution and Dispersal   *53*
  Climate and Hominin Origins   *57*
  Climate and Hominin Dispersals   66

4. Barriers to Movement   77
  Barriers Limit the Movement of Primates   *78*
  Barriers Limit the Movement of Humans   *79*

**PART TWO. ENVIRONMENTAL INFLUENCES ON
HUMAN NATURE, DIVERSITY, AND NUMBERS**

5. How Are We Adapted to Our Environment?   *89*
  Climate Affects Form and Function   *90*
  *Elevation (Altitude) and Regional Variation in Anatomy
  and Physiology*   *120*

Latitude, Climate, and Human Cultural Diversity   *128*
Latitude, Climate, and Population   *146*

6. Use of Area   *155*

Area Available and Area Used   *156*
Area and Diversity   *178*
Area, Environment, and Density   *185*

7. A Biogeography of Human Diet and Drugs   *193*

Milk Consumption and Lactase Production   *194*
Starch and Salivary Amylase   *198*
Only the Japanese Can Digest Seaweed   *199*
Alcohol Enzymes and Alcoholism   *199*
Two Enzymes, Codeine, and Antidepressants   *201*
The High-Protein, High-Fat Eskimo/Inuit Diet   *202*

PART THREE. INTERACTION AMONG CULTURES AND SPECIES

8. We Affect Our Biogeography   *207*

Deliberate and Inadvertent Extermination   *208*
Range Change in the Face of Competition   *209*
Competition and the Move out of Africa   *210*
Disappearing Cultures   *211*
Cooperation and Its Biogeographical Consequences   *212*

9. Other Species Affect Our Biogeography   *215*

Disease and Human Biogeography   *216*
Disease and Conquest   *218*
Disease and Human Population Density   *221*
Competitors, Predators, Prey, and Human Biogeography   *221*

10. We Affect Other Species' Biogeography   *225*

Humans vs. Other Homo Species   *226*
Late Quaternary and Recent Extinctions   *228*
Climate or Humans?   *230*
If Humans, How?   *233*
Reduction of Geographic Range   *235*
Change in Community Structure   *239*
Humans Benefit Many Species   *240*

References   *245*
General Index   *303*
Author Index   *313*

Acknowledgments

I thank Robert Bettinger, Chris Darwent, Victor Golla, Frank Marlowe, David G. Smith, and Tim Weaver for information or unpublished data, and for very helpful discussion.

I am very grateful to David G. Smith, Teresa Steele, John Terrell, and Tim Weaver for their commentary on several chapters. They educated me in areas not my own. I thank also Mark Lomolino for expert and detailed editorial and biogeographical criticism that enormously improved the result.

Margot Wood and Snigdha Garise helped with collation and analysis of data, and Alexandra Towns and Keir Keightley provided GIS analysis and map production. I thank them. Thanks also to Ives Goddard and Laurie Burgess for permission to reproduce the book cover's map.

The book would not have been possible without the support of my wife, best friend, severest critic, and scientific partner, Kelly Stewart. She read the whole text and hugely improved it.

A fair amount of the analysis and the occasional idea and suggestion is mine. Otherwise, the book is a synthesis of the field of human biogeography, in other words an analysis of the work and ideas of many others. They should be considered as coauthors, insofar as I have interpreted their work as they would wish.

Alexander Harcourt
Anthropology; Ecology
UC Davis, CA 95616
ahharcourt@ucdavis.edu

1

Biogeography and Humans

An Introduction

Ethnology is the science which determines the distinctive
traits of mankind; which ascertains the distribution of those
traits in present and past times, and seeks to discover the
causes of the traits and of their distributions.

—modified from T. H. Huxley, 1865, "On the Methods and Results
of Ethnology," *Fortnightly Review*

This book is about how and why our species, *Homo sapiens sapiens,* is
distributed around the world in the way it is—why we are what we are
where we are. It is therefore both anthropology and biogeography. Anthropology is literally the study of humans, and it is nicely summarized
in Thomas Huxley's description of what he termed ethnology. Biogeography perhaps needs a bit more explanation.

One part of the field of biogeography investigates why organisms are
where they are. How did they get there? Why are they where they are
and not somewhere else? Why is this sort of organism here, and that sort
of organism there? Scientists and others have been asking these questions
for centuries. "God" used to be the answer, and still is for half the American population. But in science, biogeography took off with the explora-
tions of the 18th century [104; 463, ch. 2; 464]. The early work can be
seen as culminating in Alfred Russell Wallace's two-volume *The Geographical Distribution of Animals* [803; 804].

A modern example of biogeography is the quantitative, statistical
approach of phylogeography, a subfield of biogeography that relates
data on the distribution of organisms to data on their evolutionary trajectories, their phylogeny [26; 27; 74; 348; 463, ch. 11, 12; 646; 663;

1

668; 784]. For anthropology, phylogeography is exemplified by the large amount of work on the origins of native Americans.

A major branch of biogeography is now often known as macroecology [101; 253; 463, ch. 13–15], although other terms have been used for all or part of the discipline, such as areography or geographical ecology [469; 629]. Macroecology as a branch of biogeography relates diversity (e.g., number of species or cultures in a region) or density (e.g., individuals per square mile) or the traits of organisms (e.g., body size) to various quantitative measures of space on a large scale. For example, why do taxonomic and cultural diversity vary with latitude? Do animals and human cultures at higher latitudes have larger geographic ranges? Why are organisms at higher latitudes stockier than those from the tropics? Do large islands have more species or cultures than do small islands?

A perusal of this book's table of contents will give an idea of the range of biogeography as a discipline. Mark Lomolino et al. provide in chapter 1 of the 4th edition of their great synthesis of biogeography a detailed account of its questions and its integral relation to so many other disciplines [463]. Their chapter 2 is a nice introduction to the history of biogeography, a fuller and engaging account of which is provided by Janet Browne's *The Secular Ark* [104].

WHY HUMANS?

In many respects, humans are biogeographically just another species. Why then concentrate on humans? In brief, I hope that both biogeography and anthropology can benefit from a more specific and extensive concentration on the topic of the biogeography of humans than so far attempted [757; 758].

What can anthropology contribute to biogeography? The answer is the huge amount of data and understanding that anthropology has of regional variation, both past and present, in our own species and its close relatives. Is there another species on which we have as large a sample of relevant data as we do for humans? Sample sizes of hundreds of individuals and scores of molecular markers are commonplace in studies that directly or indirectly concern the distribution of humans. Globally, the sample is 10 times as large and increasing all the time [125; 529; 627; 762]. Some of the medical anthropological data is barely available for most other species, let alone in the detail that anthropology has. We even have a biogeography of the distribution of bacteria on the human body [154].

On top of its enormous amount of data, anthropology can use areas of evidence not available to biogeographers and macroecologists of other taxa. Linguistics is an example [124; 125]. Yes, the calls of bush-babies (*Galagonidae*) can differ regionally and inform us about taxonomic relationships [554]. But no bushbaby taxonomist has anywhere near the detail available to historical linguistic anthropology [124; 125]. And while there is some nice prehistorical analysis of pack rat middens [763], it is, of course, utterly negligible by comparison to the information that archeology can provide.

Yet as far as I can tell, little of this anthropological knowledge reaches the biogeographical literature. For instance, although the title of Ian Simmons' *Biogeography: Natural and Cultural* [705] sounds as if it fully integrates the two disciplines, in fact Simmons treats only the topic of chapter 10 of this book, the effect of humans on other species' distributions. In biogeography's bible, Mark Lomolino et al's *Biogeography* (now in its 4th edition), only 15 of its 760 text pages treat human biogeography [463]. Yes, we are just one of thousands of species, but does not the amount we know about us deserve more than the 2% of space devoted to us?

An additional benefit to the field of biogeography of attention focused on human biogeography might be that because of our intrinsic interest in ourselves, the topic of biogeography as a whole would receive more than the minor amount of attention it does in the popular press [441].

What can biogeography contribute to anthropology? Biogeography in all its forms has exploded in the last half century, and perhaps especially in the last two decades [463]. It seems to me that anthropology has yet to fully embrace that immense increase in knowledge and understanding.

It is not the case that biogeography is ignored in anthropology [224; 228; 435; 757; 758]. Rather, its treatment is scattered through anthropological works and usually addressed under headings such as human origins, or human variation or adaptation [757; 758]. Despite the quote from Huxley at the start of this chapter, which is effectively an explicit statement that much of anthropology is the biogeography of humans, it is not treated as such. Consequently, relevant biogeographic patterns and principles developed largely by analysis of nonhumans have perhaps not been as fully integrated into our understanding of humans as they could have been.

Advantages of integration of other disciplines with anthropology are, as always, to increase understanding of, on the one hand, the biological heritage and influences that we have in common with so many other

TABLE I.I

TAXONOMIC TERMINOLOGY OF HUMANS, THEIR ANCESTORS,
AND CLOSEST PRIMATE RELATIVES

Taxonomic level	Taxon name	Taxa included
Superfamily	Hominoidea	*Hylobates* (gibbons), Fossil apes (e.g., *Kenyap-ithecus, Proconsul*), Hominidae
Family	Hominidae	*Pongo* (orangutans), Homininae
Subfamily	Homininae	*Gorilla, Pan* (chimpanzees), Hominini
Tribe-1	Hominini-1	*Pan*, Hominina
Tribe-2	Hominini-2	*Homo sapiens* (humans), and ancestors (i.e., *Homo, Australopithecus, Paranthropus*). *Ardipithecus, Orrorin, Sahelanthropus* still being debated.
Subtribe	Hominina	As for Hominini-2

species and, on the other hand, to help delineate what is unique about the human species.

HUMAN TAXONOMIC TERMINOLOGY

Hominoid, hominid, and hominin are terms still sometimes used interchangeably to describe humans and our ancestors. Table 1.1 shows the proper taxonomic divisions and indicates continuing debate—for instance, whether *Pan* (chimpanzees) should be included in the same tribe as *Homo* [841]. The term hominin in English is confusing because it could mean either Hominini or Hominina. Not only that, but whether the chimpanzees, *Pan*, are included in Hominini is debated. In this book, I use hominin to mean Hominina, i.e., only humans and our direct, non-ape ancestors. By human, I mean *Homo sapiens*. I need to say that because some authors use the word human to mean any *Homo*.

Colin Groves provides an expert yet readable account of taxonomic rules in what some term the bible of primate taxonomy, his *Primate Taxonomy* [292]. Richard Klein provides an exhaustive description of all of the hominids, including debate about their nomenclature, in the 4th edition of his *The Human Career* [425].

For want of another place to say it, let me add here that I use the terms gene and allele more loosely than some might like. In my defence, precise jargon meanings of the words differ, and have changed over time, and considerable debate exists over the units of selection [694, ch. 7].

HUMAN VARIATION

The observation that individuals differ is one of the foundations of the theory of evolution by natural selection [165]. Not only do individuals differ, but so does the average person in different parts of the world. Humans from different parts of the world differ morphologically, molecularly, genetically, and culturally, and some of the differences have clearly existed for millennia. The causes of those differences are the topic of this book, and of much previous work in anthropology [124; 125; 241; 332; 530; 540; 640; 722].

A focus on differences between humans from different parts of the world raises the topic of race and all its unpleasant connotations. Humans are distressingly good at not only detecting differences between "us" and "them" but also inventing differences and then at stigmatizing those unlike ourselves [622]. The contrast between us and them, whether real or perceived, is the basis of much that is worst in human nature. People from many parts of the world, even now, perceive people from different parts of the world as inferior in one or more traits. Unfair discrimination does not, however, remove the fact that regional differences exist.

It can be medically vital to recognize regional differences [61; 115; 304; 708; 722; 771; 817; 857]. Several contrasts in susceptibility to various aspects of the environment exist between peoples from different regions of the world. Do not force milk products on adults from regions where adults are not used to drinking milk: they cannot digest the milk, and they will get sick from it. This is not an artificial example: western famine-relief programs used to ship milk powder as food aid, which at best was useless but at worst could have exacerbated malnutrition [193, ch. 5]. Know that if a child in New York is sick and its parents came from west Africa, or Greece, or lowland New Guinea, or Ferrara province in Italy, the child could well have one of a variety of hemoglobin variants that seem to provide protection against the consequences of malarial infection in the parents' country of origin [115; 124, ch. 2; 540, ch. 4; 640, ch. 7; 771]. Know that codeine might be less likely to cure the headache of someone from Europe or Africa than from Asia [817]. It is not racist to recognize these differences. Understanding their causes can be helpful.

Nevertheless, between a Maasai herdsman and the average Scandinavian farmer a continual gradation of skin color exists. It would be impossible to say where African stops and European starts [640, ch. 5].

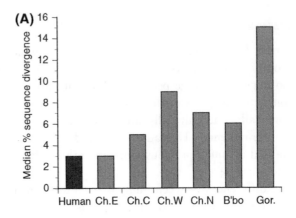

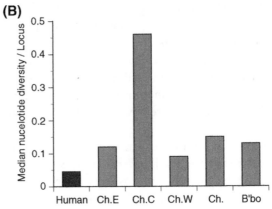

FIGURE 1.1. Genetic diversity is less in humans than in great apes. Diversity is measured in two ways: percent of sequence that is different and diversity per locus within human and ape populations. Ch. = chimpanzee, *Pan*. Ch. E, C, W, N = eastern population, central, western, northern chimpanzee populations. B'bo = bonobo, *Pan*; Gor. = gorilla, *Gorilla*. Data: (A) from [247], (B) from [730].

As Darwin wrote, "It may be doubted whether any character can be named which is distinctive of a race and is constant" [166, part I, p. 225].

Indeed, we know that more variation exists genetically within any one population of humans than between populations from different parts of the world [665]. Not only that, but a large number of studies from a variety of labs and continents on a variety of bits of the genome all agree that more genetic variation exists within chimp subspecies, even a single population of chimpanzees, than within the whole global population of humans (Fig. 1.1) [247; 730]. In other words, for a species spread over the whole globe, humans are extraordinarily homogeneous genetically.

In sum, spatial gradation in traits and extraordinary genetic homogeneity make "race" as the word is used in normal discourse a sociopolitical concept, not a biological concept.

I suggested at the start of the explanation of why I have written a book about human biogeography that biogeographically humans are just another species. That suggestion of identity will for some social anthropologists still raise the red flag of genetic determinism. Such anthropologists have always made the mistake of confounding genetic influence with genetic control. Let me emphasize that much of the biogeography that I write about has to do not with genes determining human behavior, or even influencing it, but with humans' reactions to the environment. There should not be any hard-core anthropologists remaining who deny any influence at all of the environment on human physiology, behavior, society, culture. But if there are, they need to produce alternative explanations than the ones from physical anthropology and biogeography that I review here to explain why biogeographically we seem to do the same as so many other species.

ORGANIZATION OF THE CHAPTERS

I have organized the book into three parts. The first considers why humans are where we are. It is historical biogeography, phylogeography. The second relates our distribution to the nature of the environment. The topic is adaptation, or functional biogeography, including macroecology. And the third addresses the biogeographic effects of interactions between cultures and between species.

The biogeographic history of human origins and our spread around the world from Africa is the first topic (chapter 2). People from different regions are different. In part, that is because their origins and history of settlement were different. The similarities and differences in the biological and cultural traits that lead to this inference allow us to track the spread of humans, first from Africa, and subsequently to the rest of the world.

Change of climate has been a main correlate of major evolutionary events since life began. The same is probably true of human origins and our dispersal around the globe (chapter 3). Quantitatively substantiating the connection between climate and human history is a different matter, however. Except for climate's inevitable role in delaying human migration into and through the arctic north, a lot more data and statistical sophistication are needed to prove climate's influences.

Barriers stop movement of individuals and therefore help maintain different sorts and variety of organisms on either side of the barrier (chapter 4). The most common sort of barrier is a hard geographic barrier, such as water or high mountains. However, the behavior of individuals also can be a barrier to movement. A contrast in cultures can be an extremely effective impediment to the movement and mating of individuals; hence, ghettoes in cities.

Even if people from different regions did not have different origins, and even if there were no barriers to movement, people in different parts of the world would still be different, because of adaptation to different environments (chapter 5). For instance, people with short limbs are probably better adapted to cold environments than are those with long limbs. It is not just the average trait of this sort that sometimes varies with the environment in apparently some sort of adaptive way, so also does the diversity of traits within a region. Thus, in just the same way as diversity of many organisms is greatest in the tropics, so is diversity of human cultures. We can use our understanding of the causes of the gradient in cultural diversity to test hypotheses for the biodiversity gradient, and vice versa. And of course, human numbers vary with the nature of the environment: there are not many of us in Antarctica or the Sahara.

The size of areas available to organisms and species, or used by them, is a major explanatory and organizing principle in biogeography (chapter 6). The area available affects the number of individuals that can inhabit an area, and hence it affects both the sort of species and culture in a region and the diversity of species and cultures. The area used by species and cultures, their geographic range, varies with latitude, probably in part via latitudinal contrasts in productivity of the land and thus the size of area needed for persistence of a species or culture.

Different cultures and people from different regions exploit the environment in different ways—for example, by having different diets. Different diets sometimes correlate with a different physiology, and so we get regional differences in physiology, a perhaps too-little-investigated aspect of biogeography (chapter 7).

In the final three chapters, I consider the biogeographical effects of competition within and between species. Competition between humans from different parts of the world affects human diversity across the globe (chapter 8). Human nature and diversity is affected by other species. Because the nature of these other species varies regionally, human nature and diversity varies regionally (chapter 9). And, of course, humans affect other species, driving many to extinction but increasing the numbers and geographic ranges of many as well. Which species are affected and how depends on the nature of the species, on where they are, on regional differences in what sorts of humans arrived, on when and where humans arrived, and in what numbers humans arrived (chapter 10).

To illustrate general understanding in the field of biogeography, I start a number of chapters and sections with some consideration of the biogeography of primates. The taxonomic order, Primates, is the taxon most closely related to humans, and therefore a taxon likely to be biogeographically similar to us, other things being equal. Indeed it is. In many respects, we are simply another primate.

SOURCES AND METHODS

Much of chapters 5 and 6 involves comparisons between regions and latitudes of the variety of cultures within them. Problems of definition and the variable quality of data attend all comparative analyses. One person's language is another person's dialect; one person's statement about a societal trait is based on 1 year's study, another's on 10 years of study. And so on.

Regarding differing quality of data, I have made no attempt to distinguish what I think might be data of good quality from those of poor quality. Instead, I work on the assumption that inaccurate or imprecise data are more likely to obscure a relationship than produce one.

With respect to defining "culture," considerable debate surrounds the issue, because the concept of culture is complex [122; 325; 326; 477; 819]. Consequently, anthropologists (and practitioners of many other disciplines) use a variety of definitions and measures [819].

I use language as one definition of culture, as have many others. For the quantitative analyses that I conduct, I take data from the website on the world's languages, *Ethnologue*[a] [277; 455]. Another definition I use is a hunter-gatherer society, as defined by Lewis Binford in his extraordinarily detailed collation of an amazing variety of information on hunter-gather

a. The latest *Ethnologue*, the 16th edition [455], was published only after I had collated almost all the data from the 15th edition [277]. The differences are minor enough for the 15th edition's data to be valid. For instance, in over half the countries the number of languages listed did not change, and for those that did, the average difference was less than two, or less than 10%. However, I used the latest version for my discussion of endangered languages.

cultures [62]. And I use without questioning them the various combinations of the more complex concepts used by others, such as Elizabeth Cashdan's "ethnic groups" [122].

Those more immersed in the analysis of cultures than I am will with justification object to my gross simplifications (e.g., language = culture). I have two forms of justification. One is that sometimes gross simplification can expose regularities that continual attention to complexities would hide, as is the case in almost all the work on human biogeographical variation that I review throughout the book. The other is that the concept of species is also complex, much discussed, with many definitions used, and yet for the most part, comparative analyses in biology are highly productive.

In my analyses of the *Ethnologue* and Binford compilations, I omit much of their data in order to avoid artificially inflating sample sizes in statistical tests. For instance, the peoples of the adjacent small central African countries of Burundi and Rwanda are quite similar. Counting data on them as two samples is in effect to count the same data twice. A variety of complex statistical means of avoiding this spatial autocorrelation problem exist. I have used a simple, more or less arbitrary method.

For the *Ethnologue* data, I use only states whose centers are more than 10° apart (i.e., about 1,000 km apart). For Binford's data, I use in global analyses the first listed society per state; for analyses within regions, I use societies farther apart than the sum of the diameters of the area used by potentially adjacent societies—societies that had less chance of influencing one another than if their ranges touched or overlapped.

Even so, two languages or societies in Africa are more likely to be similar to one another than either is to an Australian or European society. This effect of common origin, or phylogeny, I have mostly ignored, not because it is unimportant—it is very important—but for lack of an easily available, accepted global inheritance tree of human cultures, which takes account of both similarity by descent (phylogeny) and similarity by adoption. Future work will need to incorporate both sources of similarity, as well as of developments in statistical means of accounting for both spatial autocorrelation and correlation among the variety of potential biogeographical influences on human variation [559].

A few other rules operate too. In regression analyses I omitted influential, outlier points to avoid producing or hiding a relationship amongst the majority of the data. I did not use *Ethnologue* data from what are essentially political enclaves, such as Gaza, Gibraltar, Hong Kong, Singapore, Vatican. In analyses concerning the size of an area, I often used only mainland countries, not island countries, or presented the analyses separately for mainland and island. I made this separation because what happens biogeographically on islands can be different from what happens on mainlands (chapter 6) [463, ch. 13, 14].

I have put the results of statistical tests into footnotes in order to make the text easier to read. If I do not state the source of the data, then it is mine, in the sense that I or my research assistants collated it from a wide literature. With the statistics, I have largely used parametric tests if samples sizes are ≥20, and nonparametric tests if they are <20. The most common nonparametric test that I use is the Spearman correlation, signified by r_s. In multivariate tests, the statistical program that I use, JMP 8.0 [679], chooses its best-fit minimum models on the basis of the Akaike Information Criterion, AIC_c [112]. P values are two-tailed.

In chapters 5 and 6, I test for environmental correlates of the distribution of cultures. The data come from Binford's compendium [62]. I entered up to 10 environmental variables into the analysis and obtained best-fit models for a maximum of three variables—viewing models with more than three variables as too complex to be useful in the context. The 10 variables were chosen to indicate (a) average environment (mean annual temperature, rain, productivity) (b) extremes (mean min., max. temperature, rain in coldest, hottest, driest, wettest month), and (c) seasonal variation (mean number good growing season months).

CONCLUSION

An analytical synthesis of a field, in this case of human biogeography, is of necessity effectively the work of others. The writer brings their work together, trusting that the collation, integration, some new analysis, a more distant or different perspective, and the occasional slightly original idea can help development of the field. My hope is that my synthesis will make anthropologists and biogeographers more aware than they might otherwise have been of each other's ideas and data. Finally, and emphasizing that a book such as this necessarily depend on the work of others, "we" in the book means humans, or those scientists, anthropologists, and biogeographers studying the topic that I happen to be writing about.

Why and How Are We Where We Are?

Historical Biogeography of Humans

2

Origins and Dispersal

SUMMARY

Many of the traits that distinguish humans in different parts of the world, especially genetic traits, have little to no relationship to the environment. If the differences are not due to the environment, then they can sometimes tell us about evolutionary history, i.e., where the people originated, when they originated, and how they got to where they are now.

The human line (hominins, *Homo,* and eventually humans) all originated in Africa, probably sub-Saharan Africa. A small group of humans left Africa 65,000–45,000 years ago (65–45 kya). They migrated rapidly to eastern Asia and were in Australia by at least 45 kya. Our African origin and subsequent dispersal is strongly indicated by the facts that human genetic diversity is greatest in Africa and genetic distances between populations are proportional to geographic distance from sub-Saharan Africa. There is some agreement that the first rapid easterly movement was along the coasts. Whether humans also went east through central Asia is still debated. The alternative is that central Asia was populated from the east. Whichever, it looks as though Europe was populated from western Asia. Although humans were in far northeast Siberia perhaps 30–25 kya, it seems that it was not until 15 kya that humans crossed to America. Within about 2,000 years, humans had spread throughout the Americas, all the way down to southern South America. The Pacific was the last great region that humans arrived in, with some of the central Pacific islands not reached until less than 1,000 years ago.

Most of the evidence for dispersal comes from analysis of genetic similarities and differences among human populations. The distribution of commensals, animals that live and travel with us, either as stowaways or as domestic animals, is another source of information about human movements across the globe. The human diaspora is also evident in the extinctions of large mammals concomitant with our first arrivals in a region.

In considering the distribution of organic beings over the face of the
globe, the first great fact which strikes us is, that neither the
similarity nor the dissimilarity of the inhabitants of various regions
can be accounted for by their climatal and other physical conditions.

[165, Darwin, 1859, ch. 11, p. 346]

After several examples to illustrate his statement, Darwin goes on to argue that to understand the similarities and dissimilarities that are seemingly independent of the physical conditions, we need to know the history of dispersal of the organisms to explain their distribution. Organisms are similar or different, not because the environment is similar or different, or not solely because the environment is similar or different, but because they have got where they are from the same or different places. The traits in question then tell us about the history of the organisms. Eugene Harris provides a detailed account for humans [329].

Such similarities and differences in traits of organisms, genetic or morphological, that have nothing to do with adaptation to different environments (nothing to do with climatal or other physical conditions) are crucial for determining taxonomic relatedness of organisms and for elucidating the timing and routes of movement of those organisms around the world. The presence of the same trait in two regions can indicate common descent, as opposed to convergent evolution in response to similar environments. Conversely, different traits in two regions can indicate different origins, as opposed to divergent evolution in response to dissimilar environments. In both cases, though, the possibility of chance similarity needs to be kept in mind, as does the possibility that although we cannot see an adaptive function of the trait now, or an adaptive explanation for any difference, one might be discovered in the future.

We have two main ways to quantitatively test ideas about origins and dispersal. One is to use the quantitative phylogeographic methods that I mentioned in chapter 1: the taxon's phylogeny (evolutionary tree) is statistically compared to its geographic distribution now and in the past. The second way relies more purely on genetics, or the equations and models of genetics. Geneticists have good measures of what they call "founder effects" and "drift" [137; 445]. Founder effects result from the fact that a population's current genetic profile is determined by the subset of individuals that founded the population. Drift describes the way in which the population changes through time as a result of mostly nonadaptive, neutral modification resulting from chance consequences of mutation, breeding, and survival.

Assumptions about the nature of founder effects and drift allow specific, quantitative predictions to be made about the nature of the variation that will be seen within and between populations, or taxa [137; 445; 639; 812; 813]. Comparison of observations with predictions then allows the determination of whether traits of populations, especially differences between them, are a result of adaptation or not. If not, if the variation is as would be expected from founder effects and drift, then the precise pattern of similarities and differences allows inferences about origins and dispersal of populations. By contrast, deviations from the patterns expected if only founder effects and drift were operating can be interpreted as possible evolved adaptations. Although the models were initially formulated by geneticists, they can be applied to any trait that can be quantitatively measured.

In the following section, I briefly describe some of the many differences that we see between peoples from different regions that appear to be the result of history, of founder effects. In the second section, I concentrate on inferences from the distribution of the traits about exactly where and when we originated and when and how we dispersed across the globe. I close with very brief considerations of differences between the sexes in dispersal, and of the use of other species, as opposed to traits of humans, to tell us about our origins and dispersal.

REGIONAL VARIATION AS EVIDENCE OF ORIGINS AND DISPERSAL

Genes and Blood Groups

Human populations from different parts of the world have on average different proportions of individuals with any one type of genetic complement, as exemplified by, for example, blood groups. One of the most magisterial treatments of this topic is Luigi Cavalli-Sforza et al.'s *The History and Geography of Human Genes* [125], from which the data in Fig. 2.1 are taken.

As can be seen in Fig. 2.1B (top left), Native Americans, especially those in South America, are very unlikely to be of blood group A (or B). Instead, they are highly likely to be blood group O (Fig. 2.1B, bottom left). The average sub-Saharan African is highly unlikely, less than 10% probability, to have the Rh C blood group allele, compared to more than 30% in most of the rest of the world and about 90% in New Guinea

(A)

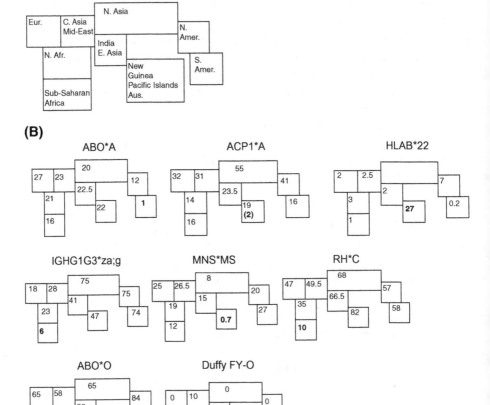

(B)

FIGURE 2.1. Frequencies of genes in human populations across the globe. (A) Schematic map of continents. (B) Distribution of some genes. Values are proportion of population with the named variant. Region with unusual frequency shown in bold. "(2)" for ACP1*A is frequency in Australia. Data from [125, ch. 2].

(Fig. 2.1B, mid-right). Many blood group variants exist in addition to those at the ABO and Rhesus loci that most of us are used to. For instance, in regions of southern Canada, Greenland, and northwestern South America, the average person is 80–90% likely to have the M blood group allele, while in southwestern Australia less than 20% have this allele [332, ch. 11; 530, ch. 4]. Other examples are in Fig. 2.1B.

Some of the global variation in blood groups and other aspects of physiology is tied to the environment [302; 303]. Associations between a person's blood group and the probability of their having one disease or another are among the most obvious examples of an environmental effect [124, ch. 4; 125, ch. 2; 530, ch. 4; 540, ch. 3]. The best known association of this sort is that between sickle-cell hemoglobin and malaria (although the sickle cell variant is not a blood group) [640, ch. 7]. Individuals heterozygous for hemoglobin S withstand malaria better than do those without the S allele, and the S allele is correspondingly prevalent in areas with falciparum malaria, such as western Africa. In the absence of malaria, S allele individuals do not do so well, and if they are homozygous for S, they die early from complications of sickle cell anemia. And so we find hardly any S individuals in regions without falciparum malaria.

With respect to blood groups themselves, people of blood group O appear to be more resistant to syphilis than those of other blood groups, as might be expected if syphilis originated in the Americas, where O is so prevalent. Blood group O people might be more likely than those of A blood type to suffer from cholera and plague. Blood group A people might be more likely to suffer from smallpox and bronchial pneumonia. And so on.

However, enough comparisons can produce chance associations. In the case of blood groups and diseases, scores of blood groups and hundreds of diseases exist, providing thousands of opportunities for a link to appear by chance. It turns out that not only are many of the associations between blood types and diseases quite nebulous, but some anomalies exist. An example is a gene in sub-Saharan African populations that protects against a form of malaria that does not exist there now [530, ch. 4].

If our blood groups are sometimes not obviously associated with disease, what about associations of other traits with other environmental factors? It turns out that connections are still not frequent or obvious. One example will suffice. Luigi Cavalli-Sforza et al. collated a sample of 122 genetic markers, along with 10 climatic measures distributed across 1473 meteorological sites [125, ch. 2]. From this sample, they found sufficient data to make 366 comparisons between the distribution of the genes and of climate. Only 19 of the comparisons showed what might be a statistically significant correlation. Nineteen of 366 is exactly the number expected by the standard 0.05 (1 in 20) statistical threshold of getting a match by accident.

In sum, if the environment does not explain why people from different regions have a different genetic constitution, we are in quite a strong position to use the differences to make inferences about ancestry as an influence. Almost all native South Americans are blood group O, not because people of blood group O do better than others in the (extremely varied) South American environment, but because native South Americans originated from one small founding population (chapter 2) [206].

Admittedly, though, we are quite ignorant of past environments and of the geographical distribution of a myriad of factors to which one population or another might have adapted.

Morphology

We all know that people of different biogeographic origin often look different. Some of the variation exists because people with one trait survive or reproduce better in a particular region than do those with another trait. I will come to such traits in chapter 5.

One of the odder examples of variation among humans, even if it is not exactly a morphological trait, is the nature of their earwax, or cerumen. In the laboratory, earwax can inhibit the growth of various bacteria and fungi [467]. However, while one study indicated that wet wax was more likely in hot, moist climates, its global distribution appears largely independent of climate or disease risk [530, ch. 7]. In general, wet earwax is characteristic of Africans and Europeans, but dry earwax is characteristic of Asians, especially eastern Asians, and native peoples of both North and South America. Yet only 50% or so of Ethiopians and Mozambiquans are wet, and Melanesians, Micronesians (western Pacific Island region), and especially the Ainu of Japan are wet, as are several Native American peoples. For what it is worth, our closest relative, the chimpanzee, also has a mix of wet-wax and dry-wax individuals [530, ch. 7]. The apparently random geography of the two forms of ear wax in humans is what we would expect of a trait that has little or nothing to do with the environment but instead reflects the effects of history of origins and dispersal, and of founder effects and drift.

Several anatomical features that have no apparent benefit or cost distinguish people from different parts of the world. Asians and Native Americans are more likely to have whorls in their finger prints (c. 25%) and less likely to have arches (< 5%); natives of southern Africa, the Khoisan, and pygmies of the eastern Congo Basin are more likely to have arches (c. 15%) and less likely to have whorls (c. 10%); and peoples from other parts of the world, including east Africans, are somewhere in between [332, ch. 12]. Roughly 90% of Chinese and of native North Americans have shovel-shaped incisors, compared to less than 15% of Americans of European or African descent [530, ch. 11]. While ridged as opposed to smooth skin might have some function, it is difficult to think of environmental explanations of the regional differences in patterns of the ridges.

Head shape has long been of interest to anthropologists, because it varies regionally. Some suggest that higher-latitude peoples have rounder heads, as an adaptation to reducing heat loss [530, ch. 10; 640, ch. 8]. However, the spread of points in the head shape by latitude graph is large [540, ch. 5], and in at least one sample of head shapes, no regional or latitudinal differences exist[a] [332, ch. 12]. Instead, the farther apart populations are, the greater the difference in the shape of their head, specifically the cranium, a biogeographic pattern expected from the effects of drift [638; 639; 641].

With far more detailed measures of head shape than just breadth as percent of height, John Relethford, and later others, found that most differences between modern human populations could be explained by phylogeography—by history, not by environmental adaptation [60; 638; 639; 813]. For instance, Weaver et al. found that about 80% of the differences among over 2,000 humans from 30 populations could be due to founder effects or drift, or both [813]. Of the remaining 20%, they could not say what the adaptation might have been nor with what aspect of the environment any of the traits correlated, because all they were looking at was deviations from a distribution of measures predicted on the assumption that there was no adaptation. Environment probably does in fact have some influence on head shape, as implied by the fact that head shape of offspring of immigrants can be different from the head shape of their parents [278].

Weaver et al. measured not just humans but also Neanderthals in order to test suggestions that some of the well-known differences, such as the heavier features of the Neanderthal skulls, were caused by Neanderthals using their teeth so much to tear meat from skin. With a sample of 20 Neanderthals to compare to their 2,000+ modern humans, they found no evidence of adaptation. All the differences measured were entirely due to

---

a. Head shape by region and latitude. Head shape = breadth/height, the cephalic index (CI). CI by latitude, $N = 30$, r^2 adj $= 0.09$, $F = 3.9$, $P > 0.05$; CI by region, counting only regions with $N \geq 5$, $N = 4$ regions, $r^2 < 0.6$, $P > 0.3$. Data from [332, ch. 12].

drift [813]. Moreover, the characteristic Neanderthal features are present in immature individuals too, which also makes lifetime adaptation to the environment an unlikely explanation for them [811].

Nevertheless, as Weaver et al. were at pains to point out, their conclusion concerning the dominant role of drift applies only to the features that they measured [813]. As I will describe later, the evidence that Neanderthal body proportions are adaptations to cold seems quite good (chapter 5).

ORIGINS AND GLOBAL DISPERSAL

Primate Origins and Dispersal

Most of those working directly on the biogeographic history of the primates are fairly explicit about our lack of information on the topic and about the extent of active debate [500]. Not only are there abysses in the fossil record, but a huge difference exists between timing of origin indicated by fossils and timing indicated by molecular analyses [95; 385; 497; 500].

Molecular data indicate a date of origin of primates that is at least 50% older than the date indicated by fossils, maybe more. The earliest fossils of primates date from around 55 million years ago (mya), whereas molecular dates of origin are between 80 and 100 mya. Getting the date of origin right is crucial, because it will strongly affect hypotheses about place of origin and routes of dispersal from that place [51; 57; 343; 385].

Fossil dates of origin are likely to be too young, because the chance of finding the originating population is so small. Conversely, molecular dates of apparent origin can be older than the real date of origin of a species, because any particular part of a genome could differ across populations within a species long before the speciation event.

If we accept paleontological evidence for times and places of origin of the various branches of the primate tree, primates are remarkable transoceanic sailors [668]. They had to have crossed several seas and a major ocean, the Atlantic, to end up where they are now. On the face of it, that is a highly unlikely scenario.

In brief, we do not yet know on which continent primates originated [348]. Many lines of evidence indicate Asia, but some argument can be made for Africa and even Indo-Madagascar [497]. But too much is unknown, including of the geology, and there has not yet been enough application of modern statistical methods [784]. The consequence is a far-too-voluminous debate for even a brief cited summary here.

With regard to apes, most current opinion has them originating in Africa in the late Oligocene, around 25 mya [216; 226; 334; 726]. They radiated rapidly, taxonomically and geographically, so that by the middle and late Miocene, 16–5 mya, many genera occurred throughout Europe and Asia, as well as Africa. The hominids—the family that led to the gorilla, chimpanzee, and eventually humans—also probably arose in Africa, around 8 mya.

But new finds continually change the story. I will close this section by pointing out that we have as yet no fossils from the 8 million years of African ape evolution that has ended with the gorilla and two species of chimpanzees. If creationists want evidence of God's hand and the sudden creation of life forms, they should use the gorilla and chimpanzees, not humans. For humans, the gradual evolutionary change from an ape-like ancestor is now extraordinarily well documented.

A Single Origin of Humans—In Africa

On these views, it is obvious, that the several species of the same genus, though inhabiting the most distant quarters of the world, must originally have proceeded from the same source, as they have descended from the same progenitor . . . Nevertheless, the simplicity of the view that each species was first produced within a single region captivates the mind. He who rejects it, rejects the vera causa of ordinary generation with subsequent migration.

[165, Darwin, 1859, ch. 11, p. 351–352]

We are naturally led to enquire where was the birthplace of man at that stage of descent when our progenitors diverged from Catarhine stock . . . as these two species [gorilla and chimpanzee] are now man's nearest allies, it is somewhat more probable that our closest progenitors lived on the African continent.

[166, Darwin, 1871, part I, p. 199]

When Darwin speculated in 1871 that Africa was the birthplace of modern humans, no fossil hominins had yet been discovered, and the world did not even know of genes. Since then, our African roots have been confirmed and reconfirmed again and again. Thomas Huxley pointed out nearly 150 years ago humans' many anatomical similarities to the gorilla in particular, but he could have equally used the chimpanzee [387, ch. 2]. Darwin built on Huxley's work to speculate that we originated in Africa [166, ch. 6], and now an immensity of information has confirmed both Darwin's contention that species arise in only one place and his speculation that humans originated in Africa [116; 124, ch. 3; 125, ch. 2; 165, ch. 11; 181; 221; 424; 425, ch. 7; 443; 456; 577; 589; 737; 776; 812].

Ordinarily, a long list of citations to substantiate a single origin of a species would not be necessary. However, in human historical biogeography, an evolving theory over the last three decades or so has suggested that in various ways the current single species of *Homo sapiens* is an amalgamation of separately evolved populations. The multiregional hypothesis is the best known term for the variants of the idea, which is mostly associated with Milford Wolpoff. I will return to the hypothesis after I present the evidence that humans, like all other species, "proceeded from the same source."

A wealth of evidence supports the idea that humans (i.e., *Homo sapiens sapiens*) originated in just one place from just one small population.

For instance, a single origin fits best the fact of humans' extraordinary genetic homogeneity. Humans across the globe are less genetically diverse than are some single populations of chimpanzees in Africa (Fig. 1.1) [247; 730]. The most likely explanation for such homogeneity is that we evolved from one small population.

Additional evidence for a single origin is that the farther apart two human populations are geographically, the more different they are morphologically, molecularly, and genetically (Fig. 2.2) [124, ch. 2; 125, ch. 2; 638; 639]. That pattern also is difficult to explain in any way other than with a single origin and dispersal from it. If humans had originated in more than one place and spread over the world, then neighboring populations could end up being the most different—people of African origin, for example, meeting people of Asian origin in, say, Afghanistan.

Along with the observation that genetic differences increase with geographic distance between any two populations is the finding that, for a given geographic distance, people from almost all regions are genetically farther from sub-Saharan Africa than they are from any other areas (Fig. 2.2; Fig. 2.3). Figure 2.3 emphasizes this point by showing the regions separately and plotting genetic distance to each other region, with geographic distance accounted for. Sub-Saharan African populations are genetically the farthest from all others, and hence with little doubt, sub-Saharan Africa is the place of origin of modern humans.

Morphological and physiological differences show the same pattern [124, ch. 3; 125, ch. 2; 126; 530, ch. 12; 627; 782].

Studies of the genetic diversity of people in Africa by comparison to the rest of the world provide another line of evidence demonstrating that modern humans originated in sub-Saharan Africa. Genetic diversity should be greatest where a population has existed for the longest and so had time to build up diversity through, for example, genetic mutation

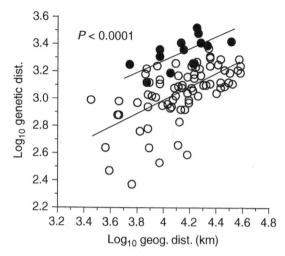

FIGURE 2.2. The farther apart populations are geographically, the more different they are genetically; populations are more genetically distant from sub-Saharan Africa than they are from other regions.* Genetic distance by geographical distance globally. • = distances of all regions from sub-Saharan Africa; ° = distance among all regions excluding sub-Saharan Africa. Data from [125, table 2.3.1A].

*Genetic distances by geographic distances, region as fixed effect. Whole model GLM, $N = 91$ distances between 13 regions (listed in Fig. 2.3), r^2adj. = 0.62, $\chi^2 = 100$, $P < 0.0001$, AICc = -82; partial, geog. dist., $\chi^2 = 50.4$, $P < 0.0001$; region, $\chi^2 = 70.4$, $P < 0.0001$; sub-Saharan Africa vs. rest (residual from overall regression), $t = 6.3$, $P < 0.0001$. Data from [125, table 2.3.1.A].

(other things being equal). And where is human diversity among native peoples at its greatest?—in sub-Saharan Africa (Fig. 2.4) [116; 124, ch. 3; 301; 456; 530, ch. 12; 771].

Not only are all other populations elsewhere in the world less genetically diverse than are sub-Saharan African populations, but the farther in geographic distance human populations are from Africa (more precisely, northern Ethiopia), the less diverse they are [456; 458; 615]. The pattern is true of not just genes but also of language sounds, phonemes [24].

Two processes could be at work, both related to the African origins of humans. The farther other populations are from Africa, the less time they have been in existence, and so the less time there has been for diversity to increase from the original small founding populations [812]. Additionally, if each more distant population was founded from only a

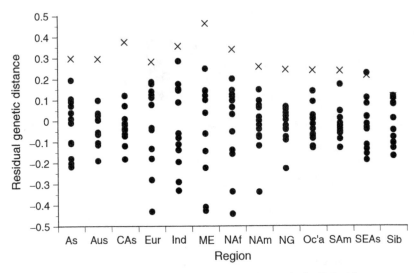

FIGURE 2.3. Populations of almost all regions are more genetically distinct from sub-Saharan Africa (x) than they are from other regions* (•). Residual genetic distances (i.e., geographic distance accounted for by taking residuals of genetic distance from geographic distance) per region. Regions are Asia mainland (China), Australia, Central Asia, Europe, India, Middle East, North Africa, New Guinea, Oceania, Southeast Asia, Siberia. Data from [125, ch. 2].

*Genetic distances, Africa vs. others. Africa residual vs. highest other residual, $N = 12$, $t = 5.8$, $P < 0.0001$, Paired t test. Data from [125, ch. 2].

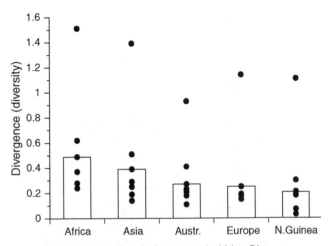

FIGURE 2.4. Molecular diversity is greatest in Africa. Divergence measures for seven mitochondrial DNA alleles in five regions (Austr. = Australia). Data from [116].

subset of individuals in the parent population, diversity would have been lost as humans expanded across the globe [24; 456].

The loss is obvious at two places in our diaspora, the Middle East and, subsequently, in Beringia [13]. The first bottleneck from Africa into the Middle East, both a genetic and a geographic one, could have been constricted even further, given that the Middle East is difficult to leave, except by narrow coastal routes. North and east are the Taurus Mountains of Turkey and the Zagros Mountains of Iran. Roughly 40 ky later, the Americas were populated from what must have been a very sparse and therefore small population from eastern Siberia, or Beringia (below and chapter 3).

Finally, old and more recent alleles can be distinguished—and while old ones are present in African and non-African populations, it is recent ones that are specific to continents [180], as expected if we originated in Africa, and continued to intermarry there, but went our separate ways genetically in the rest of the world.

As well as knowing from where humans originated, geneticists consider that they know also the rough size of the population from which modern humans outside of Africa originated. Calculations based on the genetic diversity of the present human population indicate a founding population of just a few thousand breeding individuals (somewhere between 1,000 and 20,000), or a total population of perhaps 50,000 people[b] [327; 328; 404, ch. 6; 458; 860].

The term "bottleneck" is one that frequently appears in the literature to describe the small size of the *Homo sapiens* population in Africa that was the source of all people outside of Africa. In other contexts, the term can mean reduction to a small population size. That is presumably why some write as if the whole human population in Africa were reduced to one small population of a few thousand individuals.

However, instead of humans outside of Africa originating from a single, small remaining population, we could well have arisen from one of several subpopulations scattered through Africa [443; 612]. Genetically, the two scenarios would produce the same result.

---

b. The calculations are complex, based as they are on estimates and assumptions about rates of mutation, the pattern of expansions and contractions (fast or slow), extent of subdivision of the population, whether mating was random or not within the subpopulations, and so on. Hence the variation in estimated size of the breeding population. In brief, the extreme genetic homogeneity of the human population now indicates a small originating population. Henry Harpending et al. provide a summary of some of the methods [328].

In support of the idea of a demographic bottleneck, some archeological evidence indeed indicates near disappearance of humans from southern and northern Africa just before 50 kya [424], i.e., around the time when humans finally expanded out of Africa. If the disappearance were continent-wide, the single-population scenario would be supported. But no widely accepted evidence yet exists to indicate that the eastern African populations that were the likely origin of the global diaspora were suffering at the time. Indeed, genetic and archeological evidence indicates an expanding population in Africa 80 to 50 kya[c] [238; 443].

Additionally, the fact that genetic diversity within Africa is so much greater than everywhere else in the world, and that genetic divisions within Africa are so deep, indicate that the human population in Africa was not reduced to anywhere near as small a size as the population that gave rise to the global diaspora [115; 125, ch. 3; 770; 771; 860]. Some analyses indicate a total African population of perhaps twice the size of the subpopulation that was the origin of the human global diaspora [770; 860].

The genetic diversity within Africa and its deep roots indicate also that the Africans who populated the rest of the world did not replace all humans in Africa [115; 125, ch. 3; 771]. If a multiregional hypothesis of assimilation works anywhere, it is in Africa. Africa, not the United States, was the original melting pot.

Alternative Explanations. An alternative explanation for the fact that Africans are more genetically diverse than are non-Africans is that until very recently, Africa's human population was larger than the rest of the world's [301; 812]. It obviously was originally, but with the spread of humans outside of Africa driven in large part by demographic expansion, it is unlikely that the African population remained the largest for any length of time after the exodus that peopled the rest of the world [812].

Another explanation for Africa's greater diversity is more mixing among all the other continents than between any of them and Africa [530, ch. 12]. The Pleistocene climatic fluctuations could have produced some such mixing in Eurasia, with people moving south during glacial maxima and expanding north again during minima (chapter 3). But Africa's climate too changed greatly over the same time—arid at glacial

---

c. The genetic analyses are of the same sort as those described for estimating initial size of the out-of-Africa population. Archeologically, the evidence comes from extent and concentration of remains.

maxima, wetter at glacial minima (chapter 3)—and the genetic and archeological records certainly indicate major movements of peoples [238].

Additionally, while all populations globally have some African genetic representation, Oceania and the Americas have some forms of genes that are not present in other regions, i.e., they cannot have been involved in the mixing—and yet they fit the pattern of less genetic diversity outside of Africa [664, fig. 2]. Indeed, it looks as if Australia especially might have been separated from the rest of the world for periods of over 10,000 years at a time (see the section "Asia to Sahul," below).

Another reason to reject the mixing scenario is that the distribution of alleles within the single continent of Eurasia does not show more mixing across that continent than between Eurasia and Africa; eastern Asia, for instance, is quite well separated genetically from the rest of Eurasia [664, fig. 2].

A variant of the multiregional hypothesis that incorporates mixing is Alan Templeton's suggestion of at least two major dispersals of modern humans out of Africa and subsequent interbreeding with preexisting populations in other regions [752]. However, his method ("nested clade phylogeographical analysis," to give it its full name) has some major problems [427; 589]. For instance, it appears to be largely untested, and when tested, to have an extremely high error rate [427]. Other multiregional models have problems too, such as the serious mismatch between genetic and geographic distances that I have mentioned [589].

With regard to the fact that geographic distances match genetic distances, environmental influence is at first sight a reasonable explanation [639]. The closer together two populations are, the more likely they are to share the same environment; the more distant they are, the less similar their environment is likely to be. However, the hypothesis does not work [639]. For instance, were environment an influence, relationships of measures of biological diversity of humans with north–south distances should be stronger than with east–west distances, given that environment usually changes more with latitude than with longitude. They are not [639].

. . .

In sum, even if the out-of-Africa hypothesis is not completely free of problems [589; 776], we modern humans are all Africans at origin, specifically sub-Saharan Africans, or more specifically, East Africans. However, recent confirmation that human remains and sophisticated (Aterian) artifacts in Morocco are up to 145,000 years old indicates North Africa as another potential source [44; 652]. Additionally, greater-than-expected variability in the shape of the cranium of early modern humans could indicate origins from more than one African population in more than one exodus [296]. Nevertheless, the genetic data still mostly support an origin somewhere in sub-Saharan eastern Africa.

Dates of Human Origins in Africa

Humans originated in Africa long before they finally left to populate the rest of the world. The exact date of origin of modern humans depends on one's definition of *Homo sapiens* [227]. However, genetic, paleontological, and archeological evidence indicate that the first more or less modern human, i.e., *Homo sapiens*, arose between 200,000 and 120,000 years ago [124, ch. 3; 180; 227; 443; 509; 589; 654]. Currently, the oldest identified humans of largely modern form are from Ethiopia in strata dated to 195 kya [509]. Wherever in Africa humans originated, they were throughout Africa by approximately 125 kya, from Algeria to the tip of southern Africa [227]. These early humans were anatomically not quite like fully modern humans, whose body form took another 75,000 years or so to evolve [425, ch. 6, 7; 654].

Dispersals within Africa

Hominins and humans have, of course, been moving within Africa for longer than they have been moving out of Africa [228, ch. 9; 733]. For instance, *Homo habilis* (c. 2 mya) lived in both eastern and southern Africa [733], and about 1.8 mya *H. ergaster* was in Algeria and Kenya at sites about 5,000 km apart, and soon thereafter in South Africa too [221; 425, ch. 5].

Some evidence exists for an expansion of modern humans through Africa south and west from the Ethiopian region, perhaps 100 kya [238; 771]. A subsequent expansion from the same area might also have occurred at around the time of the roughly 55 kya exodus from Africa [771].

Two relatively recent centers of dispersal within Africa are once more the Ethiopian region, and also Nigeria and Cameroon [115; 124, ch. 4; 125, ch. 3; 212; 351; 404, ch. 10; 634; 683; 771]. Ethiopia has already featured in the account of human origins. One study, though, indicates that Cameroon could be a center of origin of humans [627]. Be that as it may, the modern agricultural, iron-working Bantu peoples of much of Africa apparently originated from the Nigeria-Cameroon region and expanded south and east, starting about 3,000 years ago, with the eastern peoples eventually turning south too. Nilotic pastoralists from northeastern Africa were expanding west and south at roughly the same time, reproducing with Bantu as they went, or "admixing" in genetics termnology.

Much of the data demonstrating these movements is genetic. As Luigi Cavalli-Sforza has argued for Europe [124; 125, ch. 4], so for Africa in the last few thousand years [683]: maps of languages correlate closely with maps of genes, indicating that culture and people usually move together. Exceptions exist, of course. Thus African pygmy peoples, genetically close to South African Khoisan "click" speakers, have lost any original language that they might have had and are linguistically close to their mostly Bantu neighbors [683; 768].

Dates and Routes of the African Exodus

Prehuman Hominin Dispersals. Modern humans were not the first hominins to leave Africa [169; 172; 220, ch. 3, 7; 221; 226, ch. 17; 382; 424; 425; 443; 499; 525; 576; 580; 653; 750]. Before *Homo sapiens,* other species of *Homo* had spread at various times across western Europe, even up to southern Britain, across southern Eurasia, over to eastern Asia, down to the Sunda region, and maybe even east of the Wallace line if signs of hominins in Flores from nearly 100 kya are accepted [103; 548; 549].

Homo ergaster/erectus left Africa a little under 2 million years ago, soon after it first appeared in the fossil record. It either left as Ergaster, subsequently evolving in Asia into *H. erectus,* or it emigrated as Erectus from Africa. Erectus fossils have been found from northeastern China down to Java, the earliest dating to about 1.8 mya.

Whilst emigration of Erectus/Ergaster from Africa is the majority view, a minority argues immigration into Africa from Asia [172]. The earliest fossils of both Erectus and Ergaster are close enough to each other and to 2 mya for provenance to be uncertain. If Erectus/Ergaster arrived from Asia, it would have joined the greater number of bovids (buffalo, antelope) that moved from Asia to Africa in the Plio-Pleistocene than moved in the other direction [795].

Wherever or whenever *H. heidelbergensis* originated (Eurasia or Africa), it was in Europe by about 500 kya—unless *H. antecessor* is an early Heidelberg, in which case it was in Europe by at least 800 kya and in southern Britain during a mild period by about 800 kya [580]. A dispersal of African mammals into the Levant at this time strengthens the possibility of a hominin dispersal event then too [498]. Some have suggested that Heidelbergensis reached China, but it is possible that the Heidelbergensis-like remains there are in fact Erectus. Neanderthals (*H. neanderthalensis*) might have originated around 500 kya, either from Heidelbergensis in Europe or from something else in Africa. Definitely in Europe by about 300 kya, they also expanded to southern central Eurasia.

Several other *Homo* species might have existed, such as *H. cepranensis, georgicus, helmei,* and *rhodesiensis,* along with the newly discovered Denisovans of western Asia [431; 637]. Wu Liu et al. present evidence for modern humans in southern China at 100 kya [459]. However, the fact that the minimal remains had some archaic features, along with the discovery of Denisova in Asia, raises the possibility that these Chinese hominins might not in fact be modern humans. Of course, if they are modern humans, the account of the timing of our dispersal out of Africa needs to be rewritten [459].

Debate occurs over not only the timing of the origins, the departures, and the arrivals of all the species but even over the taxonomic status of many of them, including those such as *H. ergaster* and *H. heidelbergensis,* about which I have written as if their taxonomic status is accepted [425, ch. 5; 840]. Suffice it to say that several hominins have been dispersing within and out of Africa over more or less large areas of the globe from close to the time that *Homo* first appeared. We modern humans are not different in the fact of our origin in Africa and dispersal from it. It is the extent of our dispersion that distinguishes us.

Modern Humans out of Africa. H. sapiens expanded within Africa from about 200 kya. A little over 100 kya, our species got as far as the Middle East but then disappeared from there about 70 kya. Then some time between 70 kya and 45 kya, modern humans finally expanded out of Africa through the Middle East and on to the rest of the world [220, ch. 3, 7; 221; 424; 443; 525; 715].

For brevity's sake, I will write of this exodus as occurring at roughly 55 kya, but the 10–15 kya margin of error in that date must not be ignored. Nor must quite a bit of disagreement about this fairly simple scenario [598]. And nor must continual new finds, such as stone tools of early modern humans at about 125 kya in eastern Arabia, beyond the Levant [22]. And finally, while I write of "the exodus," debate exists on the nature of the exodus—a single population expansion, waves of expansion, or a steady trickle?

The different disciplines involved in analysis of our biogeographical history produce different dates for the exodus and dispersal. As was the case for the contrast between molecular dates and paleontological dates of the origin of primates, and for the same reasons, molecular dates of human expansion out of Africa are about 50% earlier than good archeological dates of this second dispersal, the earliest of which are from around 45 to 40 kya [607; 776]. The molecular dates often have wide margins of error, though, because different genetic molecules appeared at different times [770].

Some would have modern humans leaving Africa to populate the rest of the world as early as 80 kya [576; 577]. Archeological evidence for this early migration is, in part, the presence of sophisticated stone tools in southern India across ash layers from the well-dated Toba volcanic explosion of 74 kya [597]. However, whether the tools were made by modern humans or earlier hominins such as Neanderthals is an open question [597].

The fact that Neanderthals are not presently known to have moved as far south in Asia as anywhere in India, let alone southern India, tips the evidence in favor of modern humans as the makers of the Indian tools. The problem is that no other evidence on the ground exists for humans in the region at that time. Nevertheless, some genetic data support the idea of an early arrival of humans in India [577], and the finding of potentially yet another *Homo* species extant in Asia at around this date [431; 637] opens the doors to even more speculation. But until bones associated with the southern Indian tools are found and identified, the issue will remain open. At present, the earliest uncontested signs of modern humans in India date to around 45 kya [607].

It looks as if the modern humans that left Africa roughly 55 kya went by two routes (Fig. 2.5) [182; 215; 221; 442; 443; 471; 576; 577; 607; 783]. One, perhaps the later one, was via the northern end of the Arabian Peninsula and then on east into western Asia. The other, perhaps earlier, was along the southern end of the peninsula, to India, and on into southeastern Asia. From there, the people would have gone north up through eastern China and south down into New Guinea and Australia. The southern route was possibly a coastal movement all the way [215; 443; 607].

If the southern route left Africa via southern Arabia, the crossing at the mouth of the Red Sea would have been far easier then than now, because the sea level was so much lower [215; 238; 607]. Indeed, so shallow

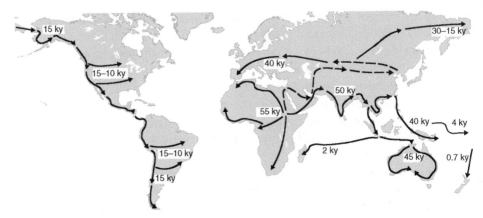

FIGURE 2.5. Approximate global dispersal routes and dates of modern humans. Dashed lines = more uncertainty than for other routes. Sources in text.

is the isthmus now that it might just have been marsh then. The migrants would have moved round the coast of Arabia to the mouth of the Euphrates and then along the coast to India. Alternatively, the initial migrants could have departed overland via the Sinai. A simulation accounting for a coastal preference, topography, and sea level suggests that from the Sinai, the likely route would not have been down the east coast of the Red Sea but rather across to the Euphrates, down to its mouth, and then east along the coast [215].

Whether or not the initial migrants used boats or rafts of some sort at the start of the exodus from Africa, the humans who finally arrived in Australia had to have used sea-worthy vessels. And they would have had to have set out with no sight of their destination, because even with the lowered sea level then, the shortest single crossing was at least 100 km, whether people arrived directly from Timor (as indicated in Fig. 2.5) or via New Guinea [147; 577; 463, ch. 16].

Much of the evidence for two routes out of Africa lies in the nature of the diversity of DNA of present-day populations along the way. European mtDNA diversity is different from Asia's and Sahul's (i.e., New Guinea and Australia), and each has in their mtDNA genome some parts that, while common in their own region, are rare or absent in the other [115; 182; 221; 238; 301; 471; 577]. At first sight, the recent finding that a southwestern Siberian population of hominins, the Denisovans, might have interbred with the dispersing modern humans [431; 637] lends support to the northern route out of Africa. However, the geographic distribution of the Denisovans is currently unknown, meaning that the signs of interbreeding allow no inferences about early dispersal routes.

Along with genetic evidence for two routes, archeological finds indicate that almost immediately after leaving Africa, differences in morphology of European, Asian, and African populations are detectable, as if different subpopulations within Africa followed different routes after their exodus [442; 443].

Marta Lahr and Robert Foley have argued that the southern, coastal dispersal event was earlier than the northern one [442; 443]. Their reasoning was based on the fact that the human fossils along the way had a morphology more similar to earlier than later humans, in agreement with some of the genetic and archeological evidence [442; 443]. For others, the argument for an earlier departure from Africa of the southern Asian and Sahul peoples is necessary because of dates of as far back as 60 kya for the earliest humans in Australia. However, confirmed evidence for dates of first arrival in Australia does not yet allow anything much earlier than 50–45 kya [80] (see the section "Asia to Sahul," below).

Back into Africa. If humans could get out of Africa, they could presumably also return. A section of the Y chromosome that seems to have arisen in Asia about 33 kya can be found in African populations, as can a mtDNA type that might have arisen about then in India [181; 238; 301]. However, the interpretation that the data show an ancient return to Africa is contested, and the fact that there are common and widespread genes outside of Africa that have not been found in any African population indicate that if there was back migration, it was relatively rare [404, ch. 9; 683].

Movement between Other Continents

Evidence for movement between other continents is much the same as the evidence for movement out of Africa. For instance, as already mentioned, genetic distance matches geographical distance across all continents, and between all continents (Fig. 2.2) [116; 124, ch. 2].

While the correlations are not perfect, the imperfections are themselves useful information. As one example, although Asia is geographically closer to Europe than it is to the Americas, and certainly has a broader geographical region of connection, eastern Asia is genetically closer to the Americas. The historical biogeographical explanation for that apparent anomaly is, of course, that the Americas were peopled by Asians, specifically Siberians.

An additional sort of evidence for human dispersal into other continents, which is possibly unique to humans, is the extinction of large-bodied fauna shortly following the date of the humans' arrival indicated by molecular and archeological evidence. I will address this issue in more detail in the final chapter on the impact of humans on the biogeography of other species (chapter 10).

To Asia. As I indicated when writing about the exodus of modern humans from Africa, at least two routes to Asia, a coastal and an inland, are likely, and several routes are a possibility (Fig. 2.5) [182; 221; 442; 443; 471; 576, ch. 5; 577; 607]. The inland route through central Asia could have been along what became the Silk Road, now pretty much along the Mongolia–China border, and also a later route through southern Siberia.

The same two routes, coastal and Silk Road, might also have been used by *Homo erectus*. We know that by 1.8 mya, Erectus was at Dmanisi, half way between the Black and Caspian Seas, well inland of a coastal route[d] [169]. After that, two independent populations of Erectus in Asia are a discussed possibility [424]. Erectus was in tropical Java by 1.8 mya, and in temperate eastern China by 1.7 mya [384; 747; 861]. Between the two populations was a lot of thick, subtropical forest [144]. If that forest was unfavorable habitat for Erectus, and it did not therefore move north from the tropics, even up the coast, then it had to have reached northern China by going east through central Asia, presumably roughly along the route of the Silk Road [144; 576]. However, a northern movement up the east coast of China seems more likely than a route through inclement central Asia.

Returning to modern humans, evidence for two routes to Asia was that southeastern Asians appeared to be genetically distinct from northeastern Asians [125, ch. 4; 576, ch. 5]. Indeed, the data indicated that southern East Asians might have been as different from northern East Asians as they were from Middle Easterners and Indians [125, ch. 4]. However, a recent analysis of nearly 2,000 individuals from over 70 Asian peoples indicates, first, that all Asians share a common ancestry, in other words that they are the result of a single event of migration from Africa. Second, the data indicate that eastern Asians arrived from the southeast [762]. Evidence for the inferred dispersal pattern comes from

---

d. Dmanisi hominin. Potentially a new species, *H. georgicus* [169].

regional genetic comparisons, including to non-Asians, and from the decline in genetic diversity of the Asian peoples as one goes north from southeastern Asia to eastern Asia.

Climatic evidence against a northern route to eastern Asia is that although central Asia is in places a rich grassland [440, ch. 1], perhaps full of large mammals [220, ch. 5], nevertheless, even in the present-day, clement interglacial central Asia is hardly a welcoming environment. There is good reason why, even now, the region is largely empty. Average winter temperatures are well below freezing, summer nighttime temperatures can drop below freezing, the winds are continuous, and the average annual temperature now on a glacier at half the elevation of the Swiss Alps is more than 10°C below the Alps' annual temperature [687].

Modern humans had reached southern Siberia by about 40 kya [272; 607]. They got to central eastern Siberia and Japan, probably up the coast from southeastern Asia, by about 35 kya [607]. And they had arrived in northern Siberia at the northwestern edge of Beringia by at least 27 kya [272; 273; 604]. The expansion into northern Eurasia, especially northeastern Eurasia might have come from central Asia, as opposed to up the coast from eastern Asia [782], even though eastern Asia was probably settled earlier [471]. However, some like the idea of both a northward coastal expansion and movement from central Asia [577].

To Europe. Turning west now, the date of modern humans' arrival in Europe, or any of the more northern parts of Eurasia, was, to repeat, no earlier than their/our arrival in Asia and Australia (Fig. 2.5) [222; 272; 273; 525; 576; 577; 604; 782]. Humans had moved into eastern Europe by 45–40 kya, as judged by archeological remains, and by perhaps 38 kya covered most of Europe [15; 220, ch. 7; 222; 238; 367; 525; 576, ch. 3], although dates earlier than 32 kya have been contested [776]. If the early dates are correct, then the movements occurred ahead of a major population increase in Eurasia from about 30 kya, as judged from mtDNA [238].

Some linguistic and genetic analyses indicate that these late Paleolithic dispersals into Europe originated not from the Middle East, let alone Africa, but from western Asia, perhaps originally from southern Asia, perhaps from roughly southern Pakistan [125, ch. 2; 220, ch. 3; 222; 576; 696]. Within Europe, it looks as if modern humans moved along two routes [525; 576, ch. 3]. One was north of the Danube; the other was more along the Mediterranean, via Turkey [576, ch. 3].

Subsequently in the Holocene, starting about 9,000 years ago, humans expanded from the Middle East, this time taking agriculture and livestock with them as the climate warmed. By 6,000 years ago, they had reached northwestern Europe. A huge amount of molecular, linguistic, and archeological evidence substantiates these later dispersals [84; 124, ch. 4; 125, ch. 4, 5; 297].

Debate persists about whether people moved taking the new agriculture with them or whether residents adopted the culture from neighbors, with little movement of the agriculturalists. Cavalli-Sforza has consistently argued that people moved with the culture [124; 125, ch. 4]. A recent study found so little mtDNA shared between skeletons of hunter-gatherer residents and modern Europeans that the inference must be that the agriculturalists moved with agriculture and replaced the hunter-gatherers, with little interbreeding between them [84].

And more recently still, these disciplines continue to tell us details of origins and migration. For instance, a difference in gene frequencies between western and eastern Iceland indicates that western Iceland was settled by people coming from Britain, but eastern Iceland by Scandinavians [37]. Similarly, differences in the shape of the peoples' heads across Ireland can, using equations from genetics, be related to country of origin, with people in the center of Ireland descendants of Scandinavian Viking immigrants and people in the east descendants of Welsh and English immigrants [643].

Asia to Sahul (Australia, Tasmania, New Guinea). Humans arrived earlier in Australia than they did in Europe (Fig. 2.5). It seems likely that northward expansion from Africa was inhibited by the colder climate of higher latitudes (chapter 3). The evidence indicates an almost nonstop, rapid dispersal all the way from Africa to Sahul. For instance, native Australian mtDNA and Y-chromosome DNA markers have African links, with no evidence of any other input except from some tens of thousands of years later [383].

Some molecular and archeological dates for arrival in both New Guinea and Australia are far earlier than commonly accepted for the exodus from Africa. Dates as far back as 75 kya have been suggested, but the presence of humans in Australia much earlier than 45 kya is, on current good evidence, unconfirmed [80; 264; 383; 570; 576; 607; 776]. However, a 67,000-year-old *Homo* toe bone that might be modern human has been found in the northern Philippines [532]. Even if not modern human, the find indicates that hominins by then had to have used some form of boat.

The earliest dates so far for the presence of modern humans in New Guinea are about the same as for Australia, i.e., about 45 kya, and soon thereafter humans were in New Britain and New Ireland, a few tens of km off the east coast of New Guinea [291; 425, ch. 7; 743].

As said, the migration to Australia has been described as rapid [471]. How fast is rapid? For Vincent Macauley et al., the answer is 4 km per year to cover the 12,000 km or so from the Middle East [471]. Four km per year is just 15 m a day with weekends off.

The speed might be fast if the journey were achieved by demographic expansion alone, as an advancing front of an increasing population. However, well-established optimal foraging theory indicates that abundant food produces rapid movement across the landscape as individuals move on in search of even greener pastures. Shellfish are a highly nutritious [92], reliable, easily harvestable resource that can occur at very high densities and which all members of the community can collect [64; 65]. Doug Bird and coworkers have estimated for a Torres Strait community (between New Guinea and Australia) an hourly harvesting rate of 1 kg of flesh per hour per person, including children who are far less efficient than are adults [65].

And we know that humans have been harvesting shellfish ever since humans evolved. For instance, in South Africa, coastal sites from over 200 kya to recent times have copious shellfish remains [723], as does a 1 mya Erectus site in Java [414]. As the shellfish were depleted in one place, people would have moved on to the unharvested abundance that they would have seen ahead [484]. If edible shellfish do not change much in form or abundance from Africa to India to Australia, no cultural or technological barriers need have hindered migration.

If the dispersing humans hunted game also, or instead, they would have encountered native animals naïve to their new weaponry. With information transmitted back about the abundance of easy resources ahead, migration, not just demographic expansion, could have been a large impetus to a population movement eastward. And we know how far hunter-gatherer societies can move in a year. Annual migrations of 100 km are common among traditional societies that move residence through the year [62].

In other words, within limits of current accuracy and precision, the data would indicate humans reaching Australia almost as soon as they left Africa. Therefore, dates of departure from Africa of the original Australians that are earlier than 55 kya, or even 50 kya, do not have to be postulated to account for an arrival at around 45 kya. Conversely, if humans in fact left Africa 65 kya, they would have plenty of time to reach Australia by 60 kya, if such an early date of arrival turns out to be substantiated.

Some earlier genetic findings indicated that New Guineans were as different from Australians as they are from anyone else except Africans [576, ch. 4], implying two migration events from Africa to Sahul. However, new data indicate close genetic similarities between the populations of New Guinea and Australia [383]. The kinship is not surprising given that from over 100 kya, and including the time of first arrival in Australia, people could have walked between the two land masses, because of the lower sea level then.

The peoples of the Sahul land mass are closely related also to the neighboring scatter of Pacific islands of Melanesia [383]. A variety of mtDNA and Y-chromosome markers are unique to this region, while others are unique to Australia [383]. An interpretation of these findings is that after humans first reached Sahul, they did not (for unknown reasons) enter it again until many thousands of years later.

Some argue that no other humans reached Australia for 40 kya after the first peopling at 45 kya [589; 680]. The next arrivals came with the dingo 5,000 years ago, they suggest. However, central-northern native Australians have a genetic marker indicating a separate input from New Guinea around 30 kya [383].

This central northern region of Australia also contains the majority of Australia's numerous language families, whereas the rest of Australia speaks variants of just one language family—or used to, for many are going extinct (chapter 8) [125, ch. 7]. These findings currently allow two interpretations. One is that central northern Australia is linguistically diverse because of input from New Guinea. The other is that it is the region of original occupation, from which a subpopulation expanded through the rest of Australia.

Either of those ideas, though, is complicated by the extreme morphological variability of Australian humans of the time, from robust to gracile, even within a region, such as southeastern Australia [425, ch. 7]. Drift, rapid adaptation to local environments, or other immigrations than that of 30,000 ya are possible explanations [383; 425, ch. 7].

Peopling of the Pacific. Much debate surrounds the biogeographic origins of the Pacific Islanders [421; 760]. The story is far from settled yet, both because of lack of full archeological and genetic coverage of the Pacific and because new data indicate greater complexity in the pattern of settlement than perhaps hoped.

A little bit of geography to start: the Pacific Islands are split into Melanesia, Micronesia, and Polynesia, which were settled by humans in roughly that order. Melanesia includes New Guinea and the islands immediately to the east and southeast, down and across to New Caledonia and Fiji. Micronesia is north of Melanesia and includes Guam and the Marshall Islands. The rest is Polynesia, from Tonga and Samoa in the west, Hawaii in the north, New Zealand in the south, and Rapanui (Rapa Nui, Easter Island) in the east.

In brief, humans might have reached the Solomon Islands 200 km east of New Guinea by 30 kya, roughly 15 kya after they reached New Guinea [576; 829]. There was then a long pause until about 4,000 years ago, when people began to colonize the nearer of the other western Pacific islands of Melanesia and Micronesia (Fig. 2.5) [125, ch. 7; 404; 716; 839]. Archeologically, this region and time is characteristic of the Lapita culture, which is recognized by distinctive pottery, among other cultural artifacts [421].

Then roughly 2,000 years ago, perhaps with improvements to the outrigger canoe and navigation, humans began to colonize the rest of the Pacific, including the distant northern Micronesian islands [125, ch. 7; 264; 279; 421; 530, ch. 8; 833; 834; 839]. We initially got as far east as Samoa, Tonga, and the Cook Islands about 1,000 years ago. However, the Lapitian pottery did not get farther east than Samoa, perhaps because suitable clay was scarce on the eastern Pacific Islands.

The islanders paused there in western Polynesia for a century and a half, and then over the next century, around 1250 A.D. (nowadays commonly termed CE, or Common Era) colonized all the other Pacific islands [834]. It is not impossible that at about the same time, from the same islands, Polynesians reached South America [732] (details later in the section "Asia to the Americas").

Unsurprisingly, the Pacific Island peoples are a genetic mix of the three regions into which anthropologists have divided the islands, Melanesia, Micronesia, and Polynesia, because of movement of peoples between the islands [125, ch. 7; 530, ch. 8; 716]. All of these are most closely related to the peoples of southeastern Asia, including eastern Indonesia and New Guinea [125, ch. 7; 421; 530, ch. 8; 716; 760; 839]. Even eastern Polynesians, the ones farthest into the Pacific, are most closely related to the peoples of southeastern Asia, particularly eastern Indonesia, Thor Heyerdahl and Kon-Tiki notwithstanding [125, ch. 7].

The family of languages spoken in the region is Austronesian. Its many variants are spoken in Indonesia, the Philippines, northern New Guinea and Melanesia, Micronesia, and Polynesia [125, ch. 4; 279]. It is also spoken in Taiwan. Indeed, close as Taiwan is to mainland Asia, its native peoples do not speak any eastern Asian language, but instead speak the greatest variety of Austronesian languages [279; 404, ch. 11]. It looks as if the population that founded Taiwan has disappeared from the mainland but survived and diversified on Taiwan. Wherever the Taiwanese came from, a plot of the evolutionary tree of Austronesian languages shows an initial movement from Taiwan into southeastern Asia, specifically the Philippines, then New Guinea, then into Melanesia and Micronesia as far east as Fiji, and finally across the rest of the Pacific Islands [279].

Some genetic studies substantiate these language results, for while present-day native Polynesians clearly have Melanesian and Micronesian ancestry [125, ch. 7; 774], a mitochondrial variant exists that is found in only Taiwan and Polynesia [774]. However, Bing Su et al. found hardly any Taiwanese Y-chromosome genes in Polynesia (or Micronesia) [740]. Instead, the analysis indicated southeastern Asia as the origin of Polynesian peoples.

Similarly, the characteristic mitochondrial "Polynesian motif" of Polynesia and Fiji is missing in Taiwan, the Philippines, and New Guinea but is found in eastern Indonesia, including Java [404, ch. 11; 648]. This Polynesian motif is most diverse in eastern Indonesia, implying an origin there, even if ancestral forms are in Taiwan [648]. Rough dates of origin of the motif are 30 kya in Taiwan and 17 kya for eastern Indonesia [648]. The most derived form of the motif is absent from eastern Indonesia and most prevalent in Melanesia, just east of New Guinea [716]. The distinctiveness of the mitochondrial Polynesian motif is substantiated by some Y-chromosome analysis that indicates that Polynesia can be separated genetically from Melanesia and Micronesia [160].

The overall scenario as judged from linguistics and genetics is then plausibly an origin in Taiwan, movement to eastern Indonesia, and then several thousand years later, a subpopulation of eastern Indonesians populating the Pacific, with Polynesia most likely populated from Melanesia, and maybe some movement back to Taiwan from eastern Indonesia [279; 648; 716; 839].

In detail, the situation is more complex. That is not surprising. The peopling of the Pacific was not a one-time, one-way movement from just one place: the Pacific Island peoples clearly traveled and traded widely

[421; 760], which explains why their genetic make-up indicates quite substantial mixing among the three major regions of the Pacific. Additionally, the sexes dispersed differentially (later section), which explains some of the contrasts in the mtDNA and Y-chromosome DNA results [299; 420].

Peopling of Madagascar. Large as Madagascar is, as close to Africa as it is, and despite primates and a few other organisms supposedly rafting to it from Africa millions of years ago [501; 854], nevertheless it was perhaps only a little over 2,000 years ago that humans got to Madagascar—from Borneo (i.e., not even the closest of the Indonesian islands; Fig. 2.5) [109; 125, ch. 3; 175; 279; 648]. Linguistics, genetics (e.g., the presence of variants of the "Pacific motif"), and archeology all agree on this origin.

Asia to the Americas

When we find the same [people] . . . ranging over a great area, as over America, we may attribute their general resemblance to descent from a common stock. [166, Darwin, 1871, part I, p. 240]

These days, few question the idea that the Americas were originally populated by peoples of northeastern Siberia, i.e., western Beringia (Fig. 2.5) [125, ch. 2; 273; 425, ch. 7; 576, ch. 7; 630; 689; 749; 782]. One of the stronger lines of evidence for such a precise assertion is the finding that only northeastern Siberians and native Americans, nobody else, have a particular section of DNA, an allele at autosomal microsatellite locus D9S1120, to be exact [205; 688; 689]. The obvious way that pattern could have arisen is for all Native Americans to have originated from the one region in northeastern Siberia, i.e., Beringian Siberia. Nevertheless, it is not completely impossible that the northeastern Siberian population came from America, whose native population came from somewhere else in northeastern Eurasia.

Accepting the idea of a Siberian source of Native Americans, for American origins to be so tightly tied to just northeastern Siberia means that the region was almost certainly largely isolated from the rest of Eurasia for some time [576, ch. 7]. Otherwise there would be genetic evidence of more widespread Native American origins from eastern Eurasia. I will return to the possibilities of that isolation a little later in this chapter, and in the next chapter, its climatological correlates.

Anthropology is full of discussion of the date and number of dispersals from Siberia [161, ch. 1; 273; 576, ch. 7]. We used to consider that the Clovis people (named after the site where the characteristic stone technology was first discovered) were the first into the Americas, at about 13 kya. Most now agree that people arrived earlier, possibly by 16 kya [273; 576, ch. 7; 715].

The earlier date agrees with archeological evidence, such as human coprolites from Oregon dated to about 14 kya [262], butchered mammoth bones in Wisconsin dated to about 14.5 kya [273], and the famous Monte Verde site in central Chile, where remains, including mastodon bones, are reliably dated to over 14.5 kya [179; 260; 273; 798].

Monte Verde is about 16,000 km from the Bering Strait. The great uniformity of some of the genetic types all the way from Alaska to South America and across both continents [749] supports a hypothesis that once the dispersal out of Beringia started, humans moved very rapidly throughout the continents, both by migration and by growth of population. As was argued for Australia, back-of-the-envelope calculations show that the journey could be done in just a few hundred years, accepting reasonable annual distances migrated. Thus, Lewis Binford's data on movements of over 200 traditional peoples across the globe produce a median of eight moves per year and 13 km per move, which would get the first Amerindians from Beringia to Tierra del Fuego in less than 200 years [62].

Skilled, tool-using modern humans moving into regions previously unharvested by anything remotely similar would have found abundant resources easily available to them. Game would be less wary, oysters and clams larger, berries denser on the bushes, grass greener. Both optimal foraging theory [432, ch. 3] and personal experience tell us that fast onward movement is likely in such a situation.

Nevertheless, we did not spread across the continents and fill them by just continual movement to richer pastures. We also dispersed by expanding demographically. Jonathan Wells and Jay Stock include humans' unusually fast reproductive rate (by comparison to their closest relatives, the great apes) among the traits that predispose us to be a successful dispersing and colonizing species [818]. Help from community members in raising the offspring probably allows that fast rate [340; 379]. Todd Surovell [744] calculated with modeling and real values on reproductive output, population increase, costs of carrying children, and number of dispersals per year that the Americas could have been peopled from Beringia in less than 2,000 years. Such calculations are just possibilities, though, without more well-dated archeological finds [798].

If archeological evidence indicates a date of 16 kya for the original dispersal into North America, some genetic evidence could indicate a far earlier date [124, ch. 3; 452; 576, ch. 7; 716]. For instance, a crude comparison of genetic distances between continents with archeological evidence of first dispersal produces ratios of date to genetic distance of around 0.2[e] [124, ch. 3]. To get a similar index from a match of genetic distance of eastern Asians to Amerindians, a dispersal date of around 30 kya has to be postulated [124, ch. 3].

That date happens to be about the time that humans arrived in northern Siberia [273; 604]. If populations in northeastern Siberia were cut off from the rest of Asia then, and subsequently were the sole or main origin of Native Americans, who did not move into America until about 16 kya, the gap between the genetic and archeological dates is explained. Indeed, the disparity in dates then substantiates the idea of a several-thousand-year isolation of the western Beringian (northeastern Siberian) peoples from the rest of Asia, before their expansion into the Americas (above).

Regarding the number of dispersals, linguistic evidence used to be taken to indicate three main language groups in the Americas, potentially implying three major dispersals [124, ch. 5; 125, ch. 6; 126; 289; 782]. However, this tripartite interpretation is now challenged by both linguists and geneticists, and has been largely rejected [205]. In fact, from linguistics, a source of Native Americans in Beringia or elsewhere in Siberia is unrecoverable, because language evolves too fast (chapter 5).

The rejection of the hypothesis of three language groups and three migrations does not mean that there is no general structure to the American population. For instance, Erika Tamm et al. [749] suggest an initial movement into American Beringia (Alaska), a pause that produced the separation of all Amerindians from eastern Siberians and yet similarity among Amerindians, then a very rapid spread across all of the Americas (which is why founding genetic types are everywhere in the Americas), and finally some division by separation and drift of genotype and language. Subsequent migrations from Siberia to northern North America can also be detected linguistically and genetically [273; 749].

Additionally, some subdivision of the Beringian population before expansion can be related to current structure of the Native American

---

e. Relation genetic distance to date. For example, Africa to Europe, genetic dist. = 9.7 units; archeological date = 43 kya; ratio = 0.23 [124, ch. 3].

peoples. Alan Rogers et al. suggest with evidence from the distribution of mammals and regional diversity of languages that a tripartite separation could have been produced during the last glaciation if human populations persisted and diversified separately in Beringia, in the coastal northwest of North America and south of the Laurentide ice sheet [659].

Until recently, no linguistic connections existed between Siberia/western Beringia and North America that might throw light on origins of Native Americans, except ones easily explained by far more recent dispersals between the regions [124, ch. 5; 125, ch. 6]. However, an increasingly strong case is being made that Yeniseian, a language of a very small group of people (about 200 speakers) from around the Yenisey River in central Russia (i.e., western Siberia) is related to arctic North American Na-Dene [418; 785]. It might be that the two populations share a common origin, rather than that one gave rise to the other, because the Yenisey is far east of where genetic analyses indicate Native Americans originated.

Debate exists also on the exact route taken by the first immigrants into the Americas. Some suggest a mainly coastal route; others suggest the corridor through east-central Canada between the Cordilleran and Laurentide ice sheets [798]. Genetic and archeological evidence exists for both, though the corridor route would probably not have opened until about 13 kya [204; 273; 300; 368].

Movement across the north of North America had to wait even longer, and it was probably not until about 4,000 years ago that northeastern North America, including Greenland, was peopled, probably mostly from Alaska [224], but also directly from Siberia, with no Alaskan input [630].

Finally, evidence is appearing that indicates that at the same time as Polynesians reached New Zealand, about 700 years ago, they also reached South America, and indeed returned to Polynesia. A site in Chile of about 1,000 years ago has produced chickens the DNA of which has its closest connections with Polynesian chickens, and plants of South American origin (sweet potato, gourds[f]) have been found in Polynesian sites also of about 1,000 years ago [732].

In sum, whatever the exact pattern of movement into and across the Americas, whatever the number of migratory events, it looks as if all Amerindians originated from the same small region of eastern Siberia. Yet many details remain to be worked out [205; 576, ch. 7].

---

f. Yam (*Ipomoea batatas*), bottle gourd (*Lagenaria siceraria*).

SEX DIFFERENCES IN DISPERSAL

Differences between the sexes in their geographic pattern of dispersal can be detected via contrasts in regional distributions of mitochondrial DNA (mtDNA) and Y-chromosome DNA. On a global scale, genetic analyses indicate equal movement of the sexes, as might be expected if the globe has been peopled by migrating groups consisting of both sexes[g] [832]. On a smaller geographic scale, however, various studies show the genetic consequences of longer-distance movements of males by comparison to females, such as obvious presence of European Y-chromosome markers in Amerindian populations and of western Asian Y markers in eastern Europe [832]. These Y chromosomes would correlate on the one hand with emigration to America of more European males than females initially, and on the other hand with the Mongol invasions.

Differential assimilation of the sexes by marriage or mating between neighboring societies is also evident from genetic studies. For instance, we can sometimes see more mtDNA of hunter-gatherers in neighboring pastoralist and agricultural Bantu populations than vice versa [621; 832]. Similarly, the movement of native American Athapascan peoples into the U.S. Southwest is characterized by a strong presence of mtDNA from indigenous southwest populations in the Athapascan peoples' genome, but little Athapascan mtDNA or Y-chromosome DNA in the indigenous peoples' genome [479; 480]. In the case of the Athapascans, they were the hunter-gatherers, and the residents with whom they reproduced were agriculturalists [479]. Mixing between neighbors is not inevitable, however. The Bantu expansion into Zambia appears to have been associated with little interbreeding or marriage with the resident hunter-gather peoples [168].

Between communities within a people or culture, some bias toward females moving to husband's parents' site of residence is evident [487]. It is especially evident among nonforaging peoples, as opposed to hunter-gatherers, among whom less of a sex bias occurs [358; 487]. The contrast is probably tied to land ownership by individuals or families, which usually runs in the male line, and does not exist among hunter-gatherers.[h]

g. Global study. One study reviewed showed females moving farther than males, but it did not take the sexes' samples from the same populations [832].

h. Place of residence jargon. Patrilocal, virilocal = males stay, females move, i.e., females move to husband's parents' site of residence; matrilocal, uxorilocal = the opposite.

But now the geographical scale seems small enough for us to have left biogeography.

However, Per Hage and Jeff Marck argue that contrasts in apparent rate and origin of movement of peoples through the Pacific Islands— slow movement if the evidence is from mitochondrial DNA, fast if from Y-chromosome DNA—result from the fact that females tended to remain in their region of birth, while males were the sea-faring, dispersing sex [299]. Fiona Jordan et al.'s evolutionary (phylogenetic) analysis of the nature of Austronesian societies and their languages indicates the same societal influence on migration patterns [415]. In other words, the nature of the society at the local scale can correlate with its regional-scale biogeography. That being the case, and given that in most societies sex bias in local movement is evident, biogeographic inferences about peoples based on genetic indicators from only one sex should probably be treated as potentially incomplete [420].

OTHER SPECIES AS EVIDENCE OF HUMAN ORIGINS AND DISPERSAL

A theme of this book is that humans are far from being a unique species biogeographically. That being the case, the biogeography of other species that use roughly the same environment as humans should inform us about the biogeography of humans. Of particular use could be those species that use humans as their environment. Pathogens, parasites, and commensals are examples.

Origins and Dispersals of Other Mammals

If hominins and humans moved out of Africa, and we are not unique, we should find that other large mammals of the wooded savannah also moved out. They did. For instance, seven African mammalian genera (and the ostrich) appear temporarily in the Middle East at about the same time as humans first appear there, in the last interglacial at a little over 100 kya [443]. Similarly, from 3 to 1.5 mya, 16 mammalian genera moved (expanded) between eastern and southern Africa, some going south, others north [733]. Among both fossil and present-day large mammals, one-third of 170 species in eastern or southern Africa occur in both regions [779]. And nowadays, of 21 genera in southern Africa, 17 also occur in Somalia, and it is usually the same species in both [294].

The dispersal of these mammals strengthens the case for hominin dispersal. Indeed, the fact that during the Plio-Pleistocene several mammals moved from Asia to Africa hints that more hominin species than currently accepted might also have moved from Asia to Africa [172].

Origins and Dispersals of Commensals

Many species live in the habitat created by humans, such as agricultural fields and grain stores. Some live in human habitation. And some live on or in humans. As humans migrate, these commensals can be carried along accidentally, or deliberately in the case of species useful to humans.

Several species carried by humans, whether parasites, pathogens, or other forms of commensal, vary regionally. Darwin noted one example, head lice: people from different parts of the world have different head lice [166, part I, ch. 7].

The same is true of the gut bacterium *Helicobacter pylori* [212; 541]. Analysis of its genotype indicate the out-of-Africa episode, subsequent dispersal to New Guinea and Australia and then isolation there, and dispersal to Asia and thence to the Americas and the Pacific Islands. For instance, Asian genotypes of *H. pylori* were most similar to both Amerindian and New Zealand Maori genotypes. Indeed, the Polynesian and New Zealand genotypes are most similar to those from people of one part of Taiwan [541]. However, no samples came from eastern Indonesia, so that region cannot be distinguished from Taiwan as the final origin of particularly Polynesia (see section "Peopling of the Pacific").

H. pylori phylogenies provide clear evidence also of western African migrations to North America (presumably the slave trade in operation), as well as to southern and eastern Africa (presumably the Bantu expansion). And they indicate European migrations worldwide [212].

The Pacific rat (*Rattus exulans*) will live around human habitation and is native to southeastern Asia. It is now spread throughout the western Pacific. It cannot possibly have swum or rafted to reach such a large area. It had to have dispersed along with humans. The rat's mitochondrial DNA maps almost exactly onto proposed maps of human movement across the Pacific. The phylogeny shows clustering in southeastern Asia (Borneo, Sulawesi, Philippines), relatedness to Melanesia (and so presumably expansion there from southeastern Asia), and a branch leading first to eastern Micronesia and then a rapid expansion across Polynesia and to the outer, northern Micronesian islands [502; 503]. This scenario is substantiated by a small lizard, the moth skink (*Lipinia noctua*), which like the Pacific rat will live around humans. Its mtDNA shows relatively deep roots in Melanesia and Micronesia, as if humans spent some time there, and then an almost instantaneous diversification over Polynesia [25].

Rodents also feature in the recent peopling of Britain. House mouse (*Mus musculus*) mtDNA in northern and western Britain has clear similarities to the mtDNA of mice in Norway. The mice presumably indicate the Viking dispersals to Britain [692].

The commensals tell us not only about sites of origin and routes taken but also about dates. For instance, C^{14} dates from the earliest Pacific rat bones in New Zealand and from seeds gnawed by the rats show arrival of humans there only 700 years ago [264; 833]. In the case of the dingo, mentioned in the section on peopling of Sahul, the dating of the second arrival of humans in Australia came from genetic evidence indicating that the dingo of Australia separated from Asian dingoes about 5,000 years ago [680]. So genetically homogeneous is the dingo in Australia that, as for humans out of Africa, it must have originated from a very small population, probably with a single dispersal.

CONCLUSION

Dates of past events are continually being revised with new methods of dating, new information, new calculations. Take the so-called molecular clock. This clock is calibrated by taking a starting date firmly established in the paleontological or archeological record, counting the number of changes that have occurred in the DNA of the species or populations that separated near that date, and then applying the calculated rate of change to knowledge of the differences in the DNA of other species or populations that diverged over roughly the same period.

However, many factors affect the clock's calculated rate [16; 186; 202; 265; 365; 616; 724]. Generation time is one, as well as the fact that rate is different in different lineages and across different times. Some argue that so much variation in real and calculated rates might exist that few dates so far calculated from molecular clocks should be considered reliable [186; 202; 616]. A contrary and more comforting view is that because a variety of molecular analyses often combine to produce similar dates of splits among the mammalian, primate, hominoid, hominin, and human lineages, the molecular dating is robust [718]. Of course, the degree of comfort depends on the acceptable amount of disagreement among the calculated dates. Phillip Endicott and coworkers argue that the way forward is to make use of known differences in rates of change across hominin evolutionary branches, along with advanced statistical analyses that can incorporate phylogeny and uncertainty in the calibrated ages used to calculate the rates of change [202].

Dating with the molecular clock depends on well-established paleontological or archeological dates from which to start the clock. And yet not only do different currently accepted dates sometimes produce different results, but dates of events in the past are constantly being revised, whether the dates come from paleontology and the far past or archeology and the more recent past. Just this century, the age of strepsirrhines (lemurs, bushbabies) in the fossil record was doubled from 20 mya to nearly 40 mya [695]. As I write, it turns out that *Homo erectus* was at the famous Zhoukoudian site near Beijing over 200,000 years before the roughly 550 kya previously calculated [144; 701]. The age of the oldest Erectus remains in northern China is 1.7 mya, not 1.3 mya [861]. Less than a decade ago, the oldest *Homo sapiens* in Africa was redated from about 135 kya to 195 kya [509]. And so on.

In sum, we can expect change in our knowledge and understanding of human dispersal across the globe, i.e., of the evolutionary origins of "the distribution of the distinctive traits of mankind in present and past times," to quote Thomas Huxley [388]. Much of the change will be of timing of events. Less will be of the pattern, given the variety of evidence that agrees on the origins of the various peoples of the world. Nevertheless, different dates can revise ideas about sites of origin and routes of dispersal.

3

Climate, and Hominin Evolution and Dispersal

Climate appears to limit the range of many animals,
though . . . in many cases it is not the climate itself so much
as the change of vegetation consequent on climate which
produces the effect. [803, Wallace, 1876, vol. 1, p. 11]

There is little doubt that climate and environmental change
underlies much of evolution in general and hominid evolution
in particular. [231, Foley, 1999, p. 332]

SUMMARY

Climate affects the origin, extinction, and spread of species across the globe. The highly variable climate of Africa during hominin evolution provides plenty of climatic events to connect to evolutionary events. However, statistical verifications of any connections are rare, in part because of lack of precise data on regional climates, and in part because of lags in evolutionary response.

Whether climate drove humans out of Africa is an open question. Once out, modern humans expanded east fast along roughly the same latitudes as their place of origin, probably eastern Africa. Some have suggested that we could not have expanded earlier through southern Asia because of aridity associated with global cooling.

Climate certainly limited modern humans' movements north. Thus, it might not have been until the end of the last ice age, around 15,000 years ago (kya), that we managed to move into northern Europe and the far northeast of Siberia. From there we moved into the Americas, probably initially by a coastal route. If we got to Siberia before 15 kya, the Beringian ice sheets of the last age probably prevented movement into the Americas until about 15 kya.

Evidence that climate might have driven hominin evolution includes the fact that a few of the climatic changes seemingly significant in hominin evolution correlate with more or less simultaneous changes in the form and number of not only hominins but also the form and number of several other taxa too, particularly bovids [171; 228, ch. 5, 9; 795]. The significance of bovids is that they flourish in the grasslands and wooded grasslands that replaced forest in Africa during glacials and which were the habitat for hominins, especially hominins that were consuming the bovids.

Thus, major increases in the number of bovids occurred around 4 mya (at the time australopiths were appearing), 3–2.5 mya (at the time *Homo* was appearing), and 2 mya (origins of *H. erectus/ergaster*). A problem with arguments connecting the bovid and hominid events is that a burst of bovids occurred also around 1.3 mya, when nothing obvious was happening in hominin evolution, except perhaps the disappearance of southern Africa's *Paranthropus robustus* [171; 795; 840]. If bovids indicate good habitat for hominins, why would a burst of bovids be associated with an extinction of a hominin? One answer is that *Paranthropus* used more of the disappearing, closed habitat than did *Homo* (next section). However, *Paranthropus* also used grassland.

Plate tectonics and climate change are two main factors affecting the distribution of taxa across the globe in the long term, with the climate sometimes strongly influenced by the mountain-building consequences of the tectonics. Thus, it is no accident that the earth's geological eras are defined by the nature of the rocks and the flora and fauna in the rocks.

Perhaps the most famous such geological boundary is the Cretaceous–Tertiary, the K–T. The two eras are separated by an asteroid impact, decades-long global winter, major changes in global flora and fauna, demise of the dinosaurs and ammonites, and renewed diversification of mammals [63; 690; 691].

Hominids, hominins, and humans would be odd taxa if they had not responded to climate change, in both their evolution and distribution [93]. Demonstrating that they did so is another matter, however.

. . .

I have two provisos to the coming discussion of the relationship between climate and hominin evolution and dispersal.

First, knowledgeable as we are about the course of hominin and human evolution, nevertheless the paleontological record is still incomplete. Witness the post-2000 discoveries of two potentially new species of *Homo* [103; 431], the 2010 discovery of a new *Australopithecus* with more features shared with *Homo* than any other [58], and the 2011 publication of the finding of stone tools in the coastal eastern Arabia potentially indicating early modern humans there 125 kya, i.e., east of the Levant over 50 kya earlier than previously thought [22]. Consequently, any association of events between climate and hominin evolution and dispersal, even if statistically substantiated, could be negated by more data [45; 171; 220, ch. 7; 222; 823].

The paleoclimatic record for the tropics is also incomplete. For instance, Shackleton's detailed reconstruction of Pliocene and Pleistocene climates in Africa is based on oceanic sampling [121; 698]. The inferred climate must be an average for a huge neighboring area, with no possibility of estimating even large-scale variation within that area. Pollen sampling on land (e.g., in lake sediments) might enable estimation of more fine-scaled variation, but palynological samples are sparse, especially in the tropics, as are any other sort of record of the paleoenvironment [45; 76]. Even in western Europe, the paleoclimatic record is generally poor [3].

Second, so variable was the Pleistocene (and Pliocene) climate [45; 171; 183; 220, ch. 6; 443; 656; 698] that without strict statistical analysis, it might be easy to find some aspect of climate that could be inferred to be a cause of any significant event in hominin and human evolution and dispersal [538].

For instance, although Robert Foley found several statistically substantiated associations between events in hominin evolution, such as number of extinctions or number of species in any 500 ky period, he performed so many statistical tests that after correction for the number, almost no obviously significant associations remain [230; 231]. Another example is Peter deMenocal's suggestion that the times of change of periodicity of climate at around 2.6 million years ago (mya) and 1.8 mya were "key junctures in early hominid evolution" [171, p. 3]. However, I could not see in his data any obvious difference in speciation or extinction rates between these periods of change and the intervening periods.[a] Indeed, despite Robert Foley's quoted statement at the start of this section, he concluded that associations between climate or climate change and events in human evolution were weak at best [230; 231].

A recent report by the National Research Council epitomizes for me the state of the argument [149]. It suggests a connection between climate change and events in human evolution. Yet, not only does the report omit any statistical tests of any association, but a plot of its listed human

---

a. deMenocal's claim. Two suggested periods of change vs. others: seven speciations, three extinctions at 2.6 and 1.8 mya vs. six speciations, nine extinctions at other times. Data from [171, fig. 11].

evolutionary events per past time period against its listed climatic events shows a "J" shape, in other words, essentially no relationship.

Clive Finlayson disagreed, producing sophisticated statistical analyses to make the case [220, ch. 1]. He suggested, for example, that the more unstable climatic periods in the last 2 million years are statistically associated with fewer hominin species, just one species in the most variable periods [220, fig. 1.1]. However, several of the most stable climatic periods also have only one species [220, fig. 1.1].

Climate must have affected hominin evolution and dispersal, but showing that it did so, let alone how it did so, in any particular instance is going to be difficult, except perhaps where sea-level changes were involved. One problem with discerning climatic causes of hominin evolution and dispersals could be the lag between change of climate and evolutionary consequence [220, ch. 3, 7]. Whether that is a practical difficulty, though, will depend on the resolution of the analysis: an actual lag of several thousand years might appear in the record as an instantaneous response. Indeed, if resolution is poor, any detected response could be spurious, i.e., a change unrelated to the climatic event, or one related to another event.

Additionally, the frequency and severity of events could be important [220, ch. 3, 7]. A sequence of relatively poor periods can ultimately drive populations to extinction, because each affects a smaller population [220, ch. 3, 7]. If sequences of mild climatic events have to be analyzed as individual severe events, statistical verification of climatic causation becomes even more complex, especially as several of the Pleistocene's large changes in climate were not associated with any obvious change in the course of hominin evolution or in the geographic range of any of the hominin taxa.

Add to climate and its effects the many other factors affecting dispersal, such as the presence of barriers, and sometimes the best that can be achieved at present will be the exclusion through modeling of particularly unlikely scenarios [538].

Let me emphasize that with these comments, I am largely discussing the relation of climate to Pleistocene evolution and dispersal. Data for more recent times is, of course, more available, and hence change of climate can sometimes be more convincingly associated with change in technology and other aspects of culture [552].

CLIMATE AND HOMININ ORIGINS

Climate and Nonhuman Primates

We can see before our eyes now that climate affects the distribution of species. With global warming, several taxa in the northern temperate regions are contracting from the southern edge of their ranges and from lower elevations (altitudes) and extending their ranges northward and upward [547; 583; 808]. Those that cannot go farther north or up are experiencing a diminishing range, and some are on the path to extinction [547].

We are largely ignorant of what is happening currently with tropical taxa as the climate changes [583]. However, we know that in the past the largely tropical taxon of primates responded to climate change. As tropical climate expanded across Eurasia and North America during the Eocene (55–34 mya), the number of primate taxa increased, and their geographic range extended into what are now temperate regions. Then, as the world cooled through the Oligocene to the Pliocene and eventually Pleistocene, many of the primate taxa dependent on tropical forest environments went extinct (almost all the apes), and their ranges retreated back to equatorial regions [226; 397; 495].

In Africa specifically, the climatic variation of the five million years of the Pliocene and Pleistocene was accompanied by speciation and extinction of baboons. As expected from suggestions that environmental change stimulates evolution, the number of major climatic cycles per half-million-year period correlates closely with the number of species of two genera of baboons (Fig. 3.1) [229].

Correlations between Climate and Hominin Origins?

The exact ages given for the geological eras and sub-eras vary across the globe, as might be expected if climate influences the nature of the strata and associated flora and fauna that are used to determine the eras. The estimated ages also change over time, as more information comes in, and new inferences become necessary. The major geological boundaries in the period relevant to our evolution are the Miocene–Pliocene (c. 5.3 mya), the Pliocene–Pleistocene (c. 2 mya),[b] and the Pleistocene–Holocene (c. 10 kya).

b. Boundary ages. The Pliocene–Pleistocene boundary used to be about 1.8 mya; it is now dated to 2.6 mya [801].

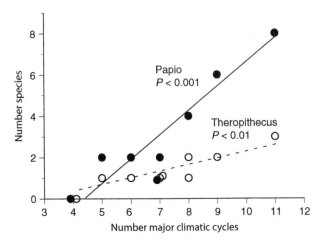

FIGURE 3.1. Climatic variation correlates with diversity of baboons.* Number species per 500,000-year period by number major climatic cycles in the period. Data from [229].

*Number species by climate change. Papio, $N = 10$, $r_s = 0.93$, $P < 0.001$; Theropithecus , $N = 10$, $r_s = 0.89$, $P < 0.01$; Spearman correlation tests, Bonferroni corrected for number of climatic variables tested. Regression line shown but not tested because small N. Data from [229].

In very brief summary of the climate over this period, the climate in Africa was relatively warm and wet between 6–3 mya. It then cooled for 2 million years, but with interglacials. The significance of glacials and interglacials is that global glacial periods are associated with extensive aridity across equatorial Africa and the replacement of forest by grassland; interglacials are wetter. Both interglacials and glacials became more extreme than previously over the 2-million-year interval, with c. 40,000-year intervals between. From about 1 mya, the glacials and interglacials became yet more extreme, and the periods between them lengthened to about 100,000 years. From 700 kya, the world experienced very regular extreme glacials and inter-glacials at 100,000-year intervals. The current peak interglacial has lasted an unusually long time, 10,000 years.

Some major events in hominin evolution happened near the geological boundaries, and at other times were associated with climatic change (Table 3.1). The lineages leading to chimpanzees and hominins separated between 7 and 5 mya [106; 135; 436; 792]. Three genera of African potential hominins appeared around 6 mya, namely, *Sahelanthropus*, *Orrorin*, and *Ardipithecus* [334; 425, ch. 4; 840]. And the first *Homo*, *H. habilis*, appeared maybe 2.3 mya [425, ch. 4; 840]. Because of the apparent associations of appearance of lineages with changing climate,

Taxa	Evolutionary event	Event date	Climate event	Climate date
Apes, hominins	Divergence	7–5 mya		
Ardipithecus	Origin	5.6 mya	? Mild interglacial Miocene-Pliocene	5.3 mya
Australopithecus	Origin	4.2 mya	? Mild glacial	4 mya
Australopithecus	More origins	4.1–2.5 mya	Cooling	3.9–3.3 mya
Australopithecus africanus, Paranthropus aethiopicus, A. garhi, A. rudolfensis, P. boisei, Homo habilis	More origins (up to 6)*	3–2 mya	Increasing variability	3–2 mya
Paranthropus	Origin	2.8 mya		
			Glacial	2.5 mya
			Pliocene-Pleistocene	2.5 mya
Homo habilis	Origin	2.3 mya		
Homo ergaster	Origin	2 mya	Mild interglacial	2 mya
A. rudolfensis, H. habilis, H. ergaster, P. robustus, P. boisei	Extinctions (up to 5)**	2–1 mya	Variable	2–1 mya
Homo erectus	Out of Africa	1.8 mya	Mild glacial	1.8 mya
			Pliocene-Pleistocene	1.8 mya
H. (denisova)			Highly variable	1 mya–0
			Interglacial	785 kya
H. erectus	In N. China	770 kya	Mild glacial	770 kya
			Glacial	720 kya
H. heidelbergensis	Origin	700 kya	Interglacial	700 kya
			Glacial	630 kya
			Interglacial	600–580 kya
	In Europe	500 kya	Mild interglacial	550–470 kya
H. neanderthalensis	Origin	500 kya	Mild interglacial	550–470 kya
			Glacial	470–410 kya
			Interglacial	400 k

(*continued*)

TABLE 3.1 (*continued*)

Taxa	Evolutionary event	Event date	Climate event	Climate date
			Glacial	350 kya
			Interglacial	330 kya
			Mild interglacial	300 kya
			Glacial	270–245 kya
			Mild glacial / interglacial	240–180 kya
H. sapiens	Origin	200 kya	End interglacial	200 kya
	Spread in Africa	200–100 kya	Glacial	180–130 kya
	Out of Africa 1	120 kya	Peak interglacial	120 kya
			Cooling to glacial	105–20 kya
	Gone from Levant	70 kya	Near peak mini-glacial	70 kya
	Out of Africa 2	55 kya	Mild interglacial	50 kya
			Peak glacial	20 kya
	Into Americas	15 kya	End peak glacial	15 kya
	Agriculture	10 kya	Interglacial	10 kya–0

Hominin events and dates from references in text. Climate events and dates from 785 kya from [443]; pre-785 kya from [76; 90; 698]. *H. floresiensis* not included, because dates not known.
* =species listed in order of origin.
** =species listed in order of extinction.

many have inferred an influence of climate on the path that hominin evolution took, at least in some times and places [76; 93; 171; 220; 238; 425, ch. 4; 698; 795–797].

However, the coincidence of era, climate, and evolution is not necessarily close. Humans and chimpanzees might have split nearly 2 million years before the Miocene–Pliocene boundary; the three genera of hominids that appear in the fossil record around 6 mya almost certainly evolved some time before the age of the current oldest discovered fossils; and *H. ergaster* appeared before the previous Pliocene–Pleistocene boundary of 1.8 mya, and well after the current date of 2.6 mya for the boundary.

Some other mismatches occur too. Why should the period from 3 to 2 mya, when the climate was variable and included an obvious ice age, be associated with a burst of origins of species (Table 3.1), but the next million years be apparently associated with a burst of extinctions, although the climate was also variable but not associated with any extreme climatic event (Table 3.1)? Would not extreme climate affect extinctions as much as originations? Why should climate variability in the last 5 million years be fairly obviously associated with number of baboon species (Fig. 3.1),

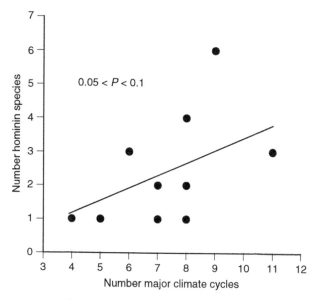

FIGURE 3.2. Climatic variation does not correlate with diversity of hominins.* Number hominin species per 50,000-year period by number major climatic cycles in the period. Data from [229].

*Homin evolution × climate variability. $r_s = 0.61$, $P > 0.05$; $N = 10$, Spearman correlation test, Bonferroni corrected for number of climatic variables tested. Data from [229]. Results do not change if *Australopithecus sediba* added [58].

but not so obviously with number of hominin species (Fig. 3.2) [229]? The answers are that we do not know, and therefore that we should be circumspect when ascribing evolutionary events to climate change.

Evidence that climate might have driven hominin evolution includes the fact that a few of the climatic changes seemingly significant in hominin evolution correlate with more or less simultaneous changes in the form and number of not only hominins but also the form and number of several other taxa too, particularly bovids [171; 228, ch. 5, 9; 795]. The significance of bovids is that they flourish in the grasslands and wooded grasslands that replaced forest in Africa during glacials and which were the habitat for hominins, especially hominins that were consuming the bovids.

Thus, major increases in the number of bovids occurred around 4 mya (at the time australopiths were appearing), 3–2.5 mya (at the time *Homo* was appearing), and 2 mya (origins of *H. erectus/ergaster*). A problem with arguments connecting the bovid and hominid events is that a burst of bovids occurred also around 1.3 mya, when nothing obvious was happening in hominin evolution, except perhaps the disappearance of southern Africa's *Paranthropus robustus* [171; 795; 840]. If bovids indicate good habitat for hominins, why would a burst of bovids be associated with an extinction of a hominin? One answer is that *Paranthropus* used more of the disappearing, closed habitat than did *Homo* (next section). However, *Paranthropus* also used grassland.

Homo *as a Generalist*

One of the suggestions to explain the success of *Homo* is that with tool use, our genus became more of a generalist and thus better able to cope with the increasingly variable environment [220, ch. 3, 5; 228, ch. 10; 601; 608; 609; 650; 818]. Bernard Wood and David Strait tested this idea with a comparison of *Paranthropus* and early *Homo* [842]. They found that *Paranthropus* was almost as much of a generalist as was early *Homo* and did not use the environment in an obviously different way than did early *Homo* [842].

A characteristic of temperate climates is that they are more variable across various time scales than are tropical climates [17; 469, ch. 6; 807]. If the argument that variable climates are associated with the evolution of variable hominins is true, then it should be the case that the more variable the climate, the more variable are the hominins. Some evidence from modern human hunter-gatherer societies matches the prediction. Temperate and arctic[c] hunter-gatherers have a more variable tool kit than do tropical hunter-gatherers (Fig. 3.3A, B) [773]. The increased variability is associated with the temperate and arctic peoples having a larger proportion of animals than plants in their diet than do tropical peoples (Fig. 3.3C) [62; 488; 773].

Hominins and Open Habitat

I started this introductory section by mentioning plate tectonics. Plate tectonics might seem more or less irrelevant to hominin evolution, because of the long time scale and huge geographic scale. However, eastern Africa was extremely active tectonically in one of the crucial eras of hominin evolution, the Pliocene [30; 584; 600]. The rifting and associ-

---

c. "Arctic" peoples. "Eskimo" used to be the term applied to arctic peoples, but while it is acceptable to some Alaskan arctic peoples, it is not, for example, to Canadian arctic peoples, who prefer "Inuit."

ated uplifts in eastern Africa could well have caused rain shadow zones east of the uplifted mountains, such as the Ruwenzoris, which would have influenced, even exacerbated, any aridification associated with global glaciation at the time [584; 600].

With aridification a major theme in the context of human evolution, it is necessary to know that during glacial periods, equatorial Africa often becomes drier and tropical forest retreats to be replaced by open woodland and savannah [45; 76; 93; 171; 795–797]. The same is true of at least southeastern Asia [86; 852]. It used to be thought that the Amazon dried during the last glaciation [298], but that idea is now disputed [145].

In Africa and elsewhere, many forest species, including primates, are limited by grassland. That limitation is one reason why so many fewer primate taxa live in temperate than in tropical regions, even if it is not a sufficient explanation [320]. More specifically as an example, chimpanzees are not found east of the savannah belt that runs down through central Uganda and east from western Tanzania, while other primate species are not found west of that savannah belt [294]. In both Africa and Asia, it looks as if formerly connected forests have been separated by aridity and consequent expansion of savannah or wooded savannah, with subsequent evolution of endemic forms in the now isolated forest blocks [86; 87; 294; 347].

Grassland and wooded grassland are argued to be good habitats for a tool-using, bipedal, running, meat-eating, scavenging, and eventually hunting hominin, and savannah might be especially good for *Homo* [85; 144; 220, ch. 3, 5; 221; 600]. The idea that savannah is good for meat-eating and scavenging and hunting hominins arises from a variety of lines of evidence. One is the apparent association of important phases in hominin evolution with aridity and the spread of grasslands in Africa and elsewhere, along with apparent adaptations for long-distance running [85] and, later, the development of refined and light-weight technology [85; 220, ch. 3, 5, 8; 600; 797]. Another is the nature of the associated mammalian fauna [635]

Thus, the fossil record indicates that from *Ardipithecus*, through *Australopithecus* and *Paranthropus*, to *Homo*, the habitat changed from woodland and grassy woodland,[d] to woodland and savannah, to savannah [45; 220, ch. 3, 5; 221; 425, ch. 4; 635; 719; 824; 825].

---

d. A technical comment in *Science* (28 May 2010) by Cerling et al. suggested more open habitat for *Ardipithecus*, but White et al. in response suggested an error in the Cerling et al. assessment [825].

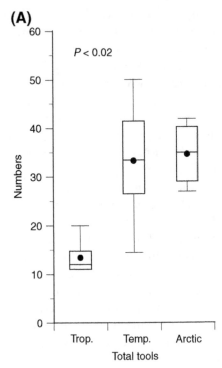

(A)

P < 0.02

FIGURE 3.3. Arctic and temperate hunter-gatherers have a more variable tool kit than do tropical hunter-gatherers and eat more animal food.* Per latitudinal zone, (A) total number tools, (B) number plant tools (i.e., tools for gathering plants), weapons (hunting), traps, and (C) percent of animals ("mobile resources," i.e., including fish) in diet. Dot = mean, line = median, box = central 50% values, lines = 10%, 90% values. Data from [773].

*Tool kit and diet by latitude. Total tools, $\chi^2 = 8.1$, $P < 0.02$; plant tools, $\chi^2 = 6.4$, $P < 0.05$; weapons, $\chi^2 = 7.4$, $P < 0.03$; traps, $\chi^2 = 10$, $P < 0.01$; % animals, $\chi^2 = 8.1$, $P < 0.02$; $N = 5$ tropical, 12 temperate, 3 arctic peoples. Wilcoxon nonparametric tests. Data from [773]. For Binford's [62] larger sample for % animals in diet, $N = 23, 37, 1, \chi^2 = 31$, $P < 0.0001$.

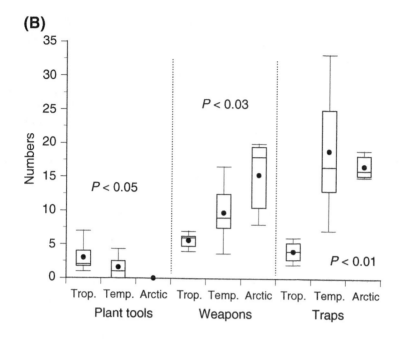

(B)

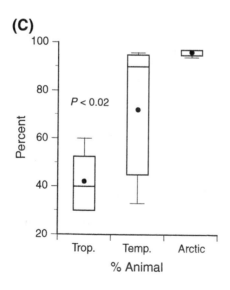

(C)

However, some evidence appears to contradict the assumption that humans liked or did especially well in open grassland or wooded grassland. For at least 250,000 years, hominins have inhabited rainforest [527]. The forest-dwelling pygmy peoples of Africa seem to do fine if left alone: they are still there, and their median density appears to be statistically the same as that of nonpygmy hunter-gatherers in Africa[e] [62]. Across the world, hunter-gatherers appear to live at about the same density in forest as they do in woodland and grassland[f] [62, table 5.10]. Whether they could do so without agriculture is debated, but it seems that they can [31]. And finally, bones do not even persist well in forests, let alone fossilize there, meaning that the density in forest of past populations of a species, including of hominins, could be seriously underestimated.

CLIMATE AND HOMININ DISPERSALS

If it is difficult to show that climate played a part in hominin evolution, what about the role of climate in our dispersal across the world? What was happening with the climate in Africa and elsewhere when hominins, as either *Homo ergaster* or *H. erectus*, first left Africa 2 mya? Can we tie climate to human (i.e., *H. sapiens*) movements out of Africa and across the globe, or even within Africa? The answer is that climate can be tied to some of the dispersal events, but that as with the linkage of climate change to hominin evolution, many discrepancies exist, and a lot more clarification is needed.

Out of Africa

The movement of Ergaster/Erectus out of Africa at perhaps a little after 2 mya (Table 3.1) was accompanied by the movement between Africa and the Middle East (or the Levant) of a number of mammals in both directions [778]. The implication is that something about the environment (drying and cooling and a spread of wooded savannah and grassland perhaps), instead of something intrinsic to hominins, allowed or

e. Pygmy vs. nonpygmy density: $N = 5, 8, z = 1.8, P > 0.1$, Wilcoxon nonparametric test. Data from [62, table 5.01].

f. Human hunter-gatherer population densities by habitat: testing separately in tropics and temperate regions (because densities decrease with latitude), no differences between the habitats, $N = 5–10$ depending on habitat classification, $P > 0.5$; $N < 5$ for tropics, so untested, Wilcoxon nonparametric tests. Data from [62, table 5.10].

stimulated the movement [228, ch. 10]. A peak in diversity of a variety of taxa at this time is also strong evidence of a general mammalian response to climatic change, and hence environmental change [228, ch. 9]. However, as with the timing of hominin evolutionary events in relation to climate, some data indicate quite an offset in the timing. Thus, Elisabeth Vrba has the appearance of a bovids peaking earlier, at around 2.5 mya, i.e., near the new data of the Pliocene–Pleistocene boundary [795; 796].

Heidelbergensis and Neanderthalensis apparently originated in interglacials, and Heidelbergensis moved into Europe from Africa during an interglacial (Table 3.1).

Modern humans' (i.e., *H. sapiens*') first movement out of Africa into the Middle East occurred near the start of the last peak interglacial, at about 120 kya, when northern Africa was wetter and the Sahara became grassland, even wooded grassland [21; 577; 578]. This relatively warm, wet period lasted from about 120 to 90 kya, even though the climate was cooling quite rapidly during it [607]. Here, the argument is that climate allowed movement by becoming more favorable.

By contrast, Shannon Carto et al. suggest that increasing aridity toward the end of the interglacial peak, and subsequently, forced humans out of Africa [121]. The problem with that argument, it seems to me, is that if humans expanded from eastern Africa, the aridity of northern Africa, especially the Sahara, and the Arabian peninsula would in fact have hindered their diaspora, as Carto et al. indeed suggest. Carto et al. solve the apparent discrepancy by suggesting that during the arid glacials and periglacials, humans dispersed along the expanded and more clement coastlines. Maybe, though the Levant on the east coast of the Mediterranean and the northern fringe of the Arabian peninsula behaves differently, becoming wetter, if colder, during glacials [245].

An addition to these climatic hypotheses is that intense volcanic activity in the East African Rift at around 125 kya, and the associated widespread ash blanketing and killing the vegetation, pushed humans out of eastern Africa in both directions, north (to the Middle East), and south [45].

So now we have hypotheses that allow both good and poor conditions to favor expansion out of Africa, with complexity introduced by regional variation in climate. There is nothing wrong with competing arguments, but the way forward is with arguments that are testable and which allow separation of hypotheses. I have not found much evidence of such proving of the hypotheses.

Despite the seemingly favorable conditions of the interglacial, the evidence until recently indicated that humans apparently did not get farther than the eastern Mediterranean, or the Levant as the region is also termed [576, ch. 1]. Topography could have been one barrier to movement out of the region [778]. Northern movement could have been prevented by a mountainous barrier, the Anatolian plateau, stretching from the Taurus (or Toros) mountains on the northeast corner of the Mediterranean across eastern Turkey to especially the Caucasus chain between the Black Sea and the Caspian Sea, which even now is glaciated. And the Zagros Mountains of present-day western Iran could have prevented eastern movement.

However, if hominins were at 2,300 m in Ethiopia 1.5 mya [366, ch. 2], why could humans not have crossed the Anatolian plateau, half as high? Additionally, the coastal route east subsequently used (chapter 2) would presumably have been available, as would a coastal route around the Mediterranean. And indeed, a recent find indicates that humans got as far as at least eastern Arabia [22].

The cooling of the northern hemisphere stopped from about 90 to 80 kya [443; 700]. Indeed, Greenland ice cores and other evidence indicate that the climate around 80 kya was briefly nearly as warm as it is nowadays [183; 607; 700]. Then the cooling resumed, and around 75 kya, increased aridity in the Levant coincided with signs of modern humans disappearing from there [183; 699; 700]. However, if the cooling and drying was associated with the spread of grassland [699; 700], why did humans suffer if we were a species that did well in grassland (above)?

With the disappearance of humans from the eastern Mediterranean, Neanderthals entered the region again [700]. It seems likely that they came from refugia in Europe or from western Asia [220, ch. 3; 699; 700]. In western Asia, they had reached southern Siberia by this time, and maybe earlier [220, ch. 3; 272].

These later Neanderthals were technologically more advanced than the earlier ones, using spears with stone flake tips, for example [699]. They were also potentially morphologically better adapted to the cold than were the early modern humans (chapter 5). For instance, Neanderthals have tibia/femur ratios close to those of present-day arctic peoples, whereas early modern humans of the eastern Mediterranean had African ratios [374; 736; 775]. Clive Finlayson argues quite strongly that the apparent adaptations of Neanderthals to cold are in fact adaptations for strength [220, ch. 4]. I address his argument in chapter 5, where I consider the general topic of human adaptations to the physical environment. Here, I will simply say that adaptations for strength need not contradict adaptations to resist cold.

Some have suggested that the eruption of Mt. Toba on Sumatra 74–71 kya, one of the larger volcanic eruptions known, caused the resumption of cooling from the peak interglacial [12; 577; 628]. However, recent evidence indicates that Mt. Toba caused less of a drop in temperature than thought, and for far less long [767]. In any case, humans apparently disappeared from the Middle East before Toba's eruption. Also, immediately afterward the climate warmed, cooled, and warmed again, with apparently no obvious massive volcanoes as a cause [183; 607]. So as with events in human evolution, a highly variable climate allows individual events—in this case, a volcano—to be linked with hominin or human history but prevents confirmation of any link.

In contrast to these various indications of cooling in the Levant, some describe the Levant climate as becoming warmer and wetter roughly 70 kya, an inference supported by a rise in sea level of the Arabian Sea and Dead Sea over a period of 2,000 years or so at this time [245; 607; 699; 700]. However, some detailed plots of chemical proxies for temperature show more a stabilization of temperature than warming in the region [700].

Further apparent evidence for amelioration of the climate is that DNA evidence indicates that human populations were increasing in density and expanding across Africa at this time [238; 526]. But perhaps that expansion was in fact associated with aridity and the expansion of the grasslands that favored *Homo,* and especially modern humans.

For maybe 20,000 years around this time 70 kya, even if humans had moved out of Africa farther east than the Middle East/Levant, it is possible that they would have been stopped by cold, dry conditions across southern Asia [607]. If so, then we have an explanation of why although there are hints (stone tools) of early modern humans in southern India at perhaps 75 kya (chapter 2) [597], we have no good evidence for a strong presence of modern humans there until after 50 kya [607].

In any case, it looks as if it was not until about 55 kya (± 5,000–10,00 yr) that humans moved out of Africa again. During that time, the climate warmed slightly, producing the Karga interglacial, or Mousterian pluvial [216, ch. 3]. North Africa will once more have been relatively wet in this period. Could these conditions have allowed humans to have again traversed this previously arid region north out of eastern Africa? Currently, we do have sufficient data on local climates to answer the question, especially given local variations [215].

A little later, about 45 kya, the cooling to the last glacial maximum resumed. Some evidence indicates that the Middle East was drying then, with expansion of deserts [700]. Others have the region remaining wet from about 70 kya to the end of the last glaciation, roughly 18 kya [245]. Whichever was the case, Neanderthals were seemingly suffering [700]. Despite their more sophisticated technology by now, their geographic range was contracting [700]. The Neanderthals' increased use of hard-seeded plant foods around this time, i.e., foods more difficult to process, also indicates that they were experiencing more difficult conditions [700]. However, the change in size of seed might have been due to the plant assemblage responding to the changing environment.

So far, I have written as if either warming or cooling were the main climatic effects on Neanderthals and humans. Clive Finlayson suggests that for Neanderthals it was in fact the variability of climate that the Neanderthals could not cope with, even though the late Pleistocene, the period of their ascendancy, is characterized by climatic variability [220, ch. 3]. By contrast, the later humans had evolved to cope with cold, aridity, and variability [700]. In the Levant, levels of the Dead Sea indeed indicate greater variability of climate 70–18 kya than in the preceding 70 kya [245].

Having presented evidence that climate could have played a role in the dispersal of hominins, especially humans, into and out of the Middle East, I have to close with a suggestion that climate was irrelevant to human movements at the time. John Shea has pointed out that the environmental changes in the eastern Mediterranean 100–50 kya were rather minor by comparison to the range of environments in which both Neanderthals and modern humans lived, and John Hoffecker emphasizes the range of climates over which Neanderthals persisted [366, ch. 4; 699]. But if climate was not the stimulus for the movements, or the limits to movements, what was, especially for retreats?

A main alternative to the idea that climate allowed and constrained the movement of hominins out of Africa is that improved technology allowed expansions more or less independently of climate. Thus, some of the early (c. 100 kya) arrivals in eastern Mediterranean produced more variable, broader, thinner flakes than Neanderthals had, and they could have had more complex cultural behaviors and cooperative abilities, if some currently not well-supported interpretations of ceremonial burials turn out to be substantiated [220, ch. 5; 699].

The idea that better technology and more cooperation allowed expansion is especially strong for the later movement out of Africa, the one

I present as occurring around 55 kya but which could have been earlier or later (chapter 2). Increasing evidence from both southern and northern Africa indicates considerable cultural and technological advances at around 70 kya, even as early as 80 kya [79; 399; 526; 550]. Indeed, the more refined aspects of Africa's Middle Stone Age revolution matches some of the technology of the European Upper Paleolithic revolution over 20,000 years later [526].

The sophistication is seen in an increased variety of small blades, tools for shaping wood and bone, and decorative art in the form of strung shells and scratches on bone and lumps of ochre [425, ch. 7]. Evidence for advanced systems of social cooperation include archeological signs of trading, such as marine shells far inland [12; 220, ch. 5; 424; 526; 699]. DNA evidence indicates the technology and cooperation were in turn associated with a local increase in population numbers and population expansion within Africa (next section).

Despite these technological and social advances, it was another 10,000 years or more before humans began their sweep out of Africa and across the rest of the world. And having left Africa about 55 kya, humans then waited another 10,000 years to move farther north. As described in chapter 2, they probably did so by expanding out of west-central Asia, not by going north out of the Middle East.

At this time of increasing cold as the last ice age approached, perhaps only further advances in technology and social cooperation allowed humans to move into the colder temperate areas that without the technology might have been too inhospitable [220, ch. 5; 366; 369, ch. 1, 3]. Once again, evidence for this proposition is the greater diversity of the tools of hunter-gatherers at higher latitudes (Fig. 3.3) [773].

But even in historic times, with of course yet more advanced technology, we know that human movements have been and are determined by climate. Southern Greenland was abandoned by Scandinavian immigrants during the Little Ice Age of the early part of the last millennium [38; 207; 220, ch. 7; 586]. The disappearance of the Anasazi peoples from northern Arizona due to drought in the 13th century is famous [56]. Drought has also been implicated in the collapse of the Mayan civilization around the turn of the 11th century [339]. And there might be a few still alive who remember the 1930s drought that caused so many to flee the Dust Bowl region of central United States [198].

Another suggestion for the 10,000-to-20,000-year delay from 70 to about 60 or 50 kya in modern humans' move back into the Mediterranean from Africa is that the Neanderthals prevented their return. I will come back to this possibility in chapter 8, where I consider the influence of hominin populations on each other's presence, distribution, and numbers.

Dispersals within Africa

Hominins and humans dispersed within Africa also (chapter 2). Sometimes climate seems to have been associated with the dispersals; sometimes not [231]. Without more paleontological data (including paleoclimatic data) and stringent statistics, I am not sure that it is currently worth trying to tie movements within Africa closely to climate. Nevertheless, I will make a comment or two.

The two moves out of Africa at roughly 120 kya and 55 kya have been associated with both expansions and contractions of the human population in eastern Africa at those times [45; 238]. As said, DNA evidence indicates population expansion and local increases in density around 80–60 kya [238; 526], but then the approaching ice age was associated with increased aridity in Africa and, presumably, a declining population [423].

A population expansion in Africa along with local increase in population density between 80 and 60 kya presumably conflicts with Stanley Ambrose's suggestion of a population bottleneck at about 70 kya, coinciding with the Toba winter [12]. Ambrose used genetic evidence to substantiate his idea of an environmentally caused bottleneck. However, as suggested in chapter 2, the genetic evidence could equally well indicate only a small founding subpopulation of a far larger total population, with any bottleneck occurring only after the emigration [13; 238; 770]. On the other hand, maybe the expansion was a response to the Toba winter, as populations moved in search of better conditions. In any case, the DNA evidence indicates that the expansion within Africa around 70 kya was from the same region (roughly Ethiopia) that produced the global expansion [238].

Dispersal across Eurasia

As described in chapter 2, it looks as if the initial expansion of humans out of Africa kept going east along essentially a southern coastal route all the way to eastern Asia [442; 443]. In effect, these dispersers stayed in roughly the latitude from which they came, and hence in the sort of ecosystem (including its range of prey) to which they had adapted in Africa [220, ch. 3]. If so, humans did what immigrant mammals and birds do, namely, move into and settle in the same latitudes as in their native continent [681]. As said (above), humans could have been prevented from this eastern movement beforehand by the cold dry climate of the region during the 20,000 years around 70 kya. By about 45 kya, however, southeastern Asia and Australia were seemingly a clement warm and wet [607].

At the same time, the northern hemisphere was rapidly approaching glacial maximum. Movement to the north in Eurasia, anywhere above about 40 degrees latitude (i.e., north of the Mediterranean or of the southern end of the Caspian, or central China) probably had to wait until cold-adapted morphology (chapter 5) and technology had developed [220, ch. 3; 272; 369, ch. 1, 3; 425, ch. 7; 650]. It is also possible that the expansion was helped by replacement of the previous grassland by shrubs and trees, which allowed the use of fire [369, ch. 1].

It is true that Neanderthal with a more primitive technology was in southern Siberia by about 70 kya, [272], but Neanderthal might have had a morphology that allowed persistence in relatively cold climates (chapter 5). Earlier, other hominins with an even more primitive technology reached as far north, or even farther north, in both Europe and eastern Asia (chapter 2). However, they did so when the climate was warmer than at the time of humans' expansion out of Africa [3; 861].

Thus, it was not until about 30–25 kya that humans were in northern Siberia, if perhaps only seasonally [273; 368; 369, ch. 1, 3; 604; 789]. Indeed, they might even have prospered then in southern Siberia as a consequence of expansion of steppe and associated high herbivore numbers [369, ch. 3; 789].

At the height of the last glacial maximum, at about 22 kya, humans disappeared from most of Siberia [238; 272; 369, ch. 3; 789]. It is not impossible, though, that a relict population persisted in the far northeast, because Beringia remained mostly ice-free during the last glacial [238; 273; 368].

If so, we would have an explanation of why the Beringians' DNA profile is different enough from elsewhere in eastern or northern Asia for geneticists to be able to identify northeastern Beringia as the site of origin of all Native Americans (chapter 2). A problem with this scenario is that the region was ice-free not because it was warm but because it was so arid [369, ch. 3]. The aridity combined with the cold reduces the likelihood that humans could have persisted in the area. However, perhaps a small population had already moved through Beringia and down south of the advancing ice sheets in America by 25 kya [238].

Whether a relict population remained or not, as soon as the climate began to warm, humans moved back into Beringia, probably from central Asia in the region of the Altai [273; 369, ch. 1, 3, 4; 789]. They must have been in Beringia by at least 15 kya (either as relicts or as immigrants) because 15–14 kya is the quite-well-established date of humans' presence in southern Chile (chapter 2). As yet, however, the earliest archeological signs of presence in northern Beringia date so far from about 14 kya [272; 273; 369, ch. 1, 3, 4; 789].

Subsequent dispersals across Eurasia are closely associated with the development of agriculture and with the continued warming of the Holocene [124, ch. 4; 125, ch. 8; 220, ch. 8]. Hence, for example, the origins of western Eurasian agriculture in the Middle East about 10,000 years ago and the slow spread of agriculturalists and their agriculture northwest to reach Britain and Scandinavia 4,000 km away about 6,000 years ago [84; 124, ch. 4; 297].

Into the Americas

Despite the severity of the last ice age, and considerable knowledge about it and about human movements at the time, debate still swirls around its influence on human movements from Siberia to the Americas and then through the Americas—as the discussion above about the occupation of Eurasian Beringia indicates.

As described in chapter 2, humans reached northern Siberia about 30 kya. Climatically, a date of 30 kya for the arrival of humans in the Americas is not impossible. As said, both western and eastern Beringia remained ice-free during the last glacial maximum, because the region was so dry [273; 368]. Soon after 30 kya, though, getting south from eastern Beringia would have involved traversing the very edge of an ice sheet for at least 1,600 km [273; 368], presumably an impossibility. However, evidence that populations of plants persisted through the last glacial maximum in northern British Columbia in regions formerly thought to have been covered by the Cordilleran ice sheet indicate that the extent of the ice might have been overestimated [490]. At the moment, though, the earliest reliable dates so far for human presence in North America is from a site in Alaska of about 14 kya [369, ch. 1, 4].

By maybe 17 kya, or 15 kya at the latest, the Cordilleran ice sheet (the western portion of the huge ice sheet covering most of northern North America) had finally retreated from the coast, so a coastal route to the south would then have been possible, especially because it would have been relatively warm and moist and demonstrably had substantial vegetal cover by at least 12 kya [273; 300; 368; 369, ch. 1, 7; 482].

Additionally, a narrow corridor between the Cordilleran ice sheet and the huge eastern Laurentide ice sheet had opened by 14–13 kya [273; 300; 368; 369, ch. 1, 7]. This corridor has been proposed as an additional route into North America. But it could not have been the route for the earliest migrants into the Americas, if they reached Chile by 15 kya, or even 12.5 kya (chapter 2). Later arrivals could have used the central corridor, though. It would have been a cold, dry, and windy route between two massive ice sheets for decades, even centuries, after its opening. Indeed, it probably would not have been traversable until about 11,500 years ago, because only then might it have supported enough biomass to support a small group of human migrants for the few weeks it might take to move from Beringia into southern North America [482].

Anthropology still debates the use of each route—the coastal route and the central corridor route—because neither route contains good evidence of passage of humans at the relevant dates [369, ch. 1, 7]. The coastal route is now under 100 m of sea. Human remains in the central corridor at 11,500 years ago will not be informative, because by that time humans could have moved down the coast and then inland south of the ice sheets, from where they could move north up the corridor [369, ch. 7].

Peoples of the far north now are on the edge of humans' survival limits, with the result that at least in northern North America, but probably all around the Arctic, an imprint of climate on humans' distribution continues to be evident [224]—for example, in the expansion of agriculture now in southern Greenland.

CONCLUSION

Climate must have had an influence on the evolution of hominins and humans, as well as on their movement around the globe. However, substantial quantitative demonstration that it did so is minimal. We need to know a lot more about variation in time and place of past climate before we can be precise about how humans were affected or how they responded. Even in the far north, with ice sheets during the last glacial maximum as an almost certain barrier, much remains to be worked out.

4

Barriers to Movement

Barriers of any kind, or obstacles to free migration, are
related in a close and important manner to the differences
between the productions of various regions . . . on the
opposite sides of . . . mountain ranges . . . and sometimes even
of great rivers. [165, Darwin, 1859, ch. 11, p. 347]

The most striking fact presented by this order [Primates] . . .
is the strict limitation of well-marked families to definite
areas. [804, Wallace, 1876, vol. 2, p. 179]

SUMMARY

Oceans, rivers, and mountains limit the distribution of primates. They define the
distribution of humans too, even now within Europe, such that different languages are spoken either side of, for example, the Alps. It could be that mountains are more effective barriers in the tropics, because organisms either side of a
mountain are not as adapted to as wide a range of conditions as are temperate
organisms, and therefore the tropical organisms are less able to cross the mountain. In addition, in humans and other animals, xenophobia, or its opposite,
preference for the familiar, is an effective barrier. Xenophobia has negative connotations, but barriers to movement and mating maintain diversity in a region.

The "strict limitation of well-marked families to definite areas" (Buffon's Law in biogeography) can occur only if there is almost no movement of individuals between the regions. With even a little movement,
differences between individuals can disappear. Interbreeding blurs genetic and morphological variation, and cultural exchange homogenizes
us. Nevertheless, the genetic, morphological, and physiological differences that still exist between humans from different parts of the world

indicate the existence of barriers to movement of individuals, even if the global ubiquity of jeans and T-shirts shows that some aspects of culture might transcend all barriers.

BARRIERS LIMIT THE MOVEMENT OF PRIMATES

Barriers can exist before the appearance of a species in the area and prevent further dispersal, or they can emerge within a species' geographic range and split the range.[a] Plenty of evidence shows that both sorts of geographic barriers exist.

Oceans are, of course, a major barrier to primates; hence Wallace's observation quoted at the head of this chapter. It is not only oceans that separate species. Within continents, rivers are important obstacles. In all four continents that primates inhabit, major rivers border the distributions of several primate taxa. In Africa, we have the Congo River [293]. In South America, the Amazon is the most obvious barrier [28; 671; 678; 838]. In Asia, an example of a barrier river is the Mekong [517]. And in Madagascar, many rivers appear to be barriers [275], so much so that we seem to be getting close to every river valley in western Madagascar having its own species of mouse lemur (*Microcebus*) [539].

Southeastern Asia is a particularly interesting example of seas separating primates, including humans. The region perfectly demonstrates a so-called vicariant event, a barrier appearing to split a formerly continuous distribution. At the peak of the last ice age (c. 20 kya), the sea level was about 120 m lower than today, and a huge expanse of now partly submerged land was exposed [347; 794]. Called the Sunda Shelf, it comprises what is currently the Malaysian peninsula and the islands of Sumatra, Java, and Borneo, as well as many other nearby smaller islands. The subsequent rise in sea level flooded the land between all of these, and by roughly 8,000 years ago, they had their current form [147]. The ocean is such a complete barrier that these separate islands are evolving their own forms of primates [88; 292].

In the subdiscipline of island biogeography, the topic of part of chapter 6, two main factors influence the number of species found on islands. One is the distance from potential sources of species. The closer an island is to a larger land mass, the more species it has. In the case of the

---

a. Biogeographic jargon. Splitting of geographic ranges is termed "vicariance." The word comes from the Latin, *vicarius*, meaning a substitute (e.g., the Pope is Christ's vicar on earth). The connection to split ranges is so nebulous that biogeography should change the jargon.

Sunda Shelf islands, the near impossibility of primates crossing water has meant that the number of species on an island bears no relation to how far the island is from potential sources of immigrants (chapter 6) [307; 345; 516; 561].

The ocean is merely an extreme form of adverse habitat. Savannah is almost as much a barrier to forest species, although primatologists are continually surprised at how far from forest they sometimes see animals that they thought were confined to forest. Nevertheless, the distribution of about 80% of primate species is quite closely determined by the distribution of tropical forest [320].

Darwin mentioned mountains as barriers (quote at head of chapter). So too did Wallace [803, ch. 1]. Dan Janzen made the interesting suggestion that a mountain could be a more effective barrier in the tropics than one of the same height at high latitudes [402]. His reasoning was that tropical animals were not adapted to withstand very cold temperatures, whereas temperate animals were if they survived winters. Consequently, tropical animals would be stopped by lower elevations than would temperate taxa. In his phrase, "mountain passes are higher in the tropics." His idea has been shown to be correct, especially for ectotherms (e.g. insects, reptiles), but also for many endotherms (e.g., mammals, birds) [261; 506; 761].

BARRIERS LIMIT THE MOVEMENT OF HUMANS

In chapter 3, I reviewed studies that suggested that climate influenced the distribution of hominins, including humans. For instance, humans initially moved out of Africa east across southern Eurasia within a climatic zone more or less similar to the one from which they originated. Only many thousands of years later did they move into northern temperate areas, probably only after they had developed a suitably advanced technology to enable them to cope with arctic winters. Are humans, like nonhuman primates, stopped by barriers other than inclement climate and associated inclement habitat?

Geographic Barriers

For a long time, seas have been as obvious a barrier to the movement of humans as they are to the movement of nonhuman primates. After the first ocean crossing to Australia, which involved one stretch of at least 100 km of water (chapter 2), nobody else arrived for maybe 20,000 years [383]; it was not until about 2,000 years ago that humans first reached Madagascar, and they did so from the direction of prevailing winds and currents, i.e., across the Indian Ocean from Indonesia, not from Africa across the far narrower Mozambique Channel [109; 125, ch. 3; 175]; it was not until about 1,500 years ago that the central Pacific islands started to be occupied [125, ch. 7; 279; 530, ch. 8]; and it was not until about 1,000 years ago that Europeans reached North America, with Leif Ericson's arrival in Newfoundland via Iceland and Greenland [806].

Even waterways far narrower than oceans are effective barriers, it seems. Eighteen of the 33 genetic and language boundaries identified in Europe by Guido Barbujani and Robert Sokal are either side of the many small European seaways, such as the English Channel [37]. Nevertheless, movement across water is so easy for humans now that the number of languages spoken in island countries bears no relation to their distance from the nearest mainland[b] [277].

Four of the 33 European linguistic boundaries are aligned with mountain ranges (from west to east, the Cordillera Cantabrica of northwestern Spain, the western and northern Alps, the northern Carpathians between Romania and Ukraine) [37]. Additionally, the major mountain ranges of the Taurus and Zagros could have been barriers to the northern movement of early humans out of the Middle East (chapter 2).

Of course, people and their cultural artifacts can move without languages moving. Plot the movement of just genes and artifacts and we find that distance appears as a predictor of similarity and difference [756; 759]. We do not yet know whether humans show the Janzen effect of mountain passes being more efficacious barriers in the tropics than at high latitudes (above). However, the equivalent of more high mountain passes in the tropics would be fewer roads across the passes, fewer vehicles on the roads, or fewer opportunities to use the vehicles.

Some evidence that difficulty of travel inhibits movement of peoples is the finding that in Oxfordshire in England, as increasingly modern forms of transport appeared, so the distance between birthplaces of spouses increased from about 1 km in the 15th and 16th centuries to about 10 km in the late 18th century to about 15 km in the mid-19th century [540,

b. Number of languages on islands by isolation. With area, latitude, and isolation, i.e., distance to nearest source, in the model, only area and latitude significant: whole model, $N = 26$, r^2adj. = 0.4, $F = 9$, $P < 0.002$; partial correlations, area, latitude, $r = 0.59$; isolation, $r = 0.1$. Data from [277].

ch. 8]. In China, too, such marriage distances have increased over time [448]. In general, the more difficult it is to move by land or water, the more diverse are the cultures of a region, globally and within regions[c] [122].

Plenty of evidence indicates that movement is more difficult in tropical countries. For instance, European countries have several thousand times more km of road per km² than do sub-Saharan Africa countries [844]. In 2001, Canada had 570 vehicles per 1,000 people, while the 11 countries in sub-Saharan Africa for which data were available had a median of only 11 vehicles per 1,000 people [844].

Add the contrasts in wealth between the average citizen of tropical and temperate countries and its effect on their ability to make use of what transport there is, and the contrast between regions in opportunity to mix cultures might be even greater. I, a Brit, and my wife, a Californian, were born 15,560 km apart. We met after a total of four plane flights, summing to well over the annual GDP per capita of sub-Saharan Africa, $1,250, compared to the United States' annual GDP of $34,000 [844].

These contrasts between tropical and temperate regions in density of roads and vehicles and ability to use either correlate with contrasts in density of cultures (chapter 5) and the geographic extent of cultures (chapter 6), as expected if lack of easy communication is a barrier to movement.

So far in this section, I have written of only geographic barriers or their equivalent. However, other species can also be barriers (chapter 9). Obvious ones are pathogenic organisms, such as bacteria and parasites. Could pathogens to which individuals of a species are not adapted act as an effective barrier to movement? For humans, it looks as though they could. Yellow fever probably kept the French out of Panama, and hence prevented them building the Panama Canal [89; 515, ch. 6], and western and central Africa are not termed the "white man's grave" for no reason (chapter 9).

Xenophobia, Xenophilia (Cultural Barriers)

Darwin considered that environment could not explain similarities or differences between peoples from different areas [165, ch. 11]. His

---

c. Difficulty of movement. Cashdan [122] did not define ease of movement; she took the measure from another study.

observations, like Buffon's for animals, were that similar peoples could live in different environments ("the Esquimaux . . . do not differ in any extreme degree from the inhabitants of Southern China"), while different peoples could live in the same environment: ("inhabitants of tropical America are wholly different from the Negroes who . . . are exposed to a nearly similar climate, and follow nearly the same habits of life") [166, part I, p. 247].

Instead, Darwin suggested that a behavioral barrier caused regional differences. Almost all organisms prefer to mate with individuals similar to themselves, and they have physiological and behavioral mechanisms to prevent cross-species mating. Put plainly, animals, including humans, prefer to mate with individuals who look like those in the general population in which they grew up. A trait, such as the red hair of some northern Europeans, arises in a region by chance founder effects (chapter 2). Because of familiarity, the trait becomes preferred in potential mates. Once the preference is in place, mating between peoples from different regions becomes less likely, and so the two populations differentiate.

Luke Premo and Jean-Jacques Hublin used this mating preference hypothesis to suggest an explanation for the fact that founding population sizes of humans in the Pleistocene were far smaller than those for chimpanzees in the same period: humans, but not chimpanzees, they suggested, preferred to mate with those similar to themselves and therefore mated with only nearby mates [612].

Of course, the preference for similarity can go only so far. Family members are the most similar to one another, and yet incest (or self-fertilization in the case of plants) can be disadvantageous. A good example is a study of 16 species of zoo primates, which found that inbred ones suffered a 50% higher rate of mortality of young than did outbred ones [626]. When I wrote about preference for familiar traits in mates, I was careful to write about familiarity "in the general population." It turns out that among mammals, including humans, individuals develop an almost total lack of interest in mating with the individuals with which they grew up—in other words, with family members [66; 703].

Irrespective of preferences, if individuals cannot speak to one another, mating is less likely, and populations are more likely to remain separate. For instance, in the Solomon Islands, southeast of New Guinea, linguistic difference was a better predictor than was geographic distance of biological differences between the populations of the islands [187]. In other words, language more than the ocean was a barrier to movement. However, the study did not account for the direction of wind and current. Those might have been more relevant indicators than distance of how seriously ocean hindered movement between islands, at least in the past.

For Europe, some nice analysis indicates that language might even now prevent movement of individuals. The study of 33 linguistic boundaries in Europe that I have already mentioned found that between nine pairs of languages and associated genotypes no discernible geographic barrier existed [37]. An example was the boundary across Italy between Florence and Rome. In the absence of any geographic obstruction, maybe linguistic difference itself was the obstacle. Such language-induced xenophobia is reminiscent of bird and insects discriminating their own from other species as mates by the sounds they make [486].

An alternative to language as the barrier is that other cultural differences, including technological differences, perhaps combined with preferences for different environments, kept populations apart. This was Cavalli-Sforza's explanation for why there seems to have been little genetic mixing between the resident hunter-gatherers in Europe and the incoming agriculturalists [124, ch. 5].

Dislike or even revulsion at other nations' diets is commonplace. The English think the French are disgusting for eating snails and frogs legs; the French think the English bestial for eating maize; and Samuel Johnson's dictionary definition of oats is famous: "a grain, which in England is generally given to horses, but in Scotland supports the people" [409].

Katie Milton has suggested that we almost consciously use dietary differences to differentiate ourselves from others, sometimes going so far as to stigmatize others for being different [536]. This was her explanation for greater contrasts in diet between Amazonian peoples than could be explained by differences in the nature of their forest environment. Milton suggested that, along with aggression between groups, the xenophobia she detected could be a mechanism to space competitors and to partition use of forest resources [536].

Corey Fincher, Randy Thornhill, and Devaraj Aran produced a similar argument to explain why linguistic and religious diversity correlate with parasite diversity, including when a variety of other potential influences are accounted for, such as latitude and land area [218; 219; 764]. In brief, the authors argue, xenophobia keeps people of different cultures apart, the benefit of which is avoidance of parasites to which an individual is not adapted. Maybe, but the correlation could as easily be a side effect of xenophobia for other reasons (above) and consequent lack of movement of individuals (and hence of parasites).

In thinking about the mechanism for the correlation, it might be worth knowing that both the number of primate species and the number of parasites per primate species are highest near the equator (chapter 5) [311; 569]. Are primates avoiding mating with other species to avoid parasites, or to avoid genetic problems of hybridization? Both would cause a correlation between primate and parasite diversity, but avoidance of genetic problems is by far the more commonly argued adaptive reason for avoidance of hybridization between species. Alternatively, we avoid people of other regions as a means of partitioning resources, as Milton argued [536], with the unintended, but maybe beneficial, consequence that we avoid each other's parasites.

Competition between peoples is an important component of several hypotheses to explain the greater cultural diversity of the tropics (chapter 5; chapter 6). The corollary of competition and associated social attitudes as a barrier between cultures is cooperation between the cultures. Daniel Nettle argued a contrast between tropical and high-latitude areas in the use of cooperation for survival [557]. The benign tropical climate allows self-sufficiency in a small area, and therefore cultures separate into distinctive populations and languages, each in a small geographic range. By contrast, at high latitudes, survival is possible only with cooperation among a large number of people, and hence geographic ranges of languages are large, and therefore diversity of cultures is low.

CONCLUSION

Even now, with our global transport system, geographic barriers separate human cultures, as they separate species. People on either side of geographic barriers are often different. Certainly, mountains and seas are less of a barrier than they were, but still the English are a different sort of people than are the French. The English Channel, or *La Manche*, is no longer an effective geographic barrier (except when a little snow paralyses the trains in the Chunnel). Instead, the relative influence of the two countries' cultural differences achieves greater prominence as a barrier to integration, even if the English now have croissants for breakfast and the French enjoy *le sandwich*. Linguistic and other cultural ghettoes within cities are overt indications of the efficacy in human populations of culture as a barrier.

Unfortunately, the ease with which we distinguish "us" from "them" can correlate with discrimination against "them" on the basis of almost no physical or cultural difference at all [622]. On the other hand, if our evolved preference for the familiar and dislike of the strange maintains diversity, xenophobia perhaps has one desirable outcome.

PART TWO

Environmental Influences on Human Nature, Diversity, and Numbers

5

How Are We Adapted to Our Environment?

[Humans' regional] differences cannot be accounted for by the direct action of different conditions of life, even after exposure to them for an enormous period of time.

[166, Darwin, 1871, part I, p. 246]

SUMMARY

Several human traits give every impression of being evolved responses to the environment, so accounting for several regional differences between human populations. Dark and light skin, respectively, prevent damage from the sun and allows benefit from it. Long limbs allow heat loss in hot climates, whereas short limbs help retention of energy in cold climates. Peoples from different latitudes respond differently to extreme cold but not, seemingly, to extreme heat, perhaps because summer temperatures at mid-latitudes and exertion at high latitudes mean that temperate and arctic peoples can experience the same high temperatures as do tropical peoples.

Although, the temperatures of high elevation can match those of high latitude, most studies of high-elevation peoples have concerned oxygen deprivation rather than cold. People resident at high elevation perform better in the thin atmosphere than do new arrivals. While acclimatization closes the difference, it does not do so completely. The data come mostly from Andeans and Tibetans, who seem to have evolved different adaptations to oxygen deprivation, though equally efficacious ones.

Some differences between the sexes match those seen between latitudes, with females equivalent to high-latitude peoples. Females have shorter limbs for a given stature and more subcutaneous fat than males have. Consequently, females are more efficient at retaining energy and probably perform better in extreme cold than do males. The greater efficiency of females is likely related to the benefits of reserving energy for care of their offspring.

Human cultures occur at greater density in the tropics than at higher latitudes: humans, like other forms of life, are more diverse in the tropics. Several explanations exist for the latitudinal gradient in biological diversity (the Forster effect). Most can explain the Forster effect in human cultures, but some do not. Those that do not work for human cultures are also invalid for several non-human taxa too. Thus, anthropology substantiates both the valid arguments for the Forster effect and the invalidity of some of the other explanations.

Even so, the best explanations for the Forster effect might differ between nonhuman species (whose ranges overlap) and human cultures, which tend to be quite territorial. One of the simplest explanations for the Forster effect among nonhuman species is that whereas the tropical climate places no limits on form, high-latitude winters severely restrict what can survive. For humans, the lesser productivity of high latitudes forces cultures to use large ranges (more in chapter 6) and hence constrains the density of cultures.

Within latitudes (for instance, within the tropics), hotspots of both biodiversity and cultural diversity exist. Explanations for the Forster effect should also explain hotspots compared to coldspots. However, the match is not good. Scale effects could be in operation; i.e., mechanisms at one scale do not operate at another. In addition, analyses of environmental explanations for hotspots, based as they are on current environmental conditions, often lack information on the vital components of history of both climate and biogeographic distributions.

Humans are distributed very unevenly across the globe. Half of the world's human population occupies less than 3% of its land area. Correspondingly, only a few cultures live at high population density; most live at low density. In general, the distribution of densities of present-day humans does not show any cline with latitude or climate. Nevertheless, extreme high and low densities correlate with good and poor climate. We are at our highest regional densities in regions of clement climate and lowest at most adverse climatic extremes. The fact that our largest cities are usually coastal indicates the importance of trade to our geographic distribution.

Hunter-gatherer cultures, more closely dependent on the immediate environment, are at highest population densities in the tropics, and at ever lower densities at higher latitudes. This gradation of density fits an explanation for the decline of cultural diversity with increasing latitude, namely, that the higher productivity of the tropics allows populations of large enough size to be viable to live in smaller geographic ranges and hence to allow closer packing of cultures.

CLIMATE AFFECTS FORM AND FUNCTION

Darwin did not consider that the conditions of life could produce the variation that he knew existed between peoples from different parts of the world (opening quote above). He suggested that instead the variation had to do with choice of mates: people from different parts of the world evolved by chance to prefer, for example, mates of one skin or hair color rather than another [166, part I, p. 250].

We now know a lot about correlates between the nature of the environment and the form and function of vertebrates [102; 514; 684; 820].

But as intimated in chapter 1, for various aspects of study of the relationship, it is humans on which we have the most data. It turns out that on this issue of adaptation to the environment, Darwin was wrong. Several human traits give every impression of being evolved responses to the "direct action of different conditions of life," so accounting for several regional differences between human populations.

Two main conditions of life that correlate with human variation are latitude and elevation (altitude). The conditions of life change with both. The sun at midday is less intense at high latitudes than at the equator. High latitudes and elevations are colder and often drier than their opposite. High latitudes are more obviously seasonal than the tropics. And so on.

Much of humans' ability to persist over so much of the earth is due to our plasticity, our adaptability, our technology, or in other words, our brains. Intelligence effectively homogenizes the environment. At the same time, it is clear that several biological traits differ between peoples whose ancestors lived in different regions of the world, and we know that the nature of some of the traits is not determined by the chance of genetic drift.

Latitude, Climate, and Color

The distribution of the variously coloured races, most of whom must have long inhabited their present homes, does not coincide with corresponding differences of climate. [166, Darwin, 1871, part I, p. 242]

Animal Colors. It is a truism that human skin and hair color vary enormously and that we pay a lot of attention to the variation. There might be some evolutionary background to that attention. Primates have the most colorful coats of any mammals. Is there another order of mammals in which a species is charcoal grey, chocolate, yellow, red, and white (Diana monkey [*Cercopithecus diana*]), or pale grey, white, black, and russet (red-shanked douc langur [*Pygathrix nemaeus*]), or whose infants are bright orange but the adults black (ebony langur [*Trachypithecus auratus*]), or the infants pure white but the adults black and white (various black and white colobus [*Colobus* sp.]) [81; 671]? Currently, explanations concerning variation across primate species in their coat colors are mostly to do with concealment, mate choice, and sharing of infants [81; 667]. Only now has climate been introduced as an explanation for contrasts between primate species in their colors [83].

Southern California's Anza Borrego State Park authorities advise that in this arid region, hikers should wear light-colored clothes in order to minimize heat stress. That seems sensible, although experiments show that Bedouin men stay the same temperature in the midday sun whether they are in black or white robes [704]. The offered explanation was that the higher surface temperature of the black robe induced greater air flow from the body outward. Why is the bright orange of ebony langur infants or the bright white of colobus infants (compared to the black of their parents) not explained in terms of heat stress?

Good evidence from ungulates shows that those with light-colored coats absorb less heat than do those with brown or black coats (Fig. 5.1) [217; 370]. That fact explains the correlation in Kenyan cattle herds between, on the one hand, the ratio of light and dark cattle and, on the other hand, elevation, potential evapotranspiration, and heat stress [217]. Light cattle do better, even survive better, under heat stress; dark cattle do better, even survive better, when it is colder and wetter. So why do humans tend to be darker in the tropics?

Constantin Gloger noticed nearly 200 years ago that animals, especially birds, tend to be darker in wetter environments [267; 463, ch. 15]. The relationship seems to work for some mammals too, including primates [83; 444]. Explanations for what came to be called the Gloger effect have not concentrated on heat stress. Instead, for birds, a suggestion is that the melanin of dark feathers resists bacteria better. A partially connected explanation is that the dark color indicates resistance to infection and is thus a signal of health to potential mates [670]. Alternatively, in more humid environments, which tend to be dark, dark animals would be better concealed than light-colored ones [463, ch. 15].

The Gloger effect has sometimes been taken to predict darker colors in the tropics, and indeed the tropical forms are darker in some mammalian taxa, again including primates, than temperate ones [83; 444; 731]. However, while rainfall tends overall to decrease with increasing latitude, some mid-temperate regions are just as wet as are the tropics, above 200 cm a year on average [469, ch. 1]. Nevertheless, explanations for the Gloger effect that postulate resistance to disease as the benefit of the dark color match the fact that parasites and pathogens of many sorts are more abundant and diverse in the tropics [569].

Latitude, Light, and Human Skin Color. When it comes to humans and skin color, explanations for variation in coat color in animals are largely ignored. That is partly because we so obviously tend to be darker, not

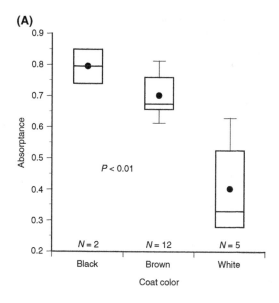

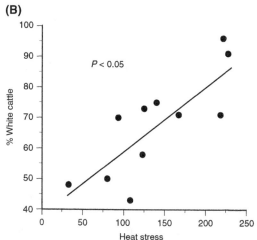

FIGURE 5.1. Coat color and temperature regulation. (A) Black absorbs heat. Absorptance by coat color. Data from 12 species of South African ungulate, with multiple measures on species with more than one obvious color, e.g., zebra.* Data from [370]. (B) Greater proportion of light-colored cattle where heat stress is higher, Kenya. Percent white cattle in herd by heat stress.** Data from [217].

*Absorptance by coat color. Four possible colors (min. of two if combine two browns and omit $N = 2$ black), $N = 6$, 6 brown; 5 white, $\chi^2 > 7.1$, $P < 0.03$, Wilcoxon tests. Data from [370].

**Color by heat stress. $N = 11$, $r_s = 0.84$, $P < 0.002$. All environmental parameters strongly correlated with one another (elevation, rainfall, evapotranspiration, radiation, heat stress); evapotranspiration (ET) correlated most strongly with % light cattle and heat stress (HS). Partial correlation of % light-colored cattle on ET and HS showed only ET significant ($N = 11$, $r = 0.69$, $P < 0.01$). Regression line for illustration only. Heat stress is a complex measure incorporating, among other factors, radiant heat, temperature, and long-wave reflectivity of the coat. Data from [217].

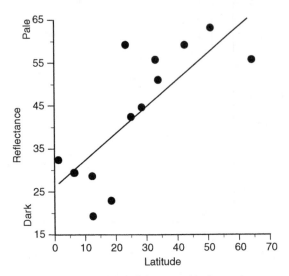

FIGURE 5.2. Human skin is lighter outside the tropics. Human skin color (reflectance) by latitude.* Data from [130; 396].

*Skin reflectance by latitude, $N = 14$, $r_s = 0.73$, $P < 0.005$, only sites $> 5°$ latitude apart, i.e., $> c. 500$ km apart. Data from [130; 396]. Polynomial fits indicating a limit to whiteness, or both whiteness and blackness, fit the data better (r^2adj = 0.64, 0.70, respectively, compared to linear fit of 0.59). However, too few data ($N < 20$) for any regression analysis.

lighter, where it is hotter (Fig. 5.2) [130; 396; 601]. Additionally, we are darkest outside of tropical forest [130; 396]. It seems that we need explanations for human skin-color variation other than those commonly offered for animal coloration.

Melanin is the main determinant of skin color. More melanin produces darker skin. Melanin appears to have two effects. It blocks light penetration, and it absorbs toxic compounds produced by photochemical interactions [396]. Consequently, dark skin protects against the damaging effect of UV radiation on folate (vitamin B_9) in subcutaneous blood vessels [396]. It also protects against sun-induced skin cancer as well as sunburn. Sunburn not only breaks the skin, but it can damage sweat glands and hence disrupt temperature regulation [396].

Protection against the damaging effects of sunlight, especially UV light, explains why people should be dark in the tropics and darker outside of forest. But which of the protective functions is the most important? Or are all the protections equally important?

Nina Jablonski and George Chaplin argue that the most important protection provided by dark skin (i.e., a high density of melanin) is the mitigation of folate damage [396]. Their first line of argument is the fact that folate is vital for the proper working of an extraordinarily wide array of bodily functions [396]. Many have to do with rapid division of cells, such as red blood cell formation. They also include several nerve functions, such as normal development of the neural tube in the embryo [396]. Second, the closest correlations between darkness of skin and UV incidence are at the UV wavelength that is the closest to absorption by oxyhemaglobin. Skin color correlates as closely with UV incidence as it does with latitude, indeed more closely at the absorption maximum for oxyhemaglobin [130; 396]. That correlation indicates that dark skin protects something in the blood against damage by UV light. And finally, folate is broken down by UV.

James Mielke and colleagues raise the possibility that at higher latitudes, where dark skin is not so much of an advantage, pale skin could be a result of genetic drift [530, ch. 11]. However, whereas pale skin is costly at low latitudes, dark skin can be costly at high latitudes, especially in the winter when not only is there little sun anyway but people are densely clothed. The cost arises because dark skin prevents people from benefiting from the fact that UV radiation stimulates the formation of previtamin D_3, a chemical vital for calcium absorption and hence bone formation throughout life, starting with the fetus [131; 396].

This cost of too little UV radiation is indicated by the higher incidence of rickets in northern European and American cities among people of African origin (15% vs. 2% among whites), female Muslims (who completely cover themselves when outside), and, for instance, immigrant Indians in the UK [396; 540, ch. 5]. Also, females are paler than are males in most populations, which could be explained by a greater need for vitamin D for their children, both while the child is a fetus and while it is being breast fed [131; 396].

These costs apply not just to clothed city dwellers [658] but also to rural people and scantily clad manual laborers [131]. It has been argued that so little exposure is needed, and that vitamin D (or precursors) are stored for so long, that a scantily clad African working outdoors at high altitudes would easily gain sufficient exposure [658]. That might be true on average over the year, but the problem is that for several months each year, such an African would have insufficient vitamin D [131].

At first sight, the fact that arctic peoples might be unusually dark for their latitude seems to contradict the vitamin D production hypothesis for light skin of high-latitude peoples [193, ch. 5]. The probable solution is that their diet of oily fish could provide more than enough vitamin D for adequate calcium absorption [193, ch. 5]. Thus, far from contradicting the hypothesis, the arctic peoples in fact support it.

Skin color is affected by a large number of genes [302]. Further evidence against genetic drift as the explanation for pale skin in northern peoples is the fact that the peoples of Europe and Asia have higher frequencies than do Africans of both the same and of different genes promoting pale skin [131; 302; 565]. The fact that a number of genes are involved and that different genes are involved makes it likely that individuals in the two populations evolved reduced pigmentation because of its advantages [131; 302].

In sum, not only was Darwin wrong about the lack of climatic correlates of skin color (accepting UV incidence as an aspect of climate), but it looks as if we know why he was wrong.

Latitude, Climate, and Size and Shape of the Body

For a century and a half now, we have been told that in many taxa the size and shape of the body varies with climate and latitude. Among birds and mammals, species and subspecies of higher latitudes and colder temperatures tend to be larger than relatives living in warmer climes [463, ch. 15]. Carl Bergmann was the first to articulate the phenomenon, over a century and a half ago [59]. Widely known as Bergmann's rule, I prefer to term it the Bergmann effect because not all comparisons show the relationship [520]. The higher-latitude taxa also commonly have shorter limbs and other extremities, such as ears. This latitudinal gradient is termed Allen's rule, after Joel Allen who described the relationship a little under a century and a half ago [7]. Allen's rule is also far from 100% applicable, and so again, I use the word "effect" instead of "rule." The Bergmann effect is about size; the Allen effect is about shape. Large, squat bodies retain heat; small, elongated ones lose it.

Derek Roberts, in one of the earlier and most detailed studies on human bodily proportions in relation to climate, pointed out that many problems exist with equating body proportions, especially stature and mass, to climate and thermoregulation [655, ch. 3]. Nutrition and disease are just two effects that could correlate with climate, and independently affect body size and proportions [655, ch. 3]. Before anyone argues that contrasts in size and shape are explained by thermoregulatory adaptations (e.g., that thin Africans are better adapted to shed heat than are squat Europeans), they would need to show that the Africans are no worse fed than are the Europeans.

Luckily for our ability to distinguish the morphological consequences or correlates of differences in nutrition from differences in temperature, nutrition affects some body proportions in ways opposite to those expected from Bergmann and Allen effects [673]. Additionally, some studies investigating one or the other of the Bergmann and Allen effects controlled for nutrition.

The Physics and Biology of the Bergmann and Allen Effects. Larger bodies lose heat slowly relative to small bodies because large bodies have less surface area for their volume, whatever the bodies are made of [82; 530, ch. 10; 673]. Of course with animals, other factors are involved such as fur and fat as insulation. Some have argued that these other influences, combined with alteration of blood flow, are so much more effective in controlling heat gain and loss that any change of size of extremities or body with latitude has little to do with thermoregulation [686].

More specifically, humans with a low skin surface area for their weight lose far less heat in humid conditions than do those with a high surface area to weight ratio: one fifth less surface area for mass correlates with four times less heat loss [241, ch. 3]. The relationship of size to heat loss might be particularly crucial for heat loss during physical work in tropical areas [655, ch. 1].

A biological consequence of the physical principle that ties size to heat exchange is that large animals can take in energy at a slower rate than can small animals or store (e.g., as fat) more of the energy that they take in; other things being equal, the large animals can therefore eat lower-quality food and can go for longer without feeding [70; 457; 514, ch. 5]. However, these apparent advantages in cold climates are in part offset by the fact that large bodies need absolutely more energy [214; 241, ch. 3; 512; 514, ch. 5].

Perhaps here is the place to point out that hominins and modern humans are very large primates (Fig. 5.3). Only one other extant species is larger than humans or any hominin ancestor. It is the gorilla (Fig. 5.3). In the past, though, a few other species, now extinct, were larger than the gorilla is now. These include *Gigantopithecus*, an ape; *Archaeoindris*, the largest of the lemurs; and maybe a hominin or two, such as *H. erectus* [226; 510; 712]. Independently of our cultural adaptations to climate, our body size alone means that we can survive in a relatively wide range of environmental conditions.

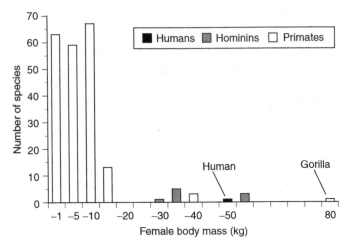

FIGURE 5.3. Hominins, humans are large primates. Body sizes of female primates. Data for extant nonhuman primates [712]; extant tropical humans [61]; hominins [510].

The explanation for the Allen effect is evident in common experience. When anyone is cold, they draw in their limbs. It is difficult to escape the physical fact that long thin bodies exchange more heat with the environment than do squat bodies of the same volume [530, ch. 10]. Perhaps because the physics is so obvious, the thermoregulatory explanation of the Allen effect in living bodies has not often been tested, it seems. But a recent study had skimpily clad volunteers sit in a cold room (22°C), and indeed those with longer legs lost more heat[a] [765].

Bergmann and Allen Effects in Animals, Including Primates. More animal studies have searched for the Bergmann effect than the Allen effect. Although Bergmann originally formulated his statement about body size and temperature as an interspecific effect and only suggested it should be seen within species [59; 400], most modern studies have tested the effect by examining variation within species. In one of the more recent more comprehensive studies, 75% of 94 bird species and 65% of 149 mammal species evinced the Bergmann effect [520]. Accepting that

---

a. Resting metabolic rate by total leg length, $N = 20$, $r^2 = 0.75$, $P < 0.001$, independently of mass of body [765]. Correlation is mainly due to thigh. As often the case [82], authors estimated heat loss by oxygen consumption, which increases with increased metabolic rate.

absence of relationships is underreported, the actual proportion of taxa showing the effect might be closer to 50% [514, ch. 5].

Latitude is a crude index of climate. Bergmann and Allen latitudinal effects are usually explained in terms of temperature and are often demonstrated by correlations with measures such as mean annual temperature. But the crucial factor could be far more precise [400]. For instance, in one test of the Allen effect, leg length among closely related gulls and terns varied with minimum summer temperature [566].

As usual in biology, only a minority of studies of the Bergmann effect are of tropical taxa [520]. Nevertheless, the presence of the Bergmann effect has been tested in humans' closest relatives, the tropical taxon of primates. It turns out that primates show the Bergmann effect. Primate taxa are on average heavier at higher latitudes (Fig. 5.4) [320]. The relationship occurs not because the high latitude taxa are larger than any low latitude taxa but because small-bodied taxa (including whole families) do not live at high latitudes (Fig. 5.4). The same triangular pattern of an array of body sizes is seen in the tropics, but only the larger sizes at high latitudes are seen within the wide-ranging cercopithecine family of Africa and Asia [320] and the genus *Papio* [190]. The genus *Macaca* shows the more normal gradual increase in body mass, such that high-latitude taxa are indeed larger than any low-latitude ones [320]. However, the very largest-bodied taxa, the great apes, are some of the most closely confined to the equator [320].

These findings match energetic explanations for the Bergmann effect [514, ch. 5]. In sum, there is little constraint on body mass in the tropics. But at high latitudes, small-bodied primates (< c. 5 kg) cannot take in energy fast enough in winter to compensate for rapid heat loss, especially as they have less energy stored as fat, and the very largest bodied primates (> c. 12 kg) cannot get enough food in total, again with winter being the likely crunch time [320]. In winter, one of the more northerly primate species, Japanese macaques (*Macaca fuscata*), are reduced to eating bark, and the high-latitude and high-elevation langurs (*Rhinopithecus bieti*) to eating lichen [1; 422].

No broad comparative study has yet been done on the Allen effect in primates. However, across eight species of macaque (*Macaca*), and within three of the species, higher-latitude populations had shorter tails relative to the length of their body (Fig. 5.5) [127; 233–235].

The Bergmann Effect in Humans. Humans show the Bergmann effect, despite our ability to regulate our own environment with fire, clothes, and

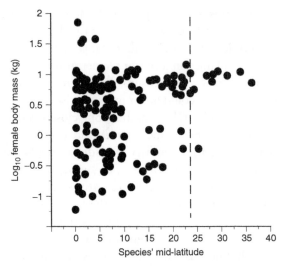

FIGURE 5.4. Primates show the Bergmann effect. Body size by species' latitude* (exc. Madagacar). Vertical dashed line = tropic limit. Data from [320].

*Bergmann effect, primates. $N = 164$, Spearman $r_s = 0.18$, $P < 0.03$, phylogeny accounted for [320]. Female body mass used to avoid confounding effects of sexual dimorphism, degree of which differs between continents [711]. Madagascar excluded because of recent extinctions of all largest species (ch. 10), plus fact that Madagascar does not reach the equator. Unusually small species outside tropics is *Leontopithecus*. Its range is the Atlantic forest of Brazil, which habitat is so reduced in extent that current mid-latitude of genus might not reflect historical mid-latitude.

shelters. Thus, in his analysis of body size of human males from all major continents,[b] Derek Roberts showed that body mass increased as mean annual temperature of the site where the person came from decreased. The relationship was true across the whole sample, as well as within regions[c] [655, ch. 1].

Christopher Ruff, among others, later confirmed the relationship for females as well as males and added the measure of mass/stature of each sex (Fig. 5.6) [673]. This mass-for-stature ratio increases with decreas-

b. Only males sampled, perhaps because in early to mid-1900s most anthropologists whose data Roberts used were males, who could courteously measure only males.

c. $N = 116$ males for mass; 285 for sitting height [655, ch. 1].

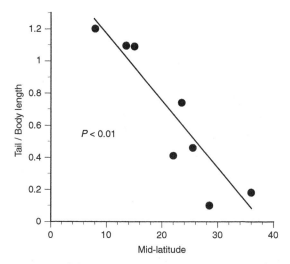

FIGURE 5.5. Primates show the Allen effect. Tail length/
body length by species' mid-latitude in macaques.* Data
from [127; 233; 234].

*Relative tail length by latitude. Spearman $r_s = 0.90$, $P < 0.01$, $N = 8$
species, 3–33 individuals per species. Data from [127; 233; 234],
with median value for four species repeated in sources.

ing mean annual temperature too, as expected if the latitudinal effect is
mainly driven by temperature and thermoregulation[d] [540, ch. 5].

Although we see the Bergmann effect despite clothing and other
external means of regulating temperature, a drop in body mass of humans, especially higher-latitude humans, that might be evident toward
the end of the Pleistocene indicates that perhaps human technology was
taking over from adaptive morphology as a means of controlling temperature [672].

In Binford's more restricted sample of hunter-gatherers, the same
relationships hold. Hunter-gatherers from the tropics weigh less, both
absolutely and for their stature, than do those from the higher latitudes[e]
[62]. Thus, the smallest peoples by both measures live within 5° of the
equator, while the heaviest are an Inuit group living close to 60°N.

d. Mass-for-stature by temperature: $N = 12$, $r^2 = 0.38$, $F = 7.7$, $P < 0.02$, Samoan outlier omitted; neither mass alone nor height varied significantly with temperature. Data
from [540, ch. 5].

e. Mass and mass-for-stature by latitude (only females analyzed): $N = 9, 8$; $r_s = 0.73$,
0.83; $P < 0.03, 0.02$. Data from [62].

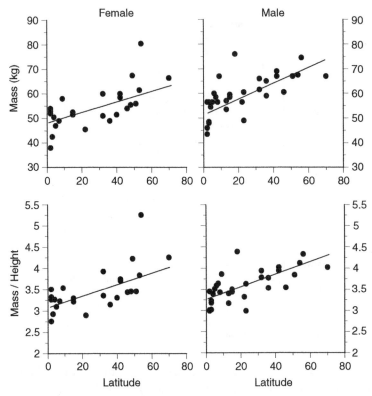

FIGURE 5.6. Humans show the Bergmann effect. Female, male body mass (kg) and mass for height (kg/cm) by latitude.* The data were collected from sub-Saharan Africa, Pacific and Australia, North Africa and Middle East, Asia, Europe, Arctic (including Aleutians). Regression lines calculated with outliers omitted. Data from [673].

*Bergmann effect in humans. Sexes separated, outliers and influential points excluded, and only one sample per country. Females, $N = 23$, R^2 adj > 0.40, $F > 16.1$, $P < 0.001$. Males, $N = 27$, R^2 adj > 0.48, $F > 26.0$, $P < 0.0001$. Data from [673]. Ruff [673] combined both sexes in his graphs and statistics, as did some others who used his data. Male and female anatomy of any one population is highly unlikely to be independent, meaning that his resultant statistical values, including any probability values, are invalid.

The smallest humans in the world are forest dwellers [125, ch. 3; 595]. (These peoples are often termed "pygmies." The word pygmy is not a genetic or regional descriptor; it simply applies to human populations whose members average under 150–155 cm). The point about small size in tropical forests is that the forests are not only warm but humid as well. Shedding heat, including by sweating (evaporative cooling), is therefore especially difficult. Thermoregulation, via a large surface area for body mass is thus one functional explanation for the forest inhabitants' very small body size [125, ch. 3; 595]. I come to another explanation later in this chapter.

It is not only adult humans who are larger at lower temperatures and higher latitudes. So also are newborns [655, ch. 3]. However, it is well known that size of newborn correlates well with size of mother, and therefore thermoregulation cannot be separated from nutrition as an explanation of the apparent Bergmann effect in newborns.

The Allen Effect in Humans. Is there anyone who has taught a course in biological anthropology that touches on human variation who has not been asked why so many of the world's top track athletes are of African origin? Humans show the classic Allen effect of relatively long extremities (i.e., arms and legs) for their body size at lower latitudes and warmer climates, and short extremities for their body size at higher latitudes and colder climates. Roberts' 30-year-old analysis showed that low-latitude men by comparison to those from high latitudes have long legs for their head–trunk length and long arm spans absolutely and relative to head–trunk length[f] [655, ch. 1]. The relationship is true within regions, as well as globally [655, ch. 1].

More recent analyses than Roberts' have usually confirmed these relationships (Fig. 5.7) [372; 673; 775]. However, authors have not always used statistical tests, or when they have, the results have not always been statistically significant. In Fig. 5.7, the tropical peoples with the same leg/trunk ratio as among arctic peoples are the indigenous peoples of the Kalahari desert—where night temperatures can drop to near freezing. The main difference in the sample as a whole is then between arctic peoples and all others, as opposed to a gradual drop in ratio with latitude (Fig. 5.7).

This arctic vs. other latitudes contrast might match the fact that summer temperatures at mid-latitudes reach tropical temperatures, but arctic ones never do. However, mid-latitude winter temperatures are closer to arctic ones than to tropical ones, and Trenton Holliday and Charles

---

f. The leg-to-head/trunk ratio that Roberts presented to argue long legs for length of body in tropical peoples was in fact "relative sitting height," i.e., sitting height/stature. Such a ratio can be interpreted either way (short trunk relative to leg, or long leg relative to trunk) but nevertheless is a common ratio used to argue Allen effects with limbs [e.g., 375].

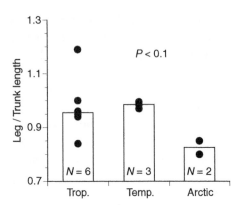

FIGURE 5.7. Humans show the Allen effect. Leg length/trunk length by latitudinal region.* Data from [673; 775].

*Leg/trunk by region. Only one sample per region (samples with smaller N omitted). Arctic vs. others, $N = 2, 9$; $z = 1.8$, $P < 0.08$; exclude Kalahari, $N = 2$, 8; $z = 2.0$; $P < 0.06$. Aleutian islands classified as arctic, because extreme high rainfall, hence low wet bulb temperature. Leg lengths from [673]; trunk lengths from [775].

Hilton argue in a detailed analysis that bodily proportions of arctic and temperate peoples are statistically indistinguishable, although both populations show the Allen effect by comparison to African populations [376]. Confirmation of the exact form of the pattern needs a larger sample, especially of temperate and arctic peoples.

Additionally, among high-latitude compared to low-latitude peoples, the distal part of both arms and legs is shortened by comparison to the near-body (proximal) part (Fig. 5.8). Both the tibia as a proportion of femur and radius as a proportion of humerus (crural and brachial index, respectively) are shorter at higher latitudes and lower temperatures and longer at lower latitudes and higher temperatures [372; 373; 673; 775]. Although the relationships of the indices with climate are clear, the difference between arctic and equatorial peoples is very slight, only about 5% (Fig. 5.8) [373; 775].

It is not obvious to me that the relationship of distal to proximal segments is predicted thermodynamically. Yes, distal parts of limbs are thinner and hence have greater surface area to volume, and in the case of the arms and legs of high-elevation peoples, the distal parts might be more likely to be in freezing water. However, the greater muscle mass of the upper parts means more heat is produced there, and one study indeed found that more heat in total was lost from the thigh than the lower leg [765]. Presumably, the mechanical benefits of powerful and hence longer proximal segments are opposing Allen effects.

While the Allen effect is usually discussed with reference to the extremities, the physics of it (less heat lost from shorter, wider shapes) applies to the trunk as well as to arms, legs, and tails. Nevertheless, although

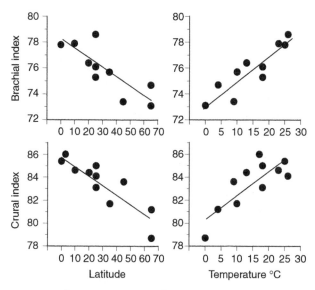

FIGURE 5.8. Relative length distal/proximal parts limbs change with latitude and temperature. Brachial (radius/humerus) and crural (tibia/femur) indices of females by latitude and temperature; all *P* < 0.05.* Data from [775].

*Allen effect. Brachial indices, *N* = 10; r_s > 0.8, *P* < 0.005; crural indices, *N* = 11; r_s > 0.7, *P* < 0.03 for the two environmental measures separately (and for males also). The relationship is stronger with temperature than with latitude, addition of the latter of which does not improve the fit. Data from [775]. Source provided estimates of mean annual temperature; I estimated latitude from the provided identities of the peoples cross-referenced to other sources, in particular [673].

humans indeed have wider trunks (measured by hip breadth) at higher latitudes (Fig. 5.9) [376; 673], including wider for length [376], they have longer instead of shorter trunks [376; 655, ch. 1], as is the case for macaques [233]. Perhaps we are seeing an interaction of the Bergmann and Allen effects here, with greater mass achieved by both elongating and broadening the trunk.

In general, for all of these measures of body parts, a ratio with another part of the body (trunk/stature, leg/trunk; brachial index) shows stronger correlations with latitude or temperature than do absolute measures of length (trunk, leg, humerus length), which often do not correlate at all. In other words, bodily proportions vary more closely with latitude and temperature than do absolute sizes [376; 655, ch. 2].

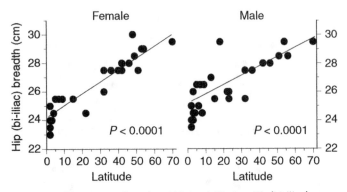

FIGURE 5.9. Humans are broader at higher latitudes. Hip (bi-iliac) breadth of individuals from different latitudes.* Data from [673].

*Hip breadth by latitude: $N = 24$ (female), 29 (male), excluding statistical outliers, $r^2 > 0.73$, $F > 79$, $P < 0.0001$ for each sex. Neither slope nor intercept differs between the sexes. Data from [673].

That should not be surprising. Although the heat that a body part exchanges with the environment is independent of the heat exchanged by another body part, the heat to be shed is supplied by other body parts. Energy to limbs comes from the trunk as well as from the limbs themselves. Heat to distal parts comes from proximal parts. In the case of the brachial and crural indices, mechanical factors presumably also play a part, as already indicated. Optimal leverage probably constrains the ratio of distal to proximal parts, which might be why the ratio varies so little with latitude (Fig. 5.9).

Mechanical ratios bring us back to the athleticism of peoples' of African origin. Could the longer leg, particularly the longer lower leg in relation to upper, provide more mechanical advantage and hence greater speed [511, ch. 5]? If it could, the question then would be whether it was speed or temperature regulation that was the main advantage causing different leg–trunk ratios at different latitudes. However, if more mechanical advantage translates into greater efficiency of locomotion, as well as greater speed, then it is the higher-latitude peoples who should have the longer legs, because both sexes move farther per day than do lower-latitude peoples[g] [488]. Endurance provided by energy stores is impor-

g. Daily distance by latitude. Females, males; $N = 6$, 8 societies; $r_s \geq 0.9$, $P < 0.01$. Data from F. Marlowe, unpublished [488].

tant too, and the heavier peoples of high latitude probably have more fat stores than do the smaller peoples of lower latitudes (next section).

Distinguishing Nutrition from Temperature as the Cause of Regional Differences in Size and Shape of Body. Better fed animals and people are heavier and taller than are less well-fed individuals. Everyone knows this. So are the Bergmann and Allen effects caused by differential nutrition, rather than different temperatures?

In the case of the brown bear (*Ursus arctos*), high-latitude and low-temperature populations are larger because of more and better food, rather than lower temperatures, as shown by the fact that distance to the nearest salmon spawning area was one of the best predictors of the average body size of different populations of bears [523]. Also, high-latitude raccoons (*Procyon lotor*) are proportionately fatter for their body mass than are low-latitude ones [514, ch. 14], as if better fed.

High-latitude peoples are larger than low-latitude peoples, and we know that arctic peoples eat more fat than do low-latitude peoples (chapter 7). Across mammal species, larger-bodied species have a greater proportion of body fat than do smaller species [685], indicating yet another thermoregulatory advantage to being large bodied. However, I have not found any information on whether the fat–body size relationship applies within species.

At first glance, nutrition might be a possible explanation for pygmy peoples' small size. All peoples currently classified as pygmies live in tropical forest, and some have argued that the tropical forest is such an impoverished environment that humans could not survive in it without agriculture [31]. That turns out not to be so, and in any case, most of the pygmy peoples use agriculture or the product of others' agriculture.

Also, if pygmies were small because of an unusually poor diet or high incidence of disease, they would grow slowly. But they do not. They grow at the same rate as any other population. Instead, pygmy peoples are small because they stop growing sooner than do others [531; 595; 802]. This finding is tied to another explanation for their small size, which is not related to either nutrition or thermoregulation.

It turns out that pygmies are likely to die at a younger age than are their nonpygmy neighbors (40% vs. 60% probability of surviving to 15 yr) [531; 802], perhaps because of a high incidence of parasites and disease in tropical forest [802]. In general, if an individual is likely to die early, it should evolve to reproduce early (i.e., before it is likely to die) [338]. One way to reproduce early is to cease early in life putting energy into growth and instead put the energy into reproduction. Pygmies seem to do precisely this, not only ceasing to grow earlier than their nonpygmy neighbors but starting to reproduce earlier too [531; 802]: pygmy women's age at peak probability of reproducing is 22 years compared to 30 years among nonpygmies [531].

A consistent additional functional explanation for the pygmy peoples' small stature has long been the advantage of small size when moving through the forest vegetation [595]. The hypothesis is largely untested. Of course, neither this hypothesis about movement nor the explanation in terms of life history tradeoffs contradicts a thermoregulatory benefit of small body size.

A common observation is that descendants of immigrants to the United States are often a lot taller than their ancestors. The rich American diet is the usual explanation. Thus, third-generation Japanese in Hawaii are taller than their grandparents [673], and Guatemalan Mayan children who grow up in the United States are taller than those who grow up in Guatemala [73]. For the Guatemala sample, we know that the difference is due to the longer legs of the better-fed U.S. children [73]. In general, the increase in stature with more food seems to result from longer legs more than longer trunks [673].

This effect of nutrition on limb length turns out to be very useful. It allows the effects of temperature and nutrition on body form to be separated. High elevations in the tropics can be as cold as high latitudes. A comparison of high- and low-elevation tropical peoples is therefore equivalent to a comparison of peoples from different latitudes, as far as temperature control is concerned. Sara Stinson and Roberto Frisancho measured bodily proportions of over 200 lowland (980 m, c. 25°C) and highland (4,150 m, c. 0°C) Quechua children and youths [729]. Importantly, the highland children, though living in a colder environment, were better fed—as indicated by the fact that they grew faster and were fatter (Fig. 5.10A) [242].

As expected from both the Bergmann effect and nutritional effects, the highland children were heavier than their lowland counterparts (Fig. 5.10C) [729]. However, whereas better feeding usually produces, as said, longer limbs, the arms of both sexes of highland children were shorter in proportion to stature than were the lowland children's arms (Fig. 5.10D) [729], in part because the trunk of highland children was longer (Fig. 5.10B) [729], as is found for people of higher latitudes (above).

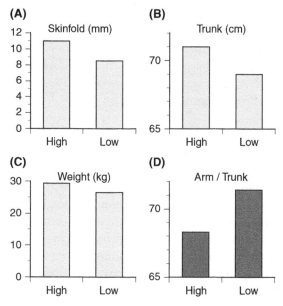

FIGURE 5.10. Highland children better nourished than lowland, yet they show Allen effect. Morphological measures of highland and lowland Andean Quechua female children: (A) Triceps skinfold thickness (mm),* (B) trunk length (cm), (C) mass (kg), and (D) arm length/ stature. Data from [242; 729].

*Skinfold. Highland and lowland males did not differ [242].

Similarly, the usual contrast in body shape between tropical peoples, especially Africans, and peoples from higher latitudes indicates the operation of the Allen effect—i.e., the effect of temperature, not nutrition. Africans are lightweight, thin, and have long legs for their trunk length by comparison to high-latitude peoples (Fig. 5.6–5.9). Plenty of evidence indicates that sub-Saharan Africans in general are undernourished by comparison to peoples from other regions[h] [845]. That impoverishment could well explain their lower weight, but it cannot explain their relatively long legs. That being so, the Africans' long legs are most likely then a result of thermoregulatory adaptation (the Allen effect).

In general, it seems that good nutrition produces both large body size and long limb proportions. Thermoregulatory effects, by contrast,

h. Africa undernourished. Sub-Saharan Africa, 29% of population; Asia, 15%; all other major global regions, < 10% [845].

produce large body size (Bergmann effect) but short limb proportions (Allen effect). The influences of nutrition and thermoregulation can therefore be separated. Wherever we see that one population is heavy by comparison to another but also has shorter limbs, nutrition alone cannot be the main explanation for the anatomical differences. Classic thermoregulatory effects are probably in play.

Other Apparent Anatomical Adaptations to Climate. Some have suggested that people of higher latitudes and colder climates have rounder heads than do those from lower latitudes, as might be expected from thermoregulatory principles [50]. However, as I described in chapter 2, genetic drift (change over time with no environmental influence) can lead to populations differing. The question should therefore not just be whether populations differ, but whether they differ more than expected by chance under neutral (nonadaptive) amounts of directions of change. John Relethford and Tim Weaver have argued that the data indicate that most regional differences in head shape are neutral [638; 639; 813]. However, Relethford recently detected greater deviation than expected in the head shape of two arctic populations, suggesting adaptation, in this case to a cold environment [642].

An anatomical feature that anthropologists commonly relate to climate is the shape of the nose. The functional explanation has to do with altering the temperature and especially humidity of incoming and outgoing air [655, ch. 4; 855]. Long, narrow noses warm and humidify better than do broad, short noses. The highest latitudes are characteristically not just cold but also dry, which might explain why long, narrow noses are characteristic of peoples from high latitudes, as well as high elevations, and deserts [120; 300; 530, ch. 10; 540, ch. 5]. The fact that desert people have long, narrow noses indicates that the humidifying function of the nose might be more important than the warming function, and indeed measures of absolute humidity appear to correlate best with nose shape [120; 655, ch. 4].

Complications exist in the nasal story, such as the fact that the difference in efficacy of temperature exchange is greatest when noses are decongested [855]. Additionally, and importantly, almost nothing is known about regional differences in complexity, and therefore surface area, of the turbinates, complex structures at the back of the nose that provide a greater surface area than does what we commonly think of as the interior of the nose [855].

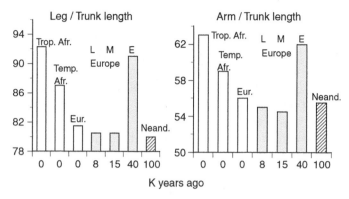

FIGURE 5.11. Limb length/trunk length of hominins of various ages and latitudes. Trop., Temp. Afr. = tropical, temperate Africa; Eur. = Europe; L, M, E Europe = late, middle, early Europe, i.e., Mesolithic, Late Upper Paleolithic, Early Upper Paleolithic; Neand. = Neanderthal. Data from [372].

Bergmann, Allen, and Inferences about Prehistoric Hominin Adaptations and Dispersals. If bodily form correlates with latitude and climate, then bodily form of extinct populations can tell us about their latitude and climate of origin [369, ch. 3; 372; 373; 588; 672; 775; 776; 810; 814]. Thus, early modern humans in Europe about 40 kya seem to have the form of modern Africans, or other peoples who live in the tropics, as would be expected if modern humans originated in Africa (Fig. 5.11) [372]. By contrast, the later, higher-latitude paleolithic populations have the squat, short-limbed proportions of present-day high-latitude populations (Fig. 5.11) [372]. So also do Neanderthals (Fig. 5.11) [372; 425, ch. 6; 672; 775].

Clive Finlayson disagrees that Neanderthal body form can be explained by adaptation to cold [220, ch. 4]. He has several arguments against the idea.

He writes that it is inappropriate to apply the thermoregulatory explanations of the Bergmann and Allen effects to differences between Neanderthals and modern humans because both effects apply to intraspecific comparisons, not interspecific ones [220, ch. 4]. But as I have already indicated, Bergmann's initial formulation was in fact interspecific [59]. Allen indeed gave mostly intraspecific examples, but he also provided an interspecific one, a comparison of six species in three genera of foxes [7, p. 113]. In other words, neither Bergmann nor Allen restricted their hypothesis to one or other of the taxonomic levels. Additionally, even if the original formulations had been purely intraspecific, if inter-specific relationships are subsequently shown to exist, as they have been [320], we do not need to stick slavishly to an original formulation.

Finlayson suggested also that it was problematical to compare species that lived in different places at different times. However, comparative biology performs such comparisons all the time, with enlightening and independently validated results.

Finally, he argued that it was unrealistic to attempt to correlate morphology with only one factor (in this case, climate or temperature) given the variety of factors likely to affect morphology. Perhaps, but climate, like body mass and latitude, correlates with so many other factors, that maybe it is in fact one of the better single indices of the variety of other factors operating [69].

Finlayson explained Neanderthal bodily proportions as adaptations indicative of strength and an active life style, not as adaptations to reduce heat loss [220, ch. 4]. Tim Weaver and Karen Steudel-Numbers disagree [814]. They prefer adaptation to cold as the explanation for Neanderthal morphology. They argue that Neanderthals' robust limb bones could indeed indicate great strength, but not an active life style.

Most importantly for their argument, Neanderthals had relatively short legs (Fig. 5.11), and therefore they almost certainly used more energy walking than did, for instance, early modern humans [814]. If Neanderthal legs were adapted to an active hunting lifestyle, then they should have been long. Also, weight of items carried, including the carrier's body weight, makes a greater difference to energy expended during walking than does a difference in length of leg [814]. Neanderthals were heavier than early modern humans, and their extra weight might have made nearly twice as much difference as did their shorter legs to their relatively greater energy expended in walking [814]. Neanderthals were also probably inefficient runners compared to modern humans [625]. In sum, as far as activity is concerned, Neanderthals' bodily proportions are a cost, not a benefit.

Weaver and Steudel-Numbers also point out that in other mobile species, such as baboons, locomotion accounts for only about 10% of daily energy expenditure, and so is anyway probably a mild influence on body shape by comparison to the influence of thermoregulation [814].

A further argument that Weaver and Steudel-Numbers made in support of the idea that Neanderthal bodily proportions by comparison to those of early modern humans can be explained by adaptation to cold conditions rather than mobility is that while thermoregulatory advantages of body shape start at birth, any influences of mobility on shape would not have started until later, after a more or less large proportion of the young had died, with the consequence that selection for mobility is weaker than for thermoregulation [814]. The same could be said, though, for all secondary sexual traits (peacock males' tails, human males' beards, phalarope females' bright colors, etc.), and yet, of course, secondary sexual traits abound.

Nevertheless, in agreement with Finlayson's rejection of the hypothesis of adaptation to cold as an explanation for the Neanderthals' limb proportions, at least one 2-year-old Neanderthal had limb proportions well within the modern, temperate human range [184], potentially implying lack of adaptation to extreme cold.

So far, I have addressed the issue of only Neanderthals' post-cranial skeleton, not their head. Finlayson (2004, ch. 4) rejected the idea that the shape of the Neanderthal nose enhanced warming of incoming air, suggesting that the very active Neanderthals would rather have needed incoming air to be cooled. In fact, it seems that the Neanderthals' nose shape is not only a mixture of low- and high-latitude features [811], but it is no different from what one would expect, given the size and shape of the Neanderthal cranium [377; 623]. The nose, then, does not help any conclusion about climatic adaptations of Neanderthals.

In addition to his arguments against the idea that Neanderthals had a body shape adapted to conserving heat, Finlayson contended that Neanderthals, in any case, did not live in particularly cold climates, and certainly not in arctic ones [220, ch. 5]. On the other hand, as Trenton Holliday pointed out, Neanderthals did not have all the cultural and technological adaptations to living in cold climates that modern humans do and therefore they needed evolved morphological adaptations for retaining heat [373].

Finally, if the Neanderthal body form was not an adaptation to cold, how did the Neanderthals manage to move back into the Levant as the climate cooled there around 70 kya, even as modern humans moved out [220, ch. 7; 699; 700]? A hypothesis of superior physical strength of Neanderthals does not seem sufficient explanation.

Latitude, Climate, and Physiology

All homeothermic species respond to thermal stress via learned
behavioural, and inherited or acquired physiological responses.
[751, Taylor, 2006, p. 91]

We know that as individuals, we sweat more when we get hotter, we get stiff fingers when our hands are cold, and we shiver (i.e., generate heat) when our bodies are very cold. We also know that tropical organisms generally cope less well with cold than do temperate organisms [261; 506; 761].

Do people whose ancestors came from different climates or who themselves grew up in different climates have different physiological traits, as well as different anatomical traits, that might enable them to perform better in their local environment? Are there physiological effects that complement the Bergmann and Allen effects? Do Europeans sweat more than Africans in hot climates, and shiver less than them in the cold?

If differences exist, to what extent and how quickly can people from one climatic regime acclimatize to another regime?

Physiological Adaptations to Heat Stress? Humans have quite variable responses to temperature [751], and hominins in at least one part of the world, northern Kenya, have had millions of years to adapt to temperatures of over 30°C [585]. Nevertheless, a difficulty with discerning regional differences in physiological adaptations to heat is that even humans from temperate regions seem to acclimatize quite quickly to high temperatures [34; 545, ch. 7; 655, ch. 2; 751]. Presumably, the reason for that adaptability is that high-latitude summer temperatures can sometimes reach tropical levels. For instance, the average July temperature at Grand Junction, Colorado, in the United States (39°N) is around 25°C, as is the year-round monthly temperature at Palma Sur, Costa Rica (9°N) [402].

Certainly, comparisons between latitudes indicate little consistency in the results, and hence the question mark of this subsection's title. For instance, humans' major way of coping physiologically with heat stress is to sweat. If there were an adaptive difference between peoples of different regions, we might expect an obvious and consistent regional difference in density or functioning of sweat glands. There seems to be none [243, ch. 3; 751].

As one example of contradictory adaptations, South African acclimatized Bantu men indeed have lower rectal and skin temperatures and lower sweat rate than do acclimatized men of European origin at the same high environmental temperature, as if they have adapted so well to heat that they can maintain low body temperatures despite not sweating much. Yet native young men of Pitjandjara in the central Australian desert have higher rectal and skin temperatures than do acclimatized European-origin men, and they sweat less, as if they have not adapted at all to coping with heat. The difference apparently contradicts adaptive explanations. And then another study's results indicated no difference between peoples of different origins: Chaamba Arab men of Algeria reacted about the same as acclimatized French male soldiers to heat stress [all examples from 243, ch. 3].

With sweating as a major means by which humans regulate temperature, a lack of a correlation between region and sweating is potentially odd. One explanation for the lack of correlation is that the Bergmann and Allen effects are operating so well that the temperatures perceived by humans are the same across all latitudes or similar enough not to produce regional differences in the sweat response.

Alternatively, tradeoffs are involved. Sweating might help shed heat, but in tropical countries the constituents of sweat (water and salt) can be in short supply [545, ch. 10]. Our craving for salt, and the large amount of money and effort that we used to expend to get it [439], indicate an evolved response to an environment chronically short of salt. In this, we are no different from many animals, as the successful use of salt licks to attract wild animals to viewing sites testifies. With respect to primates, we now know that the reason gorillas eat the dead wood of particular trees is that the wood contains enough salt for over 90% of their salt intake, even though humans cannot taste it [669].

If the operation of the Bergmann and Allen effects means that people sweat less in hot climates, then the effects have a benefit in addition to maintenance of normal body temperature: they help the body to conserve salt and water. Evidence of physiological adaptations for salt and water retention in hot environments comes from regional differences in the prevalence in populations of genes associated with affinity of the body tissues for salt and with vasoconstriction [302; 856; 857].

A high affinity for salt reduces loss of salt in sweat; an ability to vasoconstrict in response to the low blood volume that comes from dehydration inhibits deleterious effects on circulation of low blood volume, whether the dehydration is due to sweating or drought. Correspondingly, genes and alleles associated with affinity for salt and vasoconstriction are at higher frequencies in tropical populations[i] (Fig. 5.12)

---

i. Latitude and genetic variation caused by heat stress. Young et al. [857] wrote about clines in heat adaptation, although their data were in fact clines in gene frequency: heat adaptation as the environmental influence on the genes was an inference. Young et al. report strong association of one gene, GNB3 825T, and its variants with both hypertension and

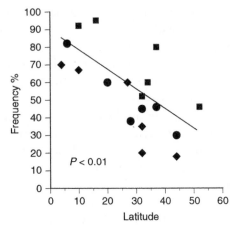

FIGURE 5.12. Stronger genetically influenced ability to retain salt, water at low latitudes. Frequency in population of three alleles that promote salt retention and via vasoconstriction counteract low blood pressure (due to low blood volume from lack of water) by latitude.* Different symbols = different alleles. Data from [857].

*Frequency three genes/alleles. I reanalyzed authors' data on multiple allele-by-latitude relationships; took the three alleles with strongest relationship with latitude, and median measures of allele frequency per continent of six continents, i.e., six allelle frequencies by latitude per allelle. Any two allelles, $N = 12$, $r_s > 0.6$, $P < 0.05$; all three alleles, $N = 18$, $r_s = 0.65$, $P < 0.01$. Six regions (and their median latitude): Africa (6), South America (16), Pakistan and Middle East—i.e., western Asia—(32), Asia (37), Europe (44). Data from [857].

[857]. As expected if the genes and their physiological effects are adaptations to mitigating the costs of sweating, the alleles are at higher frequency among the Mbuti and Biati peoples of central African forests than among open-country San—because the humidity of the forest increases heat stress [857].

Evidence of strong benefits to the relevant genes and their alleles is that tropical South American indians have the same gene and allele

---

latitude. I agree, but with their data I found the association with latitude to be the stronger: $N = 21$; whole model, $r^2\text{adj} = 0.53$, $\chi^2 = 18.0$, $P < 0.0001$, AICc = 175.5; latitude, $r^2\text{adj} = 0.37$, $F = 24.3$, $P < 0.0002$, AICc = 180; systolic blood pressure, $F = 7.4$, $P < 0.02$, AICc = 190.4. Difference in our analyses: I used only one population per nation in order to avoid spatial autocorrelation, i.e., double counting.

frequencies as do Africans, despite the fact that the native indians have been in the tropics for perhaps only 12,000 years [857] (chapter 2).

Physiological Adaptations to Cold Stress. Whereas the summer temperatures of higher-latitude regions can reach those of the tropics, temperatures of the tropics rarely reach the lows of temperate regions of similar elevation [402]. We can therefore expect latitudinal differences in long-term adaptations to cold to be more pronounced than adaptations to heat. Indeed, they might be.

Arctic and cold temperate mammals have a higher basal metabolic rate (BMR) than do tropical mammals of the same body weight, and among a variety of birds, higher-latitude taxa have higher BMRs for their body weight than do lower-latitude taxa [514, ch. 5]. The regional differences are large among the birds: the arctic species have BMRs nearly twice the average for their body mass, while tropical birds have BMRs less than expected.

In humans, comparisons at the same cold experimental temperatures (e.g., freezing point) between, on the one hand, peoples whose ancestors lived in cold, arctic or near-arctic environments, or at high elevation, and themselves spent their childhood in these cold environments and, on the other hand, peoples from temperate or low-elevation zones, indicate that where there is a difference, the people from the colder environment often have higher BMR, and higher skin, if not rectal[j], temperatures (Table 5.1) [34; 241, ch. 5; 244; 454; 545, ch. 5, 6; 655, ch. 1; 713]. For instance, a meta-analysis of the literature estimated a 4 to 5 kcal/day increase in BMR with every 1°C drop in annual mean temperature, age, sex, and body mass accounted for [244]. These latitudinal and elevational contrasts fit the fact that individuals, both nonhuman and human, seem often to respond to colder temperatures by more or less immediately increasing their BMR, with arctic peoples apparently being able to do so faster [34; 241, ch. 5; 545, ch. 5, 6; 655, ch. 1].

As is the case for heat stress, differences exist between polar and tropical populations in the frequency of genes associated with energy metabolism, for instance, some associated with mitochondrial functioning [302; 303].

In apparent exception to the generalization of higher BMR at low temperatures, Australian desert and Kalahari native peoples at around 5°C lower their BMR more than do people of European origin and have

---

j. Rectal temperature used as a practically easy index of core temperature.

TABLE 5.1
REGIONAL DIFFERENCES IN RESPONSES TO COLD (SKIN TEMPERATURE AT 0°C)

Peoples	Country	N	Latitude	BMR	Skin temperature	Rectal temperature
!Kung, White	Botswana	14, 4	22S	<	=	<
Native, White	Australia	6, 4	30S	<	<	<
Quechua, White	Peru	41, 8	50N*	>	>	>
Alacaluf, White	Chile	9, 4	55S	>	<	=
Algonkian, Chinese	C. Canada	/	55N	/	>	/
Athapascan, White	Alaska	/	63N	>	/	/
Saami, White	Lapland	9, 5	65N	<	/	<
Inuit, White	N. Canada	/	70N	>	>	/

SOURCE: Data from [241, ch. 5].
Sample size—number of individuals, native peoples first.
>, native population higher value than European origin population.
<, native population lower value than European origin population.
/ = no data given.
* high elevation (4,150m, 5°C in coldest month = c. 50°N; actual latitude, 10°S).

lower rectal and skin temperatures (Table 5.1). An explanation of the contrast is that the nighttime desert is cold, and the desert peoples are conserving energy by lowering BMR, which they can afford to do because temperatures rarely get below freezing. By contrast, high-latitude and high-elevation peoples cannot afford a low BMR, because of the danger of frostbite in their frigid environment [243, ch. 5]. This explanation fits the fact that in all water temperatures from 0°C to 40°C, arctic peoples maintain more blood flow to the hand than do people from temperate regions, perhaps in part by more responsive systolic blood pressure [241, ch. 5].

As usual, nutritional state needs to be taken into account. In the case of the Australian desert and Kalahari native peoples, the difference between the desert peoples and European-origin peoples was probably not caused by contrasts in nutrition, because if it were, the (presumably) fatter European-origin peoples should have had the colder skin temperature, if higher rectal temperature, because their fat would help retention of heat [34].

Several of the studies reported that European-origin subjects withdrew from the cold experiments (e.g., hands in ice water) before the higher-latitude peoples did so, as if the Europeans were less able to cope with the cold. However, not only could expectations have differed between the people from different regions, but acclimatization, or lack of alternatives, can considerably extend what people will withstand. Witness the appalling conditions that early British antarctic explorers put up with [136]. I say British, as opposed to Scandinavian, because the British were so much less well equipped than were the Scandinavians, who copied the Eskimo/Inuit survival techniques and technology [386].

An oft-quoted study of the effects of region of origin on ability to cope with cold comes from records of frostbite among American serviceman in the Korean War [e.g., 49]. A greater proportion of the African Americans suffered than did European Americans, leading to a quite widely expressed inference that people of recent African origin were less able to cope with cold than people from higher latitudes. However, then and later, blacks were still being lynched in the United States, and one can well imagine that the African-American soldiers were not given as comfortable billets as the European-American soldiers.

Islands, Starvation, and the "Thrifty Genotype." A major problem with life on islands is that when conditions are poor, islanders, whether nonhuman or human, cannot easily migrate to better sites. To survive, they must therefore endure the famine, rather than escape it by migrating. A consequent trait of island vertebrates is that they have various forms of efficient metabolism, so that they use less energy and therefore can survive on less food or survive for longer without food [514, ch. 10].

The island species have either a lower metabolic rate than do their mainland relatives, or, even more extreme, island organisms that are the same body size as a mainland endotherm, such as a mammal, are more likely to be an ectotherm, such as a reptile. A classic example is the 60–70 kg carnivorous Komodo dragon (*Varanus komodoensis*) of the tiny islands of insular southeastern Asia, whose 60–70 kg counterpart on the mainland is the leopard (in Asia, also called a panther; *Panthera pardus*).

An efficient metabolism appears to be characteristic of Pacific Island peoples too. Their thrifty physiology, although beneficial in the past, now brings problems of obesity and adult-onset diabetes in an environment of cheap, easily available food [61]. Whereas the average modern human in populations living in regions experiencing a mean annual temperature of about 27°C weighs around 50 kg, modern Pacific Islanders weigh about 75 kg [61].

Heavier individuals with a thrifty physiology would have been at an advantage even before they reached the islands, because on the long sea voyages, the people will often have been cold, and probably often short of food [61; 540, ch. 6]. Additionally, subsequent to arrival and for the same reasons, the island populations will have been less likely to be diluted by small-bodied people with less efficient metabolism, at least before modern means of transport.

ELEVATION (ALTITUDE) AND REGIONAL VARIATION IN ANATOMY AND PHYSIOLOGY

Most of what we know about the responses to high elevation and low air pressure comes from studies on humans.
[684, Schmidt-Nielsen, 1997, p. 209]

Two main problems of high elevation are cold and lack of oxygen. I have already discussed cold in relation to elevation. Here, therefore, it is potential adaptations to oxygen deprivation that is the topic.

By a variety of measures, populations and species of birds and mammals resident at high elevation perform better than do populations of the same species or related species of low elevation, in the sense that their physiology and anatomy appear better suited to obtaining, retaining, and using oxygen [514, ch. 8]. For instance, llama (*Lama glama*) and vicuna (*Vicugna vicugna*) have more oxygen in their blood at a given atmospheric pressure than do low-elevation mammals.

Several species of primates characteristically live at high elevation or have high-elevation populations. They include the gelada baboon (*Theropithecus*) of the Simien Mountains of Ethiopia and the snub-nosed monkeys (*Rhinopithecus*) of the eastern Himalayas in southwestern China, both of which reach around 4,000 m. Nevertheless, hardly anyone has studied primates' adaptations to high elevation. We do know, though, that geladas at high elevation are stressed, as judged by the high glucocorticoid levels of high elevation (3,650 m) compared to populations at low elevation (3,150 m) [52]. An explanation for the apparent lack of adaptation of these high-elevation animals is that climatic warming and an increasing human population in the region have recently pushed the gelada populations to higher elevations than those that they evolved in and that they have not had time to adapt either to the cold or to the depressed oxygen levels or to any of several other factors that might distinguish high and low elevation in the region [191].

Humans are just another vertebrate as regards the performance of high-elevation peoples. Almost all measures show that people born at high elevation and resident there perform better at high elevation than do people born at low elevation and not yet acclimatized to high elevation [47; 48; 241, ch. 11, 12; 243; 543; 544]. The differences between high- and low-elevation peoples have been demonstrated in Ethiopia, Peru, Tibet, and the United States, among other places.

Examples of traits for high-elevation living possessed by high-born, high-resident peoples compared to low-born, unacclimatized peoples are larger lung volume, more air moved through the lungs, and various measures of ability to deliver oxygen to the tissues, including oxygen to the fetus, which results in better fetal growth, other things being equal [48; 241, ch. 11, 12; 243; 540, ch. 6; 543; 544]

Acclimatization can more or less quickly (certainly within a lifetime) take those born at low elevation to the performance of those born at high elevation [241, ch. 11, 12; 243; 543; 544]. Those born at low elevation but who grew up at high elevation perform almost as well on most measures as do high-born, high-resident peoples [241, ch. 11, 12; 243; 543; 544]. Moreover, where high residents have been found to outperform acclimatized low-born people, it could be the case that the acclimatization did not start when the subjects tested were young, because the longer people have lived at high elevation, the more indistinguishable their abilities are from high-born natives [241, ch. 11, 12; 243].

Nevertheless, regional contrasts indicate evolved adaptation to high elevation. People have lived at high elevation for different times in different places. Tibetans have probably been high for longer than have Andeans, who might have been high for longer than the Rocky Mountain residents of North America [5; 6; 48; 543; 544; 619; 741]. High-born, high-resident Tibetans outperform high-born, high-resident Andeans on some measures, and both outperform on some measures low-born, high-resident Asians and South Americans, respectively, as well as high-born, high-resident North Americans and Ethiopians [47; 48; 543; 544].

It appears that in some respects Tibetans differ from Andeans in the apparently evolved mechanisms that allow life at high elevation in low oxygen pressure. For instance, even though high Andeans have bigger chests than do high Tibetans [290], high Tibetans can ventilate their lungs better than can high Andeans [48]. And yet the high Tibetans have hemoglobin less saturated with oxygen, lower hemoglobin concentrations in their blood, and lower oxygen concentration in arterial blood than even low-elevation peoples and far lower than high-elevation Andeans [47; 48]. Instead of high oxygen concentration in the blood, the Tibetans seem to use high blood flow as the adaptation to low atmospheric oxygen pressure, as indicated by a variety of measures, such as less constricted arteries (and hence lower blood pressure), greater blood flow to the fetus, and perhaps a greater density of muscular capillaries[k] [48].

Substantiating the claim for evolved adaptation to high elevation, a variety of studies indicate that people with high-elevation ancestry not only have more of the attributes that contribute to good performance at high elevation but also a greater incidence of genes associated with those attributes—for instance, a gene that appears to influence the amount of oxygen that can be carried in the blood [47; 48; 706; 853]. Tibetans who appear to have this gene carry more oxygen than those who do not [47; 48]. The benefits of the gene are evident in the fact that in a sample of hundreds of Tibetan women living at over 4,000 m, women likely with the active form of the gene had twice as many surviving children as did those without [47; 48].

Some of the Tibetan adaptations might have evolved within just the last 3,000 years, as judged by differences in the incidence of genes between Tibetans and Han Chinese, who diverged about 3,000 years ago [853]. Oddly enough, given findings that Tibetans have adapted by blood flow more than by blood chemistry (above, [47; 48]), the greatest difference found in this study of Tibetan–Han differences was in a gene that appears to be associated with contrasts in erythrocyte and hemoglobin concentrations [853].

An effect of elevation on birth weight was suggested by the finding that infants born at high elevation generally weigh less than do those born at low elevation [543; 544]. A problem with any inference of an altitudinal effect is that not only were the results not presented relative to mothers' weights, but also that differential nutrition, not elevation, could of course be the cause.

However, a very nice recent study allows the effects of elevation and nutrition to be distinguished [54]. Adam Bennett and his team recorded birth weight and ancestry of nearly 1,350 infants born to residents in two hospitals in the Bolivian city of La Paz, which lies at 3,600 m. Bennett et al. classified their subjects as being of high-elevation (Andean) or low-elevation (European) ancestry, or a mixture, using parental family names to identify ancestry.

k. I write "perhaps" because no statistical tests were presented [48].

They found that babies whose mothers were Andean in both parental lines were 8.5% heavier than were babies whose mothers were European in both lines, although the mothers differed in weight by a statistically insignificant less than 1% [54]. Furthermore, only 4% of the Andeans were preterm compared to 13% of the European infants, and only 10% of the Andean infants compared to 33% of the European were of low weight for their age of birth[1] [54]. In sum, people of high-elevation ancestry performed better at high elevation than did people of low-elevation ancestry by these standard measures of performance.

Sex Differences

The two sexes differ in body shape and physiology. They differ correspondingly in their ability to regulate their temperature.

From puberty, females have a lower metabolic rate than do males [249]. Females are broader for their stature and weight than are males [673], and they therefore should have a relatively lower surface area and hence lose heat more slowly. Females might also have shorter limbs for their trunk length than do males [34], but the source gave no substantiating citation or data, and I have not found any. Mass for height does not differ obviously between the sexes, because females are both shorter and weigh less than do males [673].

A potentially more important influence than shape on temperature control is amount and distribution of fat [655, ch. 2]. In general, fat people conserve heat better than do thin people [34; 241, ch. 4]. Females are not only fatter than are males (greater proportion of body mass as fat) [728], but they carry more of their fat under their skin, as opposed to internally (Fig. 5.13) [728]. Females are therefore better insulated.

And indeed, the bulk of the data on sex differences indicate that females lose heat more slowly than do males [454; 655, ch. 2]. As a particular instance, among highland Quechua Andeans sleeping in unheated houses, if under blankets, women maintained higher rectal temperature than did men, who maintained higher temperatures than did children [34].

Because females conserve heat better, they can and do have lower basal metabolic rates than do males [454], which makes the Quechua women's higher rectal temperatures particularly impressive. That it is the females' better insulation that is responsible for their lower metabolic

---

1. La Paz study. For birth weight, % preterm, and % small for age, $P<0.001$, 0.05, 0.01, respectively; $N = 170$ (both parents Andean), 61 (both parents European) [54].

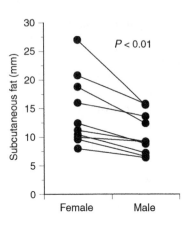

FIGURE 5.13. Females fatter than males. Subscapular skinfold thickness of young females and males in ten societies.* Data from [728].

*Sex differences in subcutaneous fat. $N = 10$, all females > males, $P < 0.01$, binomial test. The most extreme difference was Samoans; the least was the Chachi peoples of Ecuador. Data from [728].

rate is indicated by the fact that if body fat is accounted for, sex differences in BMR disappear [454]. The females' body fat and lower metabolic rate then allows them to take in less energy than do males, at least in the lower latitude countries of Africa, India, New Guinea, and South America [33].

Whatever the causes of any differences in tolerance to cold—and surely all operate (body size and shape, metabolic rate, fat amount and distribution)—three episodes demonstrate that the differences might affect survival when starving in cold temperatures.

The first episode is the famous Donner Party [282; 283; 508]. Trapped in unseasonal snow in the Sierra Nevada Mountains of northern California in the winter of 1846/7, they resorted to cannibalism. Sex appeared to have affected survival, as did age and traveling with family members [282; 283; 508]. Accounting statistically for age (young and old died) and family membership (lone migrants died), males were twice as likely to die as were females [508]. In more detail, if only families are counted (because all lone individuals, who were all males, died), 30% of family females died compared to 50% of family males, or a median per family of 25% of female members but 50% of male members (Fig. 5.14) [282]. Also, of a rescue party of 10 men and 5 women who set out from the camp to get help, 6 of the 10 men died of one or more of cold, hunger, and exhaustion (2 others were murdered); none of the women died [282; 283].

In the second episode—another tragedy in circumstances similar to those of the Donner Party—again females were more likely to survive than were males [284]. Brigham Young, the Mormon leader, funded immigrants from Europe to travel to Salt Lake City. However, the funds

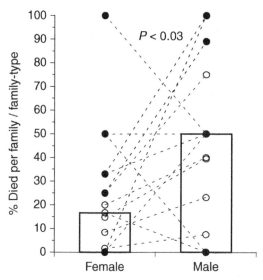

FIGURE 5.14. Males more susceptible to cold than females. Sex differences in death rates per family in Donner Party (•) and Willie Handcart Company (○) trapped over winter in Sierra Mtns., USA.* Lines connect families (DP) and family types (WHC). Data from [282; 284].

*Sex difference in death rates. Donner Party and Willie Handcart Company, $N = 19$, $t = 2.50$, $P < 0.03$ (combined data). Data for DP are family name; for WHC are family type, e.g., mother–child, siblings, singleton. However, WHC data are divided, e.g., by family type or age, sex difference in mortality rate is significant ($P < 0.01$). Data from [282; 284].

were not sufficient for horse- or oxen-drawn wagons. Instead, the people pushed handcarts. Most groups survived the journey. One, the Willie Handcart Company did not. Setting off late, like the Donner party, and poorly equipped and supplied, 68 of the company's 400 people died in a month of winter conditions as they travelled 500 km across Wyoming and into Salt Lake City. Men were over twice as likely to die as were females, whether only family members or singletons are counted (Fig. 5.14).

The third episode is the Dutch Hunger of the winter of 1944/5, when Germany prevented supplies reaching Holland during an unusually severe winter, and as a result thousands of Dutch died from cold, starvation, and their effects [36]. For 6 months, energy intake was less than the calculated optimum 2,250 calories per day, and for 4 months, less than 1,000 calories per day. As for the American migrants of the first two examples, the young, the old, and men were far more likely to die, as were working-class people [36]. In total, over twice as many deaths occurred in the Hunger Winter than in the same period a year previously. In the worst months, men were over twice as likely to die as women, with the ratio highest among the working class and only a little over 1:1 among the prosperous class [36].

The contrast between the sexes in death rates in the Donner and Handcart parties is not necessarily a result of contrasts in only metabolic rate, body shape, and amount and distribution of fat. One can imagine that men worked physically harder than did women, building, hunting, trapping, and so on, among the American migrants. In the Dutch Hunger, men did not obviously work harder than women [36]. Nor is it likely that males fighting more than females is the cause, because at the start of the German proscription, mortality rates were almost equal. It is not impossible, though, that men might have been chivalrous enough to give preference to women in food and shelter. For instance, chivalry by males might explain why females were 50% more likely than males to survive the Titanic's sinking [240].

In sum, it appears that females, like islanders, have a thrifty physiology. They have evolved to use less energy than have males, to have a more efficient metabolism, and to better conserve their energy by losing heat more slowly. The evolutionary psychological literature suggests that part of the reason for the sex differences is that females compete less actively than males do for resources and mates. As importantly, lactation means that females accrue a reproductive benefit from conservation of energy that is unavailable to males.

. . .

People from different latitudes and therefore climates differ in skin color, size, and shape of the body, and in some case their metabolic rates and reactions to the environment. People from different elevations differ in shape of the body and how they process oxygen.

A fair amount of information and analysis indicates that these regional differences are adaptive. They can be explained as evolved means to prevent harm from the sun at one latitude and to benefit from it at another; to regulate temperature at all latitudes; and to cope with low atmospheric pressure and hence lack of oxygen at high elevations.

Some of the physiological and morphological differences between the sexes mirror these regional differences. Where they match what might be expected from thermoregulatory principles, they substantiate and are substantiated by the same thermoregulatory factors as explain the bio-geographic variation in the human body form and function.

Let me conclude by emphasizing three issues.

First, anatomical differences between peoples of different regions do not necessarily imply genetic differences. We all have genes that influence our response to the environment. If the environment differs, we differ. With regard to the Allen affect in mammals, excised limbs of laboratory mice grow longer in warm solutions than they do in cold solutions [697]. I do not know whether human limbs respond in the same way, but we do know that humans adapt quite well and quite quickly to temperatures and elevations different from those in which they grew up. Of course there are limits to an individual's adaptability, and some of the regional differences must arise because of genetic differences between the peoples of the different regions.

Second, a developmental response to the environment, such as growing longer limbs in hotter temperatures (as opposed to simply inheriting long limbs from parents) does not negate a thermoregulatory function of long limbs: the ability to respond to the environment can itself be adaptive. I present this point separately from what I have just written, in reaction to the common attitude that if a trait is not fixed, then it has not evolved. A long limb in hot environments and a dark skin in bright sun are beneficial whether the long limb or the high melanin count was inherited or developed in response to the environment. The ability to respond adaptively to the environment can evolve through natural selection as much as can fixed traits.

Finally, the fact that regional differences in nutrition can explain some differences in bodily form does not mean that thermoregulation is not also an explanation. People with different nutritional regimes might need different thermoregulatory traits. Well-fed people are probably more active than are poorly fed people, in which case they will have greater need to shed heat. Hence, better nutrition, as well as a hot environment, is associated with longer limbs. And the converse for poorly fed or ill people.

In sum, in the case of human size, shape, and skin color, it looks as if Darwin was wrong. Humans' regional differences can be accounted for by the direct action of different conditions of life, both in the lifetime of an individual and over evolutionary time.

LATITUDE, CLIMATE, AND HUMAN CULTURAL DIVERSITY

It appears that the rigorous frost in the antarctic regions almost
precludes the germination of plants; that the countries in the
temperate zones . . . produce a variety of plants; and lastly, that the
tropical isles derive a luxuriance of vegetation from the advantage of
climate . . . the animal world, from being beautiful, rich, enchanting,
between the tropics; falls into . . . poverty [at higher latitudes].

[237, Forster, 1778, p. 119, 128]

The biodiversity of the tropics relative to high latitudes is a truism. Although a few taxa are more diverse outside the tropics, such as aphids and bees [159, ch. 3; 259] (and of course, penguins), most taxa are more diverse in the tropics [251; 359; 463, ch. 15]. We have known about the taxonomic richness of the tropics compared to the taxonomic poverty of high latitudes for over two centuries, as the quote by Johann Forster at the start of this section shows [237].

For brevity's sake, I have termed the latitudinal gradient of diversity the Forster effect [320]. It has previously been termed Wallace's rule [601]. However, Forster (who travelled with Capt. James Cook on his second voyage south) published nearly a century before Wallace, in 1778 compared to Wallace's 1876 [237; 803]. In addition to Forster's priority, Wallace noted so many biogeographical patterns that naming even one of them after him might not be helpful.

An immense amount of thought and ink have been spent on the topic of why the tropics are more taxonomically diverse than are higher latitudes [68; 359]. For instance, Mark Lomolino et al.'s exhaustive list of hypotheses cites about 65 studies, and several of those are reviews [463, ch. 15]. Many of the arguments can be summarized as variations on the idea that tropical stability, clemency, and productivity allow more varied forms, in part by not restricting form and in part by allowing persistent, large populations, which are less liable to extinction. The corollary is that high-latitude conditions restrict viable forms and cause more frequent extinctions.

In this section, I describe the latitudinal and other regional variation that exists in the diversity of human cultures and discuss possible explanations for the variation. Human variation has featured hardly at all in any of the discussion of the causes of the Forster effect. An aim of this section, therefore, is not just to explain geographic variation in human cultural diversity but to use knowledge of the biogeography of human cultural variation to test the hypotheses for the Forster effect among nonhumans. If the hypotheses explain human variation, they are strengthened; if they do not, they are weakened.

The argument has been made that nobody nowadays should be analyzing diversity in relation to latitude [342]. Latitude should be ignored, and instead the nature of the environment should be the independent variable. A problem with that argument is that the data available for the nature of the environment are essentially for today's environment, whereas distribution is strongly determined by history—for much of which we have few details even if some generalities are known. An advantage of latitude as a measure is that it incorporates a host of potential influences, including history, and therefore can encapsulate as an explanation the totality of potential climatic influences [69].

In any case, the latitudinal gradient is the obvious pattern, even if within latitudes, hotspots and coldspots of diversity exist. To explain the latitudinal gradient, and the siting of hotspots and coldspots, we then need more precise measures of the potential environmental influences that might correlate with latitude.

Primate Diversity

Primates strongly show the Forster effect (Fig. 5.15) [74; 157, ch. 5; 309; 311; 320; 599; 803, ch. 10, 11]. The decline in diversity of primates with increasing latitude is not simply an effect of confinement to the tropics of the primates' main habitat, tropical forest, because the diversity of primate taxa decreases from the equator, i.e., from well within the tropics (Fig. 5.15) [311; 320].

Human Cultural Diversity

Definitions and measures of "culture" are varied and can be problematical, as I discussed in chapter 1. I will here simply repeat that I use the original authors' definitions of culture.

Human cultures are like most taxa of plants, invertebrates, and vertebrates in being more diverse in the tropics [122; 146; 277; 325; 473; 477; 483; 542; 557; 746]. As a specific example, 130 tropical countries across the globe had nearly three times as many identified languages (mean of 64) as did 94 nontropical countries (mean of 23) [data from 277]. The relationship is true not only globally but also within separate regions (Fig. 5.16A–E) [122; 146; 542]. Combine both linguistic and

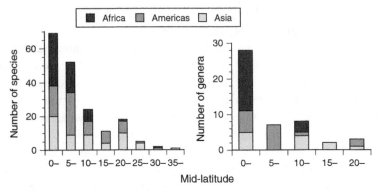

FIGURE 5.15. Primates are more taxonomically diverse in the tropics.*
Number species by 5° latitudinal band in which the species' mid-latitudes fall.
Data from [311].

*Madagascar omitted because it covers too small a range of latitudes.

taxonomic diversity, and 17 of the top 20 most diverse countries are wholly within the tropics [461].

Papua New Guinea famously has the highest density of languages [277; 461]. Part of the reason is that the rugged environment and topography makes movement difficult, and therefore cultures are extremely confined [178, ch. 15]. Additionally, though, island countries in general have higher densities of languages than do mainland countries at the same latitude. The reason for this relationship is a simple function of the fact that the slope of the relationship between number of species or cultures with area is less than 1 (chapter 6) [666, ch. 2]. With island countries mostly smaller than mainland countries, the result is inevitably a higher density of languages on islands. That is a fact. To get at the process, we need to see whether island and mainland countries differ for their area. Analysis of residuals from the line is the way to account for this effect (Fig. 5.16A, Fig. 5.16B). The regression lines for islands and mainlands turn out to be indistinguishable (Fig. 5.16B).

Globally, total diversity of hunter-gather societies does not decrease significantly with increasing latitude[m] [62]. A main reason is the very high detected density of hunter-gatherer societies in the western coastal United States [62, fig. 5.03; 268]. Omit North America, and a decline in diversity of cultures with latitude is evident (Fig. 5.16C) [62].

m. Number *hunter-gatherer societies* by 5° latitudinal band (only one society per state counted, e.g., France, Florida): $N = 14$, $r_s = 0.3$, $P > 0.1$. Data from [62].

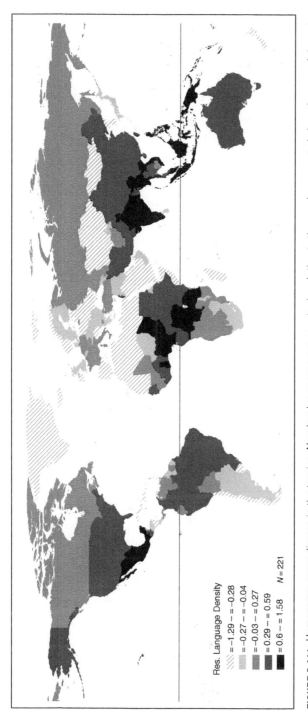

FIGURE 5.16A. Humans are more diverse in the tropics. Number languages per country in relation to the number expected given the area of the country [277]. (Residual number of languages by latitude, area accounted for. Mainland, $N = 45$, r^2adj. = 0.2, $F = 11$, $P < 0.003$; islands, $N = 26$, r^2adj. = 0.45, $F = 21$, $P < 0.0001$.)

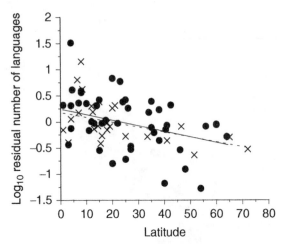

FIGURE 5.16B. Humans are more diverse in the tropics. Residual density (from area) of number languages. Mainland (• —-), island (x - - - -) countries separated. (Density languages by latitude, hemisphere ignored. Globe: Mainland, $N = 43$, r^2adj. $= 0.24$, $F = 14$, $P < 0.001$; island, $N = 26$, r^2adj. $= 0.41$, $F = 18$, $P < 0.0005$. Data from [277]. Diversity languages by latitude, hemisphere noted. Mainland countries, $N = 43$, r^2adj $= 0.17$, $F = 5.20$, $P < 0.01$; island countries, $N = 26$, r^2adj $= 0.51$, $F = 14.2$, $P < 0.0001$. Data from [277].)

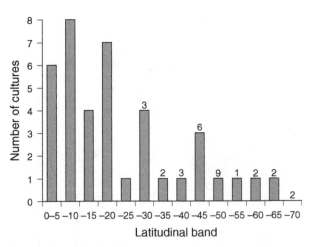

FIGURE 5.16C. Number of hunter-gatherer cultures by 5° latitudinal band, excluding North America. Numbers above bars are for North America. (Number hunter-gather societies by 5° latitudinal band (excluding North America), $N = 13$, $r_s = 0.8$, $P < 0.002$. Data from [62].)

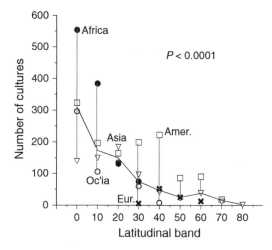

FIGURE 5.16D. Number of cultures by latitude. Separate symbols for each region, Africa, Americas, Asia, Europe, Oceania. (Linguistic diversity by latitude. Africa, $N = 4$, $r_s = 1.0$; Americas, $N = 8$, $r_s = 0.76$, $P < 0.04$; Asia, $N = 9$, $r_s = 0.92$, $P < 0.001$; Europe, $N = 4$, NS; Oceania, $N = 5$, $r_s = 0.90$, $P < 0.04$. Data from [146].)

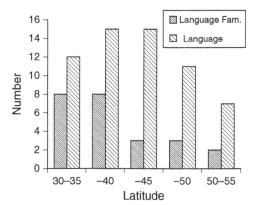

FIGURE 5.16E. Humans are more diverse at lower latitudes. Number of language families and languages of Native Americans in the Great Plains of North America. Data from [268].

Turning to the Forster effect within continents, the *Ethnologue* data set of languages had a sufficient sample for Africa and Asia. In both continents, the residual number of languages for area declined with increasing latitude.[n] The *Ethnologue* data do not allow assessment for North America, because the country is lumped as one area. However, an independent analysis, using a different source than did *Ethnologue*,

---

n. Diversity *languages* by latitude per continent. Only two continents had sufficient sample to be analyzed individually ($N > 5$ latitudinal bands), although all had fewer languages at higher latitudes: Africa, $N = 12$, $r_s = 0.7$, $P < 0.01$; Asia, $N = 14$, $r_s = 0.65$, $P < 0.02$. Data from [277].

found that there too the diversity of languages declines with increase in latitude [473].

Ian Collard and Robert Foley's [146] data on diversity of cultures by latitude allow analysis of the Americas, Asia, and Europe separately. The Forster effect operates in the first two.[o] North America peaks significantly between 35° and 45° (Fig. 5.16D) [62; 146], because those latitudes are not only where the continent is at its widest in reasonably productive climes, but they include coastal central-western United States' extraordinarily high density of cultures [62, fig. 5.03; 146; 268].

In order to test for a general relationship unaffected by outlier regions, I conducted an analysis for North America that omitted this unusual hotspot (Fig. 5.16E). I took a longitudinal band of roughly 1,000 km wide by 2,000 km long up the most homogeneous part of the continent that I could find—the Great Plains. Longitudinally, the area is between 105° and 95°W at 30°N and between 107° and 93°N at 55° N (a little north of the Gulf of Mexico to opposite the where James Bay meets Hudson Bay). The rough west and east bounds are the Rockies and the Mississippi.

In the Great Plains of North America, as in the continent as a whole, a general decrease in cultural diversity (in this case linguistic diversity) toward higher latitudes is evident, but the peak in diversity around 40° is still evident (Fig. 5.16E). In other words, the peak appears to be real; it is not a result of the width of the continent or just the central west coast's high cultural diversity. That coastal density of language families is nearly five times that of the Great Plains of North America, 29 in 1.5 million km^2 compared to 6 in the same area of the Great Plains [268].

With hunter-gatherer societies, only in South America did diversity decrease with latitude[p] [62], but then so does the area of the continent. None of the other continents showed a linear relationship.

Why Do We See Greater Diversity at Lower Latitudes?

Using Human Cultural Diversity to Test Biological Hypotheses for the Forster Effect. I here discuss in turn the various hypotheses for the Forster effect that I present in my summary (Table 5.2) of the far more detailed

---

o. Diversity *cultures* by latitude per continent. Americas, $N = 8$, $r_s = 0.8$, $P < 0.03$; Asia, $N = 9$, $r_s = 0.9$, $P < 0.0005$. Data from [146].

p. Number *hunter-gather societies* by 5° latitudinal band, per continent: S. America, $N = 12$, $r_s = 0.7$, $P < 0.02$; all others nonsignificant. Data from [62].

table in Lomolino et al.'s *Biogeography* [463, ch. 15, table 15.2]. These hypotheses were produced in the almost total absence of consideration of anthropological data on human diversity of any sort, including cultural diversity. The data on human cultural diversity and explanations for why it, like nonhuman biodiversity, shows a latitudinal gradient and thus constitute a largely independent test of the hypotheses.

The numbering of the following paragraphs refers to the numbers in the leftmost column of my Table 5.2.

The first two hypotheses address the idea that the latitudinal gradient in diversity is a side effect resulting from more or less unrelated ecological and biogeographical influences. Such models are termed neutral or null models, because they require no special biological cause.

1. Mid-Domain Effect

The Forster effect exists because species/cultures with large latitudinal extents are more likely to overlap in the center of the globe than elsewhere (think pencils of different length in a pencil box). This mid-domain effect does not explain the Forster effect in human cultures, I suggest. Nor does it for at least some nonhuman species (see Table 6.1 and discussion).

2. Neutral Theory

The stochastic birth, death, and dispersal rates of individuals within populations produce more or less stochastic speciation and extinction of species, which happens to result in more speciation and less extinction in the tropics because of environmental differences across latitudes [53; 380; 381; 555]. Stephen Hubble is one of the stronger proponents now of the idea. He and others discuss it mostly in the context of the representation of species within communities, especially the number of individuals of each species. However, add latitudinal contrasts in birth and death rates, and the result is latitudinal contrasts in rates of speciation and extinction that can produce the Forster effect. Whether or not neutral theory of this sort can explain the Forster effect in human cultures will not be known until we have data on the relevant parameters.

The next three explanations of the Forster effect provide specific mechanisms for how diversity at one level of the system can explain diversity at another level. They indicate how diversity might be maintained or constrained. To some extent they can be seen as begging the question of the origins of diversity.

TABLE 5.2
HYPOTHESES FOR THE FORSTER EFFECT, THEIR EXPLANATIONS, AND
CONCLUSIONS REGARDING THE VALIDITY OF THE HYPOTHESES FROM THE
GEOGRAPHIC PATTERNS OF HUMAN CULTURAL DIVERSITY

Hypotheses for Forster effect	Explanation regarding tropical diversity	Relevance of human cultural diversity
Null/Neutral models		
1. Mid-domain effect	Ranges (pencils) overlap more at center of available space (pencil box).	No. Diversity of cultures within each polar hemisphere is usually highest near tropics, not at mid-latitudes.
2. Neutral theory	Stochastic demographic effects produce more speciation, less extinction in tropics.	Unknown?
Biology		
3. Diversity of parasites, competitors, predators	High diversity prevents a minority of species dominating.	Yes.
4. Diversity of hosts, resources	A diversity of hosts produces or allows a diversity of dependents.	Maybe. If different cultures use the environment in different ways.
5. Environmental heterogeneity	Heterogeneous, patchy environment promotes diversity.	Yes.
6. Energy available, productivity, population growth rate	High energy, productivity in tropics = larger populations = more speciation, less extinction.	Maybe, but many complications.
7. Rapoport effect	Smaller geographic ranges in tropics allows closer packing. (But many large geographic ranges could overlap).	Yes. Especially as human cultures tend not to overlap.
Physical environment		
8. Stability over short and long time periods	Tropical stability allows/ promotes specialization and speciation; less change means less extinction.	Yes, e.g., correlation of cultural diversity with climatic extremes; less specialization (more varied tool kits) at high latitudes.

(*continued*)

TABLE 5.2 (*continued*)

Hypotheses for Forster effect	Explanation regarding tropical diversity	Relevance of human cultural diversity
9. Environmental clemency	Tropical clemency allows specialization, prevents generalization.	Maybe. But note NW coast cultural diversity in North America.
10. Intermediate disturbance	Intermediate disturbance frequency in tropics allows high biodiversity.	No evidence of intermediate disturbance frequency in tropics?
11. Energy, 1	More energy allows high metabolism, higher reproductive rate, and hence higher diversity.	No. No evidence for the relationships.
12. Energy, 2	More energy causes higher mutation rates.	No. No evidence for the relationships.
13. Area	Greater area of tropics allows more species (No. number species more than expected for area].	No. Residual number of languages for area higher in tropics.

SOURCE: Based on table 15.2 in [463, ch. 15].

3. Diversity of Parasites, Competitors, Predators

Robert MacArthur noted that for many temperate species in North America, the climate becomes more equable south of their present range [469]. That immediately raises the question of what prevents them extending farther south. MacArthur suggested that it was competition with the great diversity of tropical taxa. Pathogens and parasites are also more diverse in the tropics [122; 219; 295; 569], so these, as well as competitors, could prevent not just movement south but movement within the tropics, so keeping ranges small and hence maintaining diversity. This argument matches the explanation for why some tropical species do fine in temperate regions, namely that the normal suite of competitors, pathogens, and parasites are absent [101, ch. 6]. Additionally, if actual geographic range and potential geographic range are compared (potential calculated from the nature of the physical environment of actual range), high-latitude mammals fill more of their potential range than do low-latitude mammals, indicating that something other than the physical environment is limiting the distribution of the low-latitude mammals [551].

Given that parasite diversity correlates with cultural diversity globally, including when a variety of potential intervening factors are accounted for [122; 219], the effect could be operating in humans. Also, the high tropical-pathogen loads (as opposed to their diversity) might be the factor preventing a minority of cultures becoming dominant [122]. Of course, the argument would work the other way round if some other influence caused the cultural diversity, and the diversity was accompanied by relatively little contact between cultures. I discuss further in chapter 6 the issue of contact between cultures and its connection to diversity via size of geographic range.

4. Diversity of Hosts, Resources

The argument that a diversity of hosts or resources allows or produces a diversity of consumers (e.g., hunter-gatherers) is supported by the finding that the diversity of both the human cultures and the resources, such as plants, that the people use are higher in the tropics [477].

The explanation will work best for specialist species or cultures, those that live on only a minority of hosts. Humans in any one place use many resources. Nevertheless, the great regional (beta^q) diversity of the tropics [463, ch. 15] might prevent any one culture from expanding if people do not recognize or accept resources to which they are not accustomed. The effect would be especially powerful if the otherwise perfectly acceptable alternative resources used by neighboring cultures are demonized, in the way that English-speaking nations demonize frogs' legs as food [536] (chapter 4). If tropical cultures move less (chapter 4), they interact less. Small differences in language and other aspects of culture can then arise, and these differences can grow by drift to become large differences (see chapter 2), such that eventually the languages become unintelligible to each other.

5. Environmental Heterogeneity

The idea that environmental heterogeneity begets diversity is similar to the previous argument. The finding that diversity of habitat matches cultural diversity, including when both latitude and area available are accounted for, is evidence for this argument [122; 473].

---

q. Diversity jargon: alpha diversity = diversity at any one local site ("local site" more or less undefined); beta diversity = difference in diversity between sites, sometimes measured as the increase in diversity with distance from a starting site or the proportion of the regional number of species found at one site; gamma diversity = diversity in a region, i.e., a combination of alpha and beta diversities; delta diversity = same as beta, but across regions.

Mountains correlate with environmental heterogeneity in two ways. They act as barriers, especially in the tropics (chapter 4), and they cause abrupt change in the environment over short distances (chapter 4). Thus, the lower slopes of the Andes are a strong hotspot for many taxa, including primates [308; 675]. Three major hotspots of cultural diversity are also topographically varied (New Guinea, east-central Africa, and central-western North America), although the influence might not be wholly environmental heterogeneity but also difficulty of movement (chapter 4).

Hypotheses 6–13 explain the maintenance of diversity and its origins.

6. Energy Available, Productivity, Population Group Rate

More energy is available year-round in the tropics than at high latitudes,[r] and with adequate rainfall, productivity is greater [35; 251; 826], including within hunter-gatherer ranges [542]. The productivity allows larger populations, which correlates with more opportunity for speciation. This is, effectively, the environmental gradient part of the relevance of hypothesis 2, neutral theory, to the Forster effect. Whether the argued mechanism works, though, even in nonhumans, is another matter. I discuss the issue at length in a later section.

7. Rapoport Effect

Geographic ranges of species and cultures are often smaller in the tropics, a biogeographic pattern termed the Rapoport effect. Small ranges mean that more species and cultures can be packed into a given area and hence that diversity is greater in the tropics [146; 725]. In chapter 6, where the topics are the biogeography of area and range size, I discuss in detail the Rapoport effect, the potential relations between the Forster effect and the Rapoport effect, and the role that information on human cultures might have in substantiating the explanations and mechanisms.

8. Stability over Short and Long Time Periods

9. Environmental Clemency

Tropical stability and climatic clemency allow specialization and hence diversity. Additionally, if specialists can outcompete generalists in a stable and clement environment, tropical stability and clemency prevent generalization.

---

r. Energy available. Measured by, for example, potential evapotranspiration, which is affected by among other factors, rain and temperature.

I find it easier to think of the constraints of the high latitude winter preventing diversity than tropical stability and clemency (or productivity, hypothesis 6) causing it. In the beginning of this chapter, I suggested that humans from any latitude might be almost equally capable of adapting to extremes of heat, because the highest temperate temperatures reach tropical temperatures. By contrast, the lowest high-latitude temperatures are far lower than the lowest tropical temperatures at similar elevations [402]. High-latitude winters will severely constrain the form that organisms can take, and they will constrain the movement of species out of the tropics. Primates provide an example. Primates of all body sizes can persist in the tropics; only those of 5–10 kg can persist outside the tropics (Fig. 5.4) [320].

A finding that diversity of cultures correlates with mean daily minimum temperature in the coldest month, not mean daily maximum temperature in the hottest month [122], might be evidence that the same high-latitudinal constraints apply to human cultures. Similarly, Daniel Nettle argued that the Forster effect in human cultures was best understood by considering risk of death imposed by the environment. However, he proposed a mechanism that is unlikely to apply to nonhuman species. Nettle argued that the less likely it is that a culture could survive independently of other cultures, the more cultures would cooperate with others. The resultant homogenization would then mean less cultural diversity in risky regions. Judging risk by the duration of the growing season, he found fewer cultures (languages) in riskier environments [557].

10. Intermediate Disturbance

11. Energy, 1

12. Energy, 2

For these explanations of diversity in terms of environmental disturbance or energy in relation to metabolic rates and mutation rates, I have found no relevant anthropological data. It seems unlikely that either of the energy-related hypotheses apply to warm-blooded vertebrates (i.e., animals that maintain a more or less constant internal environment whatever the external environment).

13. Area

Michael Rosenzweig argued that one reason for the greater number of species in the tropics is that the tropics cover a wider area than does any other single vegetational zone [666, ch. 9]. In general, a sphere has more area at its equator than away from the equator. More specifically, the tropical biome covers a greater area than any of the other four biomes (subtropical, temperate, boreal, tundra) [666, ch. 9].

The argument does not work for several reasons [661]. The earth has far more land north than south of the equator, and yet diversity drops with latitude similarly in both regions. The total area of the tropical biome is indeed greater than the total area of the nontropical biome (c. 70 vs. c. 60 million sq km) [661; 666, ch. 9]. However, tundra is the second largest vegetation zone after the tropics in Rosenzweig's account [666, ch. 9], and yet it has the fewest species and cultures of any of his five zones. The number of species of many taxa, and certainly of primates, drops far faster with increasing latitude than does latitudinal area of the globe (Fig. 5.15). Thus, an analysis for one family of monkeys that took account of both area and latitude found that latitude correlated better with diversity than did area [74]. And finally, and perhaps most convincingly, when one accounts for area, the tropics have a greater diversity of languages than do higher latitudes (Fig. 5.16A, Fig. 5.16B).

In conclusion, of the ten biological explanations for the Forster effect for which information is available for human cultures, six can explain the effect in humans (hypotheses 3–8 in Table 5.2); three cannot (hypotheses 1, 2, 13), and one is questionable (hypothesis 9).

Is It Valid to Seek Common Explanations for Variation in Biological and Cultural Diversity? The fact that a greater diversity of both cultures and species exists at lower latitudes [483; 542] indicates the potential for common explanations for both. The match is far from perfect, though.

One reason for the mismatch must be that the comparison is between variation within one species (humans) and variation across not only hundreds of species but species very different from humans, such as birds and even plants. Additionally, it seems likely that human cultures can more directly and quickly influence each other, by outcompeting each other, than can species. The history of humans has been one of unending invasion, direct and indirect slaughter, and aggressive and passive assimilation of one culture by another (chapter 10).

Is any more explanation necessary than imperialism for the relative paucity of cultures and languages of higher latitudes [122; 162]? Perhaps not. Expansive imperialist cultures have swept over the tropics, but they have existed for longer at higher latitudes, and when measured by degree of political complexity, they are found more at high latitudes than low [162]. On the other hand, gradients of diversity with latitude exist within regions conquered by colonial powers [122], and given that two-thirds of the top 25 most biodiverse countries are also the most linguistically diverse [325], it seems that explanations of the Forster effect unique to humans, such as imperialism, cannot be sufficient.

Correlating Cultural Diversity and Biodiversity with Productivity. Several of the hypotheses for the Forster effect have to do with various aspects of environmental productivity (Table 5.2, hypotheses 6–9 directly, hypothesis 7 indirectly). I here discuss the relationship in more detail.

Productivity means the rate at which plant material is produced. It is assessed by a variety of measures, some quite indirect, such as evapotranspiration rate [799]. In Binford's data set for hunter-gatherers, "productivity" is the same as "evapotranspiration rate" and correlates significantly with all the other climatic measures.[s]

The relationship between biodiversity and productivity is very loose. One review found that in only two-thirds of 200 or so studies was there a relationship, and in over a fifth of those (12% of total), the relationship was negative—i.e., less diversity in more productive sites or regions [799]. Even when a relationship of biodiversity with productivity is present, a lot of scatter exists around the relationship.

The loose relationship is only to be expected given the variety of measures of productivity, and the fact that soil quality, a crucial measure, is so often missing. Across species, the lack of a close match is also expected given the different requirements and habitats of different species. Some of the scatter, maybe even much of it, is also due to the different scales in time and space at which biologists test the relationship of diversity to productivity [826]. At the local scale, the relationship is more often hump shaped than any other relationship; diversity peaks at intermediate productivities. The shape at larger scales then depends on a variety of factors, including the nature of the sampling across the region, but it is mainly at very large scales that diversity more consistently increases with increasing productivity [132; 133; 666, ch. 12; 799; 826].

A relationship of diversity with productivity has been tested for humans by correlating cultural diversity with, for example, various measures of temperature and rainfall. The results vary, depending on the definitions of culture used and the sources. For instance, Elizabeth Cashdan, and Ian Collard and Robert Foley related cultural diversity to both latitude and

s. Productivity correlations. All, $P < 0.01$–0.0001 for globe. Data from [62].

various environmental measures, but they used different sources for both sets of parameters [122; 146]. Both studies found diversity of cultures to correlate with latitude, but only Collard and Foley found diversity to correlate with mean annual rainfall [122; 146]. And Cashdan found that while global diversity of what she termed noncomplex societies did not correlate with any of 13 climatic variables that she tested, diversity of chiefdoms and states correlated with seven of the variables [122]. Cashdan states that in Africa, cultural diversity correlated with productivity, but she did not in the statement distinguish the two levels of society [122].

In the case of hunter-gatherer societies in Binford's compendium, their significant decline in diversity with latitude (Fig. 5.16C) matches a more strongly significant decline of diversity with decline of productivity[t] [62].

Ian Collard and Robert Foley suggested, following Dan Nettle, that high productivity allows a high density of cultures, because the high productivity means that a population large enough to be viable can persist in a small area [146; 557]. The argument is supported by the facts that hunter gatherers live at higher densities in the tropics, the productivity of their land is higher in the tropics, populations live at higher densities in areas of higher productivity, and they live in smaller geographic ranges when at high density and in areas of higher productivity[u] [62]. However, it needs to be noted that Collard and Foley's study included far more than just hunter-gatherer societies.

The idea that large populations are less likely to go extinct than are small ones is amply supported in the literature on biology of extinction [157, ch. 7; 306; 310; 321; 617]. However, a potential problem with the Nettle–Collard–Foley argument is that high productivity and a large population could allow a competitive culture to expand across the region and hence reduce diversity. With local communities of species, especially of plants, exactly that expansion sometimes happens, which is a reason why the relationship of productivity to diversity can be humped (above). In the Americas as a whole, it looks as if cultures originally diversified

t. Diversity by productivity, hunter-gatherers. Number of societies summed per 5° latitudinal band; one productivity value per continent per band taken; global value is average of continental values ($N = 1$–5 per band); N. America excluded because of extraordinary central west coast diversity. $N = 15$, $r_s = 0.87$, $P < 0.0001$. Data from [62].

u. High cultural diversity at low latitudes via high population densities in small ranges when productivity high. Global density by latitude, $N = 60$, r^2adj. $= 0.3$, $F = 24$; productivity by latitude, $N = 59$, r^2adj. $= 0.6$, $F = 79$; density by productivity, $N = 60$, r^2adj. $= 0.34$, $F = 31$; area by productivity, $N = 60$, r^2adj. $= 0.3$, $F = 23$; area by density, $N = 61$, r^2adj. $= 0.44$, F $= 49$; all $P < 0.0001$. Data from [62].

rapidly, but then after some time, the more successful cultures begin to expand, and the less successful to go extinct [558]. Indeed, the power of Old World cultures, as judged by their political complexity, correlates better than any environmental factor with the size of the culture's range [162].

Collard and Foley's hypothesis was slightly different from Dan Nettle's. Nettle argued that consistent high productivity (as measured by a long growing season) would allow a culture to be self-sufficient and therefore to be able to avoid interacting widely with other cultures [557]. Merging of cultures, or assimilation of one by another, would then be less likely in the productive tropics than the less productive high latitudes, where cultures need to cooperate to survive. The opposite happens with a less self-sufficient culture.

Nettle's sampling to support his argument has been criticized because, for example, he ignored temperate regions [709]. I do not see a problem with the omission. It seems to me that the more homogeneous a sample, the fewer confounds and therefore the more robust the finding and interpretation. Certainly, in Binford's sample, which extends beyond the tropics, diversity of hunter-gatherers societies correlates significantly with growing season,[v] if not as strongly as with productivity [62].

A major problem with analyses that relate environmental variables to diversity is that although past history of the environment must play a large role, the values of the measures of the environment that we use are usually one-time, recent values. So also are the measures of biodiversity in a region and counts of number of cultures, because those are all we have. Exceptions exist, such as with the marine record [393; 395], but they are few. However, we have essentially no idea of what cultures were where before the arrival of people who kept written records. Ives Goddard's map of the distributions of Native American languages is based on first-contact accounts—i.e., largely accounts of European contact [268].

Nevertheless, human history and cultural diversity in the Americas indicates a major role for the environment relative to history. Humans expanded across the Americas so quickly (chapter 2) that in effect all of both Americas was populated simultaneously. And yet America is no different from Africa or Asia in having overall not only a higher density of cultures in the tropics than outside the tropics (Fig. 5.16A) but also in showing a number of the other biogeographical patterns that we see across human cultures elsewhere in the world (chapter 6).

v. Diversity of societies with length of growing season. $N = 15$, $r_s = 0.55$, $P < 0.05$, diversity of cultures measured as number per 5° latitudinal band. Data from [62].

Interactions of Regional Diversity and Local Diversity. Biogeography and macroecology are disciplines that investigate patterns at large spatial and temporal scales. What happens in a region affects what happens locally, because the species or cultures in any one locality are likely to be a subset of the species or cultures in the surrounding region, or related to them. Thus, biodiverse regions have biodiverse local communities [463, ch. 15].

Analysis of the relationship between regional and local diversity can be informative, telling us, for example, about intensity of local competition [463, ch. 15]. If the number of species at a locality stops increasing after a certain point, even as the number of species in the surrounding biogeographic region keeps increasing, a valid inference is that competition between species is so intense that no more can fit in the locality. For reasons yet to be determined, just such a phenomenon occurs among primates in Asian forests, but not in African or South American ones [449].

In the case of human cultural diversity, the regional pool does not necessarily have to be adjacent to the local community. Empires can be far-flung, creating far-distant regions from which local communities can draw residents. India and Algeria were within, respectively, the UK's and France's colonial spheres. There are 20 times as many Asian Indians in the UK than in France, for example [357, ch. 11], whereas 25 times as many Algerians live in France as in the UK [390].

Although at some point, a locale's diversity must be affected by regional diversity, isolated species and cultures can diverge rapidly from their parent stock. The high degree of endemicity of island flora and fauna is one example [805; 827]. For human cultures, Native Americans are an excellent example. Arriving in the Americas from northeastern Siberia perhaps just 15,000–12,000 years ago (chapter 2), the diversity of native languages in the Americas is nearly as great as anywhere else in the world, a median of 7 per 100 sq km, compared to 11 in Africa and Asia[w] [277], despite the deliberate and inadvertent genocide of Native Americans of both Americas by Europeans [483; 735].

. . .

The diversity of human cultures is greater in the tropics than at higher latitudes, as is the case for biodiversity. A variety of functional mechanisms

---

w. Diversity of languages. Calculations for only mainland countries with centers more than 10° lat. apart. Data from [277].

could operate to produce this biogeographical pattern (Table 5.2). Some mechanisms proposed for biodiversity can be rejected as applicable to human cultural diversity, and so their generality might be questionable. Others fit humans and so are substantiated.

I think that it is possible that the best combination of explanations for human cultural diversity could be slightly different from that for biodiversity. Productivity, stability, and clemency are important for both cultural and biological diversity. However, for human cultural diversity, the argument that I like depends more on productivity of the tropics. For biodiversity, it depends more on the tropics' long-term environmental stability and clemency.

The tropics' greater year-round availability of energy allows consistently higher productivity than at higher latitudes. This stable high productivity then allows viable populations to persist in small ranges, which in turn allows denser packing of cultures. The same argument could apply to species, but not so strongly, because unlike the ranges of most cultures, species' ranges can overlap, meaning that the packing argument is not so crucial to explaining tropical biodiversity. Instead, a main influence is long-term clemency of the tropical climate allowing a diversity of forms to exist.

The arguments and their distinction might be more comprehensibly expressed in terms of the low productivity, inclemency, and short- and long-term instability of the higher latitudes. Low productivity there, especially in winter, forces large ranges on high-latitude cultures, which correlates with low diversity of cultures in any one spot. The same explanation might be valid for species, but in addition, the high-latitude winters restrict the forms that species can take, and the ice ages keep preventing in situ evolution.

LATITUDE, CLIMATE, AND POPULATION

I have described so far in this chapter how and why human form differs across latitudes and climates, and similarly for diversity of cultures. In this section, the topic is the distribution of population size, measured as both density and numbers of people.

Humans Are Distributed Very Unevenly

Nobody needs telling that we are distributed unevenly across the globe. Nobody needs telling that some cultures have many people, others few.

My point here is the extreme unevenness of the distribution. We occur at our highest densities regionally in central Europe, in a large area just south of the Himalayas, and in much of central China [142]. Relative to the rest of the globe, these three main regions of high population density are exceedingly small, corresponding to the fact that half of the world's human population occupies less than 3% of its land area, excluding Antarctica and boreal forests [710]. Humans are common hardly anywhere; we are rare almost everywhere.

The same highly skewed distribution is evident in the population density of speakers of each of the world's languages and the population density of hunter-gatherer cultures (Fig. 5.17A, B), as well as in the numbers of people in each culture, whether culture is defined by language or hunter-gatherer society. Globally, 5% of languages have over 1 million speakers, 75% have under 30,000 speakers, and 12.5% have less than 100 speakers [277]. For hunter-gatherer societies, the least numerous 10% numbered less than 16,500 people each, while the top 10% numbered more than 360,000 each [62]. These counts for hunter-gatherer societies come from over many decades past, and all are probably lower now.

It turns out that there is nothing unusual in this sort of uneven spatial distribution and skewed distribution of densities and numbers. Population sizes of species of diatoms and butterflies are skewed [555], as are densities of the world's primates. Of 149 primate species for which I have obtained from the literature data on their population density, the lowest 15% occur at densities of under 15/km²; the top 15% live at densities of over 130/km².

In fact, distributions, densities, and numbers of items of almost anything are distributed unevenly. In Shakespeare's works, only about 850 words appear 100 times or more; over 25,000 appear less than 10 times [197; 555]. The top 6 words in current English in TV and cinema are used more often than the next 19 of 1,000 [831]. The most common item in my kitchen cupboard was as common as the next 8 in total, of 10 categorized.

There must be some biology in the distribution of living organisms, but the fact that species, including humans, and inanimate objects have roughly similar skewed distributions indicate that the role of biology, as opposed to accident, might be minor. More or less random birth, death, and dispersal rates among more or less similar organisms will produce the skewed distribution that is so often observed [53; 380; 381; 489; 555]. Extra biological explanation is needed for only those species or cultures that are obviously more or less common than expected from

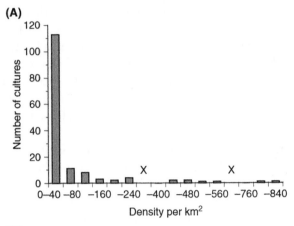

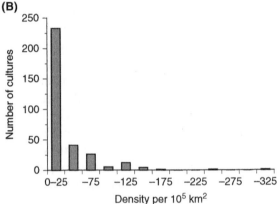

FIGURE 5.17. (A). Speakers of most languages live at low population density. Number languages by median density (#/km²) speakers per language per country (excluding N. Korea (16,500), islands, political enclaves, e.g., Gaza, Vatican). x = break in x-axis. Data from *Ethnologue* [277]. (B). Members of most hunter-gather societies live at low density (excluding islands). Number of societies by population density (# per 10^5 km²). Data from [62].

the mathematically described skew. One difficulty in estimating whether any special biological explanation is needed is that a variety of mathematical, nonbiological formulae produce somewhat different null distributions [555].

With humans, factors in addition to chance and climate play an important role in our distribution [710]. For instance, 12 of 14 of the world's cities of over 10 million people in the 1990s were coastal [571]. In general, humans live at high densities on coasts, along rivers, and at low elevations [710], although, of course, exceptions exist, such as the 8–9 million people of Mexico City at 2,250 m.

The geographic (as opposed to climatic) correlates of human density are probably a combination of ease of transport and communication by water, a food supply from both water and land, and in the case of proximity to rivers and their associated low areas, a supply of fertile alluvial soil [710]. The extraordinarily high concentrations of humans at the southern base of the Himalayas and in central China can probably persist because the year-round snow and glacier-fed river flow from the Himalayas in northern India and central China, along with the relatively clement climate in those regions, allows highly productive year-round agriculture [710]. The same explanation could apply to central Europe, the rivers of which are fed by the Alpine snows and glaciers.

Population Density

Agricultural productivity allowing a large population is, in effect, the argument in the last section about why the tropics have a high diversity of cultures. The high productivity of the tropics allows relatively large numbers of people of each culture to subsist in a relatively small area, which facilitates a dense concentration of cultures [146].

But are human populations (as opposed to cultures) more dense in the tropics? Among nonhuman primates, no relationship exists between latitude and population density of taxa[x] [309]. Likewise, among present-day humans on mainlands, no relation of density of speakers per language to latitude is evident (nor of number of speakers)[y] [277]. However, the median density of speakers per language on islands is higher at lower latitudes[z] [277].

Correlations of latitude and associated climate with human population density might be more sensitively indicated by examination of

x. Density by latitude. Primate genera, $N = 44$, r^2adj = 0 (lemurs, outliers omitted, lemurs because at low latitude and unusually high density) [313].

y. Mainland population density or size per language by latitude. Median density, number speakers, $N = 44$, NS for globe, continents separately. Data from [277].

z. Island population density speakers per language by latitude. Globe, $N = 25$, r^2 adj = 0.4, $F = 15$, $P < 0.001$. Data from [277].

peoples who depend more on their immediate environment than do so many humans currently. As described in the last section, hunter-gatherer populations indeed live at higher densities in the tropics [62]. The data presented there were for the globe. The same is true of sub-Saharan Africa, Australia, and Asia, but the two Americas show no relation between population density per culture and latitude[aa] [62]. These hunter-gatherer data, then, partially support the hypothesis that the high productivity of low latitudes could promote cultural diversity there through allowing high density and hence a viable population in a small area [146].

Latitude is an index of a variety of variously interrelated climatic variables. With regard to climate and current human density across the globe, climate influences the extremes of density. Humans occur at high density in lands where mean annual temperatures are between 10 and 20°C and mean annual rainfall is between 50 and 200 cm [710], and we occur at our highest densities where the range of annual rainfall is highest, in the monsoon areas of India and eastern Asia [710]. We are at lowest densities where temperatures or rainfall are extreme (poles, deserts).

Among hunter-gatherers, multivariate models of population density by climate produced no clean result. The temperature in the coldest month of the year was the strongest correlate for both the globe and North America, but plant biomass was the sole strongest correlate for Australia[bb] [62].

The finding that population density correlates best with temperature in the coldest month of the year fits arguments concerning environmental constraints on diversity [320]. The finding that density correlates with primary biomass fits the Collard–Foley argument relating productivity to cultural diversity via population density [146].

The relationship of high productivity and density to small geographic range in the tropics is substantiated at the smaller scale of camp moves by groups. Hunter-gatherer groups move shorter distances per move at

aa. Population density per hunter-gather culture by latitude: Globe, $N = 60$, r^2adj. = 0.3, $F = 24$, $P \leq 0.0001$; Asia, Australia, sub-Saharan Africa, $N \geq 15$, F, $r_s \geq 0.4$ (r_s if $N < 20$), $P < 0.01$. Globally, latitude had strongest effect (AICc = 89, $P < 0.0001$), but minimal models included continental effects (AICc = 85). Data from [62].

bb. Population density by temperature coldest month. Globe, $r^2 = 0.3$; North America, $r^2 = 0.4$. By primary biomass, Aus., r^2adj = 0.7. Data from [62].

lower latitudes,[cc] higher-density populations move less,[dd] and populations in areas of higher productivity and higher primary and secondary biomass (plant, mammal) move less[ee] [62]. And finally, cultures that move less far per move on average cover smaller areas,[ff] and these populations are low-latitude ones[gg] (chapter 6). The Collard–Foley explanation for high diversity at low latitudes seems to work well, even if multiple models of both comparisons (effects on distance moved, effects on area) show only latitude as a significant correlate.

I will return to the association of area covered by cultures with latitude and climate, and the relationship's correlation with the Forster effect (i.e., the latitudinal diversity relationship) in chapter 6, under the heading of the "Rapoport Effect."

Population Size

So far, the topic has been density of peoples in relation to latitude and climate. I turn now to numbers—population size. Population sizes do not necessarily show the same relationship with either latitude or climate as does density because the area occupied by species and cultures changes with latitude and climate (chapter 6). Numbers are relevant to the ability of populations to persist in the face of stochastic risk and to competition between populations (chapters 8 and 10).

Among nonhuman primates, no relationship between latitude and population sizes exists [313]. Regarding humans, we all surely know that more people live in so-called developing countries than in so-called developed countries (roughly 5.5 billion compared to 1.5 billion [844])—hence, the fairly recent terms of "majority world" and "minority world" for the two divisions. To some extent, developing and developed countries are tropical and temperate countries, respectively. However, China and India, which between them have about 2.5 billion people, are still sometimes

cc. Average distance moved per move. By latitude, $N = 43$, r^2adj = 0.6, $F = 69$, $P < 0.0001$. Data from [62].

dd. Average distance moved per move. By population density per culture, $N = 44$, r^2adj = 0.2, $F = 31$, $P < 0.0001$. Data from [62].

ee. Average distance moved per move. By productivity, $N = 42$, r^2adj = 0.2, $F = 11$, $P < 0.01$; by plant biomass, $N = 43$, r^2adj = 0.5, $F = 44$, $P < 0.0001$; by mammalian biomass, $N = 8$, $r^s = 0.9$, $P < 0.01$. Data from [62].

ff. Area used by cultures. By average distance moved per move, $N = 45$, r^2adj. = 0.2, $F = 14$, $P < 0.001$. Data from [62].

gg. Area used by cultures. By latitude, $N = 62$, r^2adj. = 0.4, $F = 40$, $P < 0.0001$. Data from [62].

counted as developing countries, but their densest regions are temperate, just north of the Tropic of Cancer.

Across all humans currently, the number of speakers per language group does not change with latitude, either globally or per continent, or whether only continental land masses were counted, or only islands [277]. Using the same source, *Ethnologue,* Currie and Mace found that number of speakers per language in the Old World increased with latitude, up to about 45° [162]. As far as I can see, they did not correct for spatial autocorrelation, and they used latitudinal bands rather than absolute latitude of each language. Nevertheless, for the same region, and with the same statistic, I get the same result with the spatial control that I used, including the peak at about 45°, if at a lower level of significance[hh] [277].

Similarly, population sizes per society of high-latitude hunter-gatherer societies were higher than those of low-latitude societies globally[ii] [62], if not significantly so within each continent, because the societies at higher latitudes inhabited larger geographic ranges (chapter 6). The relationship between numbers and latitude is so loose (only 10% of variance explained), even if statistically significant, that I will not attempt to relate numbers to any environmental factor. Instead, I will wait until the next chapter to investigate how area used by a culture might vary with the environment.

Conclusion. The population density of hunter-gatherer cultures and the number of their members seem to have been more closely tied to the climate than the densities or numbers of agricultural and urban humans. Trading systems and modern technology obviously allow most of us to be more divorced from the environment than are hunter-gatherers. Nevertheless, the densest populations of humans are still found where climate is most clement and year-round high agricultural production is possible.

Adaptive explanations for our uneven distribution across the globe and the highly skewed distribution of population density and size of

hh. Mean number speakers per latitude. $N = 56$, $r_s = 0.35$, $P < 0.008$. Data from [277].

ii. Population size of hunter-gatherer societies by latitude. Globe, $N = 61$, r^2adj. = 0.1, $F = 8.0$, $P = 0.01$. However, continental effects were strong: best model, AICc = −16.8, Americas vs. other continents; next best (AICc = −16.2), latitude, Americas vs. other continents. In four of five continents, number increases with range size (not S. America), but none significantly so. Data from [62].

each culture need to be tempered with the knowledge that random distributions are uneven and that numbers of inanimate objects are also highly skewed. Perhaps an explanation for uneven distribution of densities and numbers is something to do with the likelihood that only a few ways exist to be a generalist, and hence common, but many ways to be a specialist, and hence uncommon.

6

Use of Area

The geographical areas of distribution are the Chinese-
lantern shadows produced by the different taxa on the
continental screen: it is like measuring, weighing, and
studying the behaviour of ghosts. [629, Rapoport, 1982, p. 1]

SUMMARY

In common with a number of other large-bodied animals, primates living on islands are usually smaller than are their mainland relatives, the so-called island rule. Thus, the extraordinarily small Flores "hobbit" (*Homo floresiensis*) fits biogeographic expectations, especially if it is more closely related to *H. erectus* than to modern humans.

No other large-bodied mammalian species has as large a geographic range as do humans. But a minority of species with a large range, and a majority with a small range, is the norm for all taxa as well as for human cultures.

The taxa and cultures most likely to have a small geographic range are tropical (the "Rapoport effect"). The explanation for tropical cultures using small areas might be that a greater density of resources in the tropics allows cultures to persist in small areas at population sizes that resist extinction. Additionally, the greater climatic variability of high latitudes means that for humans to survive there, they need to cover large areas. Additionally, tools and the longitudinal homogeneity of high-latitude zones allow them to do so.

The less space available, the fewer species there are, for example, on islands by comparison to neighboring mainlands. The same is true of human cultural diversity. However, the great tropical diversity of both species and cultures is not explained by area, because in fact the tropics encompass less land than do temperate zones.

Many of the explanations for why the geographic ranges of cultures are smaller in the tropics than at higher latitudes come from our understanding of the biogeography of nonhumans, including our understanding of the Forster effect—the phenomenon of high tropical diversity compared to temperate diversity. Those explanations that also work for humans are strengthened; those that do not are weakened, unless we can show why the explanations should work for nonhumans but not humans.

Island populations of animals, including primates, live at higher densities than do mainland populations. That is not the case for humans. For primates, absence of predators and competing species are main explanations for their high densities on islands. However, humans are the top predator and competitor wherever they go. Instead, a paucity of parasites and pathogens on islands could explain humans' relatively high island densities.

If Eduardo Rapoport was measuring ghosts (chapter's opening quote), he had psychic powers, because he can be counted as one of the founders of quantitative biogeography [629]. He is particularly famous in the field for his analysis of the biology of the geographic range size of species.

As in the previous chapter, and as I hope is evident from the summary, this chapter is separated into the three themes of the biogeography of the average person or culture in a region, of the diversity of cultures, and of numbers of individuals in a region.

AREA AVAILABLE AND AREA USED

Area and Body Size: The "Island Effect"

If you want to see otherwise large mammals that are unusually small, go to a small island. Until about 10 kya, elephants as small as 1.5 m at the shoulder inhabited several Mediterranean islands [610]. If you want to see small animals that are unusually large, go to a small island. The 20 kg dodo (*Raphus cucullatus*) of Mauritius is (was) a pigeon; the 70 kg Komodo dragon (*Varanus komodoensis*) of insular Indonesia is a lizard. In general, where island species differ in size from mainland species, large-bodied species, especially mammals, become small on small islands, and small-bodied species become large, with the greatest change evident in lizards and birds [462; 463, ch. 14; 519; 522; 575]. Combining nonflying taxa (reptiles, mammals), the crossover size is between 100 g and 1 kg [462].

The effect is called the "island rule" in the literature, but as for the Bergmann and Allen rules, I use the word "effect," because the rules are far from universal. The island effect applies to both mainland–island comparisons and to small islands compared to large islands.

Explanations for the island effect combine the consequences of competition for resources and release from predation [462; 463, ch. 14; 513]. Large-bodied animals need a large quantity of food [512; 513] but benefit by being vulnerable to fewer predators [513; 707]. On small islands, where it is difficult to move away from bad conditions, the costs of needing lots of resources can outweigh the benefits of being large to escape predation, especially because small islands often have fewer predators[a] than equivalently sized areas on the mainland, perhaps even none, and where there are some, they too have evolved to be smaller [462; 519]. Small individuals of large-bodied species then do better than do large individuals, especially if on islands they have lowered metabolic rate, as they often do [513].

By contrast, small-bodied species, which hide to escape predators, no longer benefit so much (because there are fewer predators) and can instead benefit from being large in contest competition with not only their own species but other species too [239]. Of course, as both extreme sizes approach a threshold size, they will start to experience the opposite direction of island influences.

A potential confounding sampling error is present, though. The arrival of the large-bodied mammal *Homo sapiens* on islands around the world was the death knell for almost all the islands' large-bodied, eatable inhabitants (chapter 10). In other words, the upper end of the distribution of body mass of the island species has been removed. Is the detected island miniaturization therefore merely a consequence of human-caused extinction of large animals? It could be. On the other hand, the island effect includes gigantism of some smaller-bodied taxa [462; 463, ch. 14], which takes some species well into the range of body sizes that humans hunt. The Komodo dragon, which still occurs on Flores, is an example.

Nonhuman Primates and the Island Effect. Nonhuman primates ("primates" from now on) are no exception to the island effect. Primates are relatively large mammals with regard to the island effect (a median mass of about 3 kg), and where there is a difference between island and mainland forms, or between small islands and large islands, the forms on the smaller land mass usually weigh less or have shorter bodies (Fig. 6.1) [94]. The extreme examples are, for body weight, the 7 kg pigtailed langur of the Mentawai Islands (*Nasalis concolor*) compared to its congener, the 10 kg proboscis monkey of Borneo (*Nasalis larvatus*), and for body length, the 26 cm Preuss's guenon (*Cercopithecus preussi*) of Bioko Island compared to the 45 cm of the same species in neighboring Cameroon.

---

a. Few predators on islands. Predator populations are tens of times smaller than prey populations [117] and therefore far more likely to go extinct in any given area [753].

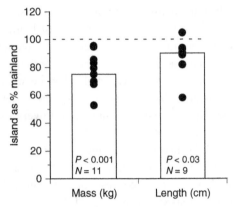

FIGURE 6.1. The island effect in primates. Primates on small islands are smaller by two measures than are their relatives on larger islands or the mainland.* Island size as a percent of mainland size, for mass and length. Data from [94].

*Island effect in primates. In all island–mainland pairs of related species (N = 11), the island form weighed less (t = 5.4, P < 0.001); in eight of nine pairs, the island form's body length was shorter (t = 2.7, P < 0.03); for weight, but not length, the larger the mainland form, the greater the decrease in size of the island form (N = 11, r_s = 0.88, P < 0.001). If each genus is counted only once (and the median species' pair's values used), all six genera for mass and all four for length of body were smaller on islands. Data from [94].

In keeping with the island effect of miniaturization of large taxa, and maybe for some of the same reasons (lack of resources keeps large animals small), larger species occur in larger land areas, whether islands or not; only small species live in small areas [225]. Within and across species on islands too, body size matches island size [344; 462].

The relationship works for primates too. In the Sunda Islands, the size of the smallest island on which a genus occurs matches body size of the genus[b] [307; 310]. For instance, while small-bodied genera of primates occur on islands of a variety of sizes, genera of over 10 kg (probos-

b. Body size of Sunda island primates by area of island. $\log_{10}$ body mass of genera by $\log_{10}$ smallest island inhabited, N = 8, r_s = 0.87, P < 0.01. Data from [310].

cis monkey, siamang, orangutan[c]) occur on only the two largest islands, Borneo and Sumatra [307]. In other words, the island effect might be, more generally, an area effect.

The mammals of Borneo are an exception to the island effect, despite the proboscis monkey, siamang, and orangutan. Borneo is larger than the neighboring Malaysian Peninsula and than the nearby other two next smaller Sunda Islands, Sumatra and Java. Its taxa should therefore be larger than their relatives on the peninsula or the smaller islands. And yet the Bornean taxa are on average smaller [521]. Of the 13 comparisons of six species of primate males and females in the sample, half are smaller on Borneo. Two explanations might work. Borneo's soils are poorer than either Sumatra's or Java's, so food is less abundant, causing animals to be smaller [521]. Also, the surprising lack of large terrestrial predators on Borneo[d] might allow smaller animals to persist [521].

I prefer the soil/food explanation. A lack of terrestrial predators does not explain why arboreal primates are smaller on Borneo. Also, the difference in size of the Bornean and other regions' forms must be too small to have made a difference to likelihood of capture: the non-Bornean form or species is never more than 10% larger than the Bornean form, a difference well within the range of sizes of prey of most predators [707].

Homo Floresiensis. The previous analysis of the island effect in mammals and especially primates brings us to *Homo floresiensis,* the miniature "hobbit" of Flores [103; 548]. *H. floresiensis,* at a little over 1 m tall and perhaps only 25 kg in weight, is far smaller than any extant pygmoid peoples (Fig. 6.2), indeed smaller than a chimpanzee, which weighs nearly 40 kg (Fig. 5.3). The smallest peoples, in the eastern Congo rainforest, are about 1.4 m tall and about 40 kg in weight [356; 595; 647]. Floresiensis' brain is also extremely small [103]: at about 400 cc it is the size of a chimpanzee's, less than one-third of the brain size of pygmies and other modern humans [510].

One of the reasons that anthropologists and biogeographers were excited by Floresiensis is that it was an obvious potential example

---

c. *Nasalis larvatus, Hylobates symphalangus, Pongo.*

d. Lack of Bornean predators. Borneo has no tigers, for instance, although tigers are present on the far smaller Sumatra and Java.

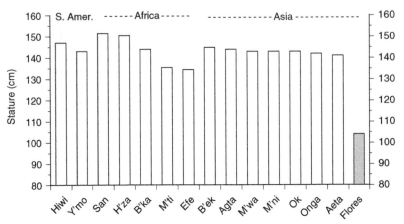

FIGURE 6.2. Stature of Flores "hobbit" compared to the world's pygmoid peoples (female only). Data from [595; 647].*

*Stature from [595] (only male given), altered to female stature with median ratio from [647].

among hominins of the island miniaturization effect [518; 563; 595; 821], and indeed of other traits of island species such as robust limbs [518]. It was yet another example that our ancestor and therefore perhaps us were following the same biogeographical trends as other mammals. The smaller the island, the smaller the animals on it, according to biogeography's "island effect." And Flores is a small island, just 13,500 km². It is about the same size as Northern Ireland (Ulster) of the United Kingdom, and a little smaller than New Jersey, the fifth smallest state in the United States. It is nearly 10 times smaller than Java, the island on which Eugène Dubois found the first *Homo erectus* in 1891.

I mention Erectus here, because from the start Erectus seemed a likely ancestor for Floresiensis [103]. The fact that the earliest hominin stone tools on Flores appear to lie in a 1 million-year-old stratum [105] substantiates the hypothesis, assuming that the tools did not descend from a younger stratum. Erectus as the ancestor appears to be the current consensus among morphologists and paleontologists, although some are suggesting *Homo habilis* as the sister taxon [4; 19; 29; 210; 211; 549].

Early hominins covered a wide range of body sizes [510]. Habilis and certainly Erectus are far larger than Floresiensis, and they have larger brains. Habilis females weighed about 30 kg and had roughly 600 cc brains; Erectus was heavier than modern humans, at a little over 50 kg, though with brains smaller than our 1350 cc, at 800–1,000 cc [510]. So whether Floresiensis is related to Habilis, Erectus or us (modern humans), it is miniaturized. To find a hominin as small as Floresiensis in both body and brain size, we have to go back to *Australopithecus afarensis*, a species alive in Africa 4–3 mya [510].

Island taxa with skull and body sizes 50% of their mainland relatives are not uncommon [462]. Of course, the degree of miniaturization varies [462; 519]. Nevertheless, while Floresiensis' stature is about 30% less than that of the average female pygmy (Fig. 6.2), the reduction is far greater than humans' closest relatives, the primates, have apparently experienced on islands, namely, a median of 10% less than mainland size (Fig. 6.1) [94].

However, that primate value of 10% comes from comparing extant island populations with extant mainland ones. Many of the extant island populations of primates have been isolated for perhaps less than 20,000 years—since rising sea levels cut off islands formerly connected to mainlands. As Floresiensis has been on the isolated island of Flores for around 90,000 years, and hominins for maybe 1 million years [105] (with proviso about evidence getting from younger to older strata), plenty of time has existed for island dwarfs as extreme as Floresiensis in body size to have evolved.

In addition to island miniaturization of hominins, another reason for anthropological and biogeographic interest in Floresiensis was that it was the first new species of *Homo* claimed for Asia since Dubois' Erectus over a century previously. It increases to three the number of hominins in the region for about 20,000 years since modern humans arrived there, around 50 kya. Erectus lasted on Java until perhaps 32 kya [748], about the same date as the demise of Neanderthal [220, ch. 7]. Floresiensis completely overlaps those 20,000 years, for its remains span much of the last 100,000 years, from roughly perhaps 95 kya to 17 kya [103; 548; 549].

Several interpretations exist other than that Floresiensis is a classic example of island miniaturization, evolved from an early hominin [18]. One is that the morphology is a result of a normal genetic mutation [647]. Another and quite common argument has been that the morphology is that of an individual suffering from a pathology, such as microcephaly [353; 398; 429; 496; 572; 816].

Suggestions that extraordinary new hominins show pathologies are not new. When the first Erectus and Neanderthal skeletons were discovered, they too were described as showing pathologies, in part because of resistance to the idea of human evolution [170]. Those suggesting pathology now for Floresiensis are of course far from creationists. Instead, they hypothesized pathology because, for example, the reduction in brain size seemed to be too great to be accounted for by the reduction in body size [496]. Subsequent studies that specifically investigated the potential for pathologies have both denied [210; 211; 416] and confirmed them [194], with continuing confirmations of the denials and denials of the confirmations.

Leslie Aiello nicely reviews this debate, along with all the other debates—including almost political ones—that surround the extraordinary Flores hominin finds [4].

. . .

If Floresiensis is a natural hominin taxon, understanding of the biogeography of the island effect can be validly applied to understanding of *Homo floresiensis*. Conversely, Floresiensis has already received more detailed anatomical analysis than almost any other island species—cranium, mandible, upper and lower skeleton, wrist, and feet have all been minutely examined [549]. That detailed analysis is going to advance understanding of island miniaturization in other species. On the other hand, if all the Floresiensis material shows signs of pathology, then our understanding and appreciation of the limits of human adaptability and ability will be shaken.

On the topic of human adaptability and ability, Flores is east of the Wallace line and has never been connected to mainland Asia. If Floresiensis is a natural hominin taxon, then its existence is evidence of overseas dispersal of hominins approaching 1 million years earlier than most accept (i.e., earlier than roughly the time of modern humans' dispersal out of Africa).

Most Species/Cultures Use Small Areas

Nonhuman Primates. I have already shown that most nonhuman primate taxa live at low density (chapter 5), an example of the fact that only a few "things" are common; most are rare. Concomitantly, most primates have small geographic ranges (Fig. 6.3) [157, ch. 5; 195; 196; 311; 413; 591; 601; 858].

Because this sort of skewed distribution is more or less universal, it could be expected that the species that are rare are a random selection of all species. That appears not to be the case for primates. The biology of the rare ones is different from the biology of the common ones. In brief,

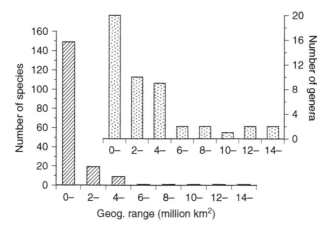

FIGURE 6.3. Most primates are rare—most have small geographic ranges. Number of taxa by size of geographic range. Data from [313].

in the case of primates, the rare ones are more specialized than are the common ones[e] [311; 314].

Humans. Most human cultures[f] too cover a small area (Fig. 6.4A, Fig. 6.4B) [473].

In the case of hunter-gatherer societies, and therefore some languages, the fact that most find themselves in only a small geographic range could be because more powerful cultures have compressed them into a small range. In other words, we are seeing an effect of politics and colonialism, not a natural phenomenon. That is not the case.

In the first place, competition between populations is very much a natural phenomenon. Individuals compete, groups compete, populations compete, species compete. So, if indeed a few cultures have competitively compressed many into small ranges, that compression is a natural phenomenon. Secondly, the fact that if we examine only hunter-gather societies (i.e., exclude colonial powers) we still see the same skewed pattern of many small ranges and only a few large ones indicates that at least

e. Before extinction, a species has to be rare. Nevertheless, the biological traits of rare primates are not the same as the traits of primates prone to extinction [311; 314]. In general, the biology of rarity is more complex than the biology of extinction [437] for as yet unexplained reasons.

f. I explain my use of "culture" in chapter 1.

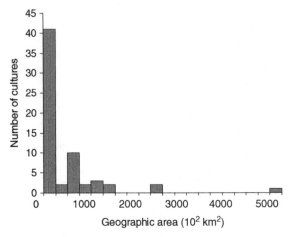

FIGURE 6.4A. The majority of hunter-gatherer societies use a small geographic range. Number of cultures by size of geographic range. Data shown for only mainland societies. Data from [62].

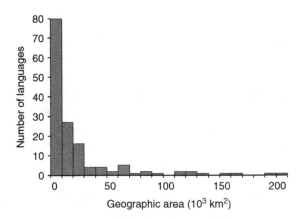

FIGURE 6.4B. The majority of languages cover a small geographic range. Number of languages by size of geographic range, calculated as mean of number languages/size of nation. Data shown for only mainland societies. (Number languages by size of range. Only mainland cultures; odd enclaves omitted [e.g., Gaza, Monaco, Vatican]; Kazakhstan and Saudi Arabia, at 350–400,000 km², omitted. Data from [277].)

colonialism is not the only cause of the pattern. Indeed, a little later in the chapter, I show that hunter-gatherer ranges are largest precisely where we might expect colonial influences to be strongest, namely, in the very continents of the colonial powers, the ones at higher latitudes.

Animals and Humans Use Less Area at Lower Latitudes

Among many sorts of organisms, those with small geographic ranges (the majority) are more often at low latitudes (tropics), while the few with large ranges are more often at high latitudes. This pattern is known in biogeography as the Rapoport rule, after Eduardo Rapoport [629, ch. 5]. Although many exceptions occur [254, ch. 3; 258], which is why from now I term it an "effect," the effect exists in all sorts of organisms— marine, terrestrial, plants, invertebrates, and vertebrates [629, ch. 5; 725].

Nonhuman Primates. Where the Rapoport effect exists, it is often stronger among temperate taxa than in the tropics [254, ch. 3; 258; 660; 674; 677; 725]. It was claimed to be largely, even solely, a temperate phenomenon, in part because environmental conditions were deemed to be fairly constant across the tropics. However, although primates are largely a tropical taxon, they too show the phenomenon of increase in size of geographic range with increase in latitude, if not consistently (Fig. 6.5) [158; 195; 196; 309; 674]. The effect is especially strong in African and Asian primates [309]. It is nonexistent in South America and Madagascar. In South America, the effect might be superseded by the strong influence of the Andes on distributions [676; 677], whereas Madagascar perhaps does not extend over enough latitudinal extent for any effect to exert itself.

The correlation between latitude and range size is often stronger with latitudinal extent as the measure of range size, rather than total geographic range size. The distinction has to do with the fact that a wide latitudinal range almost inevitably covers a wide range of climates, whereas a wide longitudinal range does not necessarily do so. High-latitude organisms, then, must cope with a seasonally variable climate, with the corollary that the necessary adaptability enables them to extend over a wide latitudinal range; conversely, tropical organisms are not adapted to change or extremes and are therefore confined to small geographic areas [725].

For primates, the climate at the higher latitudes that the order reaches is more variable than it is near the equator [158; 309], and correspondingly, primates at higher latitudes and those with greater latitudinal

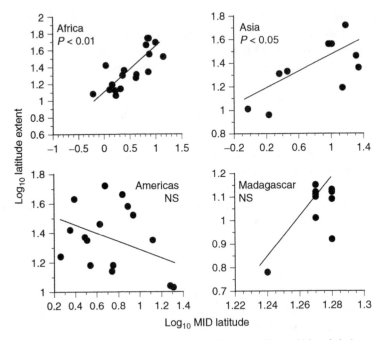

FIGURE 6.5. Some primate genera show the Rapoport effect—Africa, Asia.* Latitudinal extent of generic range by mid-latitude of genera. Redrawn from [309].

*Rapoport effect, primates. Africa, $N = 18$, Spearman $r_s = 0.75$; Asia, $N = r_s = 0.6$; Americas, r_s - 0.2; Madagascar, $r_s = 0.2$. (Outliers excluded; P value shown accounts for phylogeny with CAIC). From [309].

extent are more variable, as indicated by greater morphological variability (measured by number of species per genus[g]), broader diet, and greater variety of habitats used (Fig. 6.6) [309; 314].

There is a problem, though, with concluding that the evidence matches the explanation for the Rapoport effect. It is that correlations of latitudinal extent of taxa with latitude do not always match the variability of taxa (as measured by their dietary variety, for example) [309]. Nevertheless, the variability explanation remains the best on the table.

Humans. Human cultures too show the Rapoport effect of increased geographic range with increasing latitude [62; 146; 162; 277; 473; 629,

---

g. More subtaxa per taxon at higher latitudes. Note, this phenomenon does not negate the fact that taxonomic diversity is greater at lower latitudes: lower latitudes have a lot more taxa, even if they are not subdivided into as many subtaxa.

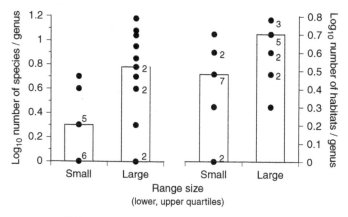

FIGURE 6.6. Primate genera with large geographic ranges (top 25% in their continent) are more variable than are genera with small geographic ranges (bottom 25% in their continent).* (A) Number of species per genus by range size. (B) Number of habitats per genus by range size. Column = median. Numbers by points = number of genera with that number of species. Data from [314].

*Range size by variability. N = 13, 13; # species/genus, z = 3.6, P < 0.0005; # habitats, z = 2.9, P < 0.004; # items in diet, z = 2.95, P < 0.005 (not shown in figure). Data from [314].

ch. 5]. For instance, area covered by languages increases with latitude globally and within several continents, accepting the crude calculation of that area (Fig. 6.7A), as does the area used by hunter-gatherer cultures (Fig. 6.7B) [62; 277]. A far more precise data set, that of the Great Plains Native American languages at the time of first reliable records, also might indicate operation of the Rapoport effect (Fig. 6.7C) [268]. As specific examples, the mean geographic range size of indigenous languages in the tropics is less than a quarter of that outside (3,000 vs. 12,800 km² [277]), and similarly with hunter-gatherer societies, which range over roughly 4,000 sq. km in the tropics compared to 13,500 sq. km outside [62].

Explanations offered for why humans show the Rapoport effect are similar to those offered for animal taxa. We need to be careful, though, when comparing phenomena within species (humans) with patterns across species (e.g., nonhuman primates).

Nevertheless, the idea that adverse conditions at higher latitudes require larger ranges for survival fits humans. Globally, range size of hunter-gatherers increases with decrease in mean annual temperature and rainfall; in the two continents with a reliably large sample, range size increases with decrease in one or more of productivity, mean minimum

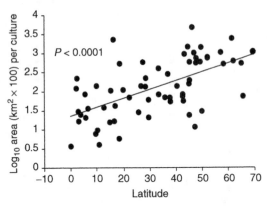

FIGURE 6.7A. Human languages show the Rapoport effect. Range size per language (mean per nation) by mid-latitude. Regression line excludes outliers. (Rapoport—languages. Globe, $N = 43$, r^2adj $= 0.24$, $F = 14$, $P < 0.001$; Africa, $N = 13$, $r_s = .65$, $P < 0.02$; N. America, $N = 2$; S. America, $N = 8$, $P > 0.1$; Asia – $N = 14$, $r_s = 0.55$, $P < 0.05$; Oceania, $N = 2$. Only mainland cultures. Regression line in graph omits outliers. Note: language area is simply area of country/number languages in country; in effect, territoriality of languages is assumed, as in other works [146]. Data from [277].)

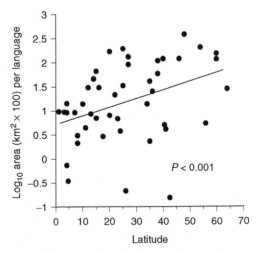

FIGURE 6.7B. Human hunter-gatherer cultures show the Rapoport effect. Range size of culture by mid-latitude. (Rapoport—hunter-gatherers. Globe, $N = 62$, r^2adj $= 0.39$, $F = 40.5$, $P < 0.0001$; Africa (sub-Sahara), $N = 14$, $P > 0.1$; N. America, $N = 75$, r^2adj $= 0.15$, $F = 8$, $P < 0.01$; S. America, $N = 18$, $P = 0.1$; Asia, $N = 15$, $r_s = 0.7$, $P < 0.01$; Australia, $N = 26$, r^2adj $= 0.3$, $F = 6$, $P < 0.01$. Regression line in graph omits outliers. Data from [62].

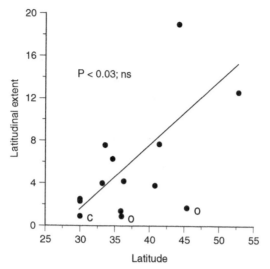

FIGURE 6.7C. Human languages (Great Plains, Native Americans) might show the Rapoport effect, depending on sample. Latitudinal extent of language family by mid-latitude. (Rappoport—Great Plains languages, North America. $N = 11$, $r_s = 0.7$, $P < 0.03$. Data are language families extending within 95–105°W, 30–55°N, i.e., a 1,000×2,650 km band up central North America, east of the Rockies and west of the Mississippi, i.e., as homogeneous an area as possible that omits, e.g., the highly diverse west coast of the United States (see chapter 5). Regression line and statistical analysis omits two language families completely surrounded by a single other language family (o), and the one coastal language family (c). Relationship not significant if (i) single language with widest latitudinal extent omitted; (ii) omitted three data included. Data from [268, map].)

temperature of coldest month, and mean annual rainfall[h] [62]. Across languages in the analysis by Thomas Currie and Ruth Mace, productivity was less of a significant correlate of range size than was a shortened growing season, although the variance explained was less than 5% [162].

It is often the case that in impoverished environments, animals need to work harder to obtain resources. For example, they must travel farther

h. Range size by environment. Globe minimum model, $N = 61$, r^2adj $= 0.4$, $F = 23$, $P < 0.0001$, AICc $= 111.5$; Australia, $N = 26$, productivity, r^2adj $= 0.6$, $F = 36$, $P < 0.0001$, AICc $= 31.4$; N. America, $N = 74$, productivity, annual rain, coldest monthly temp., r^2adj $= 0.4$, $F = 17$, $P < 0.0001$, AICc $= 138$. Data from [62].

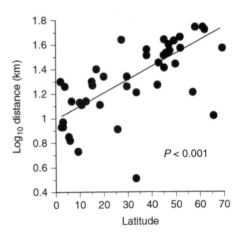

FIGURE 6.8. Hunter-gatherer groups move farther per group-move at higher latitudes. Mean distance moved by hunter-gatherer peoples per nomadic move by latitude.* Regression line omits outlier. Data from [62].

*Distance by latitude, area (N = 42). N = 43, r^2adj. = 0.5, F = 42, P < 0.0001, AICc = –6.8; with area, AICc = –4.5. Data from [62].

in a day [401]. And indeed, in the case of hunter-gatherer societies, the mean distance travelled by nomadic groups per move increases with increased latitude globally (Fig. 6.8) [62]. The minimum model for potential environmental influences globally has mean distance increasing with drop in mean annual temperature, mean monthly rain in the driest month, and number of good growing season months. Per continent, only North America has a sufficiently large sample, and for it, distance per migratory move increases with decrease in mean annual temperature of the coldest month[i] [62].

Large ranges are more difficult to defend than are small ones [465; 537]. That being the case, territoriality ought to be more evident at lower latitudes. I have not found data to test this possibility. But whether or not a latitudinal gradient in intensity of territoriality exists, plenty of evidence exists for xenophobia among traditional cultures, as among modern-day cultures. For instance, early accounts of explorers in North America consistently mention moving from one often-warring culture to another as the explorers traversed the continent [533; 644], and ethnographic evidence clearly shows that Native Americans perceive the land on which they range as owned territory [433]. Such territoriality is only to be expected within a species. Humans are so similar one to another that a main means by which cultures become and remain separate must be the near-absence of spatial mingling between cultures.

i. Mean distance travelled per nomadic move. Globe, N = 43, r^2adj = 0.6, F = 31, P < 0.0001, AICc = –13. America (N = 52, r^2adj = 0.35, F = 28, P < 0.0001, AICc = 22). Data from [62].

Conversely among humans, people at higher latitudes might survive only if they cooperate with neighbors [557]. To do so, a common language would be necessary, and that necessity plus the cooperation could lead to cultural homogenization over larger areas than in the tropics, where cultures can be more self-sufficient in a smaller range [557].

An alternative view is that socially complex societies (i.e., hierarchical societies) can control more land than can simple societies, because the complex societies are effectively larger, and can therefore conquer or absorb the simpler societies (chapter 8). Correspondingly, the best correlate in Currie and Mace's study of the size of the geographic range of a language was the degree of complexity of its political structure, rather than any of the environmental variables tested, such as productivity and length of growing season [162]. Nevertheless, the proportion of variance explained by political structure was small, only about 15% [162]. That fairly small explanatory power is perhaps not surprising, given that hunter-gatherer societies (i.e., politically simple societies) show the Rapoport effect.

Another explanation of the Rapoport effect in animals is that high-latitude taxa need to be more variable, more adaptable, in order to cope with the greater environmental variability. The adaptability then allows the taxa to range over wider areas [725]. I showed previously that among primates, taxa at higher latitudes are more variable than those at lower latitudes (Fig. 6.6).

The converse is that tropical taxa have not had to evolve the ability to cope with a range of environments and are therefore not adaptable [761]. Even minor changes in habitat then become a barrier to extension of range, and hence the tropical taxa are confined to small ranges. The Janzen "mountain pass" hypothesis (low tolerance of tropical organisms to cold means even low mountains are impassable) is a specific example of how tropical taxa might be more confined than are temperate taxa by lack of variability, in this case in environmental tolerance (chapter 4) [402; 506].

Humans too might be more variable at higher latitudes, as indicated by the greater variety of the subsistence tool kit of cultures that live at higher latitudes (chapter 3) [773]. A similar line of argument is that the modern human hunters who expanded across the plains of Eurasia as the climate cooled in the late Pleistocene, and who probably traveled long distances, had a lighter-weight, more portable tool kit (termed Aurignacian) than previously (chapter 3) [222, ch. 3; 369, ch. 3].

Regarding the Janzen mountain-pass idea, I suggested an equivalent effect in humans, namely that of differences between high- and low-latitude countries in ease of communication (chapter 4). In brief, more difficult communication in the low-latitude countries, as indicated by, for example, a lower density of roads and vehicles, prevents low-latitude cultures from having large geographic ranges.

Finally, in the long term, species or cultures that cover a large area will be more likely to find themselves in a clement-enough refugium to survive a particularly poor period [394; 666, ch. 6].

Rapoport × Forster Effects: Elucidation From and of Human Biogeography. In chapter 5, I discussed one of the major biogeographical patterns of nonhuman species and humans, namely, the great diversity of the tropics by comparison to higher latitudes. For the sake of brevity, I have termed this latitudinal gradient the Forster effect [320]. In Table 5.2, I listed the several explanations for the Forster effect, some of which work better than others.

Several explanations have been offered for the Rapoport effect too. Many of them, it turns out, are similar, even the same, as those offered for the Forster effect, even though the Rapoport effect concerns average size of geographic range and the Forster effect concerns diversity of form. Some explanations explicitly tie the two together, with the Forster explanations explaining the Rapoport effect, and vice versa.

I here use knowledge and understanding of both effects in humans to reciprocally refine explanations for both effects in humans and nonhumans (Table 6.1).

The numbering of the paragraphs below matches the numbering of the tables. I briefly repeat the explanations for the Forster effect to obviate having to switch back and forth with chapter 5. Where there are no relevant data for humans (hypotheses 2, 10–12), I do not discuss the explanation.

1. Mid-Domain Effect

If latitudinal extent of species or cultures is wide enough, then in the same way that long pencils in a pencil box must overlap more at the center of the box than at the ends, so the geographic ranges will overlap more at the equator than at higher latitudes. It is not impossible that some bird ranges might be large enough for the effect to operate [483]. However, for other nonhuman species the explanation does not work. For instance, among

Hypotheses for Forster effect	Explanation regarding tropical biodiversity	Relation with human Rapoport effect
Null/Neutral models		
1. Mid-domain effect	Ranges (pencils) overlap more at center of available space (pencil box).	No. Lat. extent of hunter-gatherers and languages too small to produce effect.
2. Neutral theory	Stochastic demographic effects cause more speciation, less extinction in tropics.	Unknown?
Biology		
3. Diversity of parasites, competitors, predators	High diversity prevents a minority of species/cultures dominating.	Yes. As indicated by smaller geographic ranges of cultures in tropics.
4. Diversity of hosts, resources	A diversity of hosts produces a diversity of dependents.	Maybe. If cultures less likely to find familiar resources (hosts) if extend geographic range.
5. Environmental heterogeneity	Heterogeneous, patchy environment promotes diversity.	Maybe. Smaller geog. ranges of tropical cultures. But high lat. ranges far smaller than size environmental zones.
6. Energy available, productivity, population growth rate	High (energy, productivity in tropics) = larger populations = more origination (speciation), less extinction.	Yes. Pops. large enough to persist can exist in smaller areas in tropics, and do. No. Pop. size of hunter-gatherers, languages not greater in tropics.
7. Rapoport effect	Small geographic range allows closer packing. But species' ranges overlap.	Yes. Especially as human cultures tend not to overlap.
Physical environment		
8. Stability over short and long time periods	Tropical stability allows more time for specialization and origination; less change results in less extinction.	Yes. Especially given potential evidence of less specialization (more varied tool-kits) in larger geographic ranges at high latitudes.

(*continued*)

TABLE 6.1 (*continued*)

Hypotheses for Forster effect	Explanation regarding tropical biodiversity	Relation with human Rapoport effect
9. Environmental clemency	Tropical clemency allows specialization, prevents generalization.	Yes. As indicated by smaller geographic ranges of tropical cultures.
10. Intermediate disturbance	Intermediate disturbance frequency in tropics allows high biodiversity.	No evidence of intermediate disturbance frequency in tropics?
11. Energy, 1	More energy allows high metabolism, higher reproductive rate, and hence higher diversity.	No. While smaller tropical geog. range might fit, no evidence exists for the proposed mechanism.
12. Energy, 2	More energy causes higher mutation rates.	No evidence.
13. Area	Greater area of tropics allows more species/cultures (No. number species/cultures more than expected for area).	No. Geog. range size tropical cultures indicates denser packing of tropical than nontropical cultures (see # 7).

SOURCE: The first two columns repeat from Table 5.2, based on [463, ch. 15, table 15.2].

mammals that inhabit only North America, the species with the largest latitudinal extents are not concentrated at mid-latitudes [629, ch. 5].

For most languages and hunter-gatherer societies, their geographic range is far too small to produce the Forster effect at the global scale. With geographic ranges calculated as circles, the maximum latitudinal extent of those hunter-gatherer societies with the largest geographic ranges is only about 1,000 km, as it is for the languages of the highest latitudes [data from 62; 277]. That distance is so much less than the roughly 6,500 km from the arctic circle to the equator that it cannot explain the Forster effect in humans. Also, within each hemisphere, diversity peaks at or near the equator, not at mid-latitudes. Where diversity does peak at mid-latitudes, as in western California, that is because of a concentration of small-range cultures there [268].

3. Diversity of Parasites, Competitors, Predators

If the diversity of almost every life form in the tropics prevents through competition, disease, or parasitism a minority of species becoming dominant, the diversity could also prevent the range sizes of most becoming large; that is, the Forster effect produces the Rapoport effect. I presented evidence for such competitive prevention of dominance in chapter 5. The match of the Forster effect and Rapoport effect in humans fits the explanation.

4. Diversity of Hosts, Resources

The argument that a diversity of hosts (e.g., resources) allows a diversity of consumers (e.g., hunter-gatherers) is analogous to the previous explanation. Of course, the question of why the tropics has a diversity of hosts is begged (as was the question of why it has so many species inimical to others). Nevertheless, the argument could explain, and therefore be supported by, the Rapoport effect in humans if cultures are less likely to find familiar resources if they extended their geographic range. Such might be the case given that regional diversity (so-called beta diversity), as well as site diversity (alpha diversity), is greater in the tropics than in temperate regions [463, ch. 15]. However, at least one study found no relation between plant diversity and range size of human language in the Old World, after accounting for a variety of other potential influences [162].

5. Environmental Heterogeneity

The argument here is not simply that tropical diversity begets diversity, but that the distribution of types of habitat (to which local nonhuman species and humans become adapted) is more patchy in tropical than in temperate regions. Therefore, species and cultures have small geographic ranges in the tropics, and so the tropics are diverse. However, the geographic ranges of even high-latitude hunter-gatherer cultures are far smaller than the extent of habitat available.

This hypothesis calls to mind the suggestions that modern humans' rapid expansion east out of Africa was aided by the fact that southern Eurasia all the way to Sahul is of a similar latitude, and hence climate, to our place of origin in Africa (chapter 3). The explanation is similar also to Jared Diamond's and others' suggestions that the great longitudinal extent of temperate Eurasia allowed a degree of cultural expansion not available to tropical cultures [178]: the steppes of eastern Europe are for a horse the same as those of eastern Mongolia; wheat grows as well in Spain (and Kentucky) as it did in Mesopotamia.

6. Energy Available, Productivity, Population Growth Rate

More energy is available year-round in the tropics than at high latitudes; therefore, productivity is greater (other things being equal), including within hunter-gatherer ranges [251; 542], and so populations are more likely to reach viable sizes in small areas. Certainly, hunter-gather societies usually live at higher population densities in smaller geographic ranges in the tropics than they do at high latitudes (chapter 5). However, range sizes at higher latitudes are so much greater than in the tropics that population sizes are higher at higher latitudes (chapter 5). Furthermore, for languages, no relation exists between latitude and either population density or population size (chapter 5), for the number of speakers per language peaks at intermediate latitudes [162].

7. Rapoport Effect

The argument is that smaller geographic ranges in the tropics allows tighter packing of species and cultures and hence greater diversity [146; 725]. I have already discussed in chapter 5 how the Rapoport effect could arise in relation to the Collard–Foley explanation for the high cultural diversity seen in the tropics. In brief, the humid tropics' high, year-round productivity (6, 8, 9) allows viable populations to persist in small geographic ranges. Later I discuss further the relationship of climate to geographic range size of hunter-gatherer cultures.

Additionally for species, if tropical organisms are not adapted to cold to the extent that high-latitude organisms are, then even low mountain ridges will be barriers to tropical organisms (chapter 4) [402]. The consequence is more barriers and smaller ranges in the tropics.

The first mechanism for the Rapoport effect does not necessarily explain the Forster effect in species, because the high tropical diversity could be produced by overlap of many species each with a large geographic range [483; 629, ch. 5]. However, it could well explain the Forster effect in human cultures, territorial as cultures are [483], except that productivity and length of growing season explain such a small proportion of the variation in range size of languages [162].

This packing explanation should work, though, to explain diversity of subspecies within species, given that subspecies are almost necessarily identified by lack of overlap of geographic range. However, the opposite seems to happen: taxa at higher latitudes have more subtaxa in their larger geographic ranges than do those at lower latitudes, in what I have termed the Darwin–Rapoport effect. The same pattern is true of human cultures (see Fig. 6.13). Part of the explanation for the larger ranges of high-latitude taxa might lie in their variability, as represented by the greater number of subtaxa [309; 311].

The second mechanism, the mountain-pass effect, could explain at least some of the Forster effect among nonhumans (chapter 4) [261; 506; 761], but it seems unlikely to work for humans, adaptable as humans are.

However, something similar to the mountain-pass effect could be operating. The lack of good means of transport (roads, cars) in the tropics could well limit movement and hence restrict geographic range there (chapter 4). Additionally, the tropical zones of South America and Africa are narrower latitudinally than the temperate zone of Eurasia, and the tropical region of Asia is markedly insular (chapter 4) [178].

8. Stability over Short and Long Time Periods

9. Environmental Clemency

Species can become specialized and diverse in the tropics because for long stretches of time, no severe periods restrict the evolution of "endless forms most beautiful and most wonderful" (chapter 5). If the specialization includes restricted environmental tolerance—for example, via the mountain-pass effect (chapter 4) [402]—then the Rapoport effect (7) comes into play and explains the Forster effect (chapter 5). By contrast, at high latitudes, the taxa that do best are those that can survive a range of conditions, which only a small subset can do so. If the members of that subset can survive a range of conditions, they can cover large geographic ranges. Indeed, to survive, individuals may have to cover large home ranges, which means that the species is more likely to cover a large range.

13. Area

The argument that the Forster effect is explained by the larger total area of the tropics does not work (chapter 5). The Rapoport effect confirms the negation of the argument, because the smaller geographic ranges of tropical species and cultures (where they are smaller) indicates not more diversity because of greater available area but because of denser packing of species and cultures.

In sum, then, of the explanations for the Forster effect in nonhuman species that could explain or be explained by the Rapoport effect in humans, three do not work (1, 11, 13), two might not work (5, 6), four work well (3, 7, 8, 9), two might work (again, 5, 6), and one works conditionally (4).

Other explanations exist for the Rapoport effect than those I have discussed here. Probably a range of explanations is needed even for one taxon, let alone for all, especially given, on the one hand, the several exceptions to the Rapoport effect [254, ch. 3; 258] and, on the other hand, the fact that within one species—humans—such a small proportion of the variance in range size is explained by any one variable [162]. Nevertheless, because several of the explanations that come from nonhuman biogeography are negated by the data on the Rapoport effect in humans, we might need to rethink those explanations for nonhuman species. Whether the connections between the explanations seem to work or not for both humans and nonhuman species, the proviso that explanations that work across taxa do not necessarily explain patterns within taxa, and vice versa, must always be borne in mind [337, ch. 6].

AREA AND DIVERSITY

The number of [plants] is likewise commonly proportioned to the extent of the country . . . Islands only produce a greater or lesser number of species, as their circumference is more or less extensive.

[237, Forster, 1778, p. 120]

Diversity Increases with Area

The more space available, the more sorts of objects can be fitted into the space. In biogeographical terms, larger areas contain more species, other things being equal. Johann Forster (see opening quote of this section) was one of the first to note the effect. This is the same Forster as noted that higher latitudes contained fewer taxa than did lower latitudes [237].

In biogeography, the relationship of biodiversity to area is known as the species–area relationship and is part of the subdiscipline of "island biogeography" [463, ch. 13; 470; 827]. The phrase applies both to real islands in bodies of water and to so-called habitat islands lying in an imaginary "sea" of unsuitable habitat. New York's Central Park is a perfect example of a habitat island. Another very common example is the fragments of a former forest surrounded now by cultivation or habitation. As habitat is destroyed, so it is often fragmented into separate patches. The effects of fragmentation are a special area of study in conservation biology [208; 331; 447], but the concepts applied are fundamentally biogeographic.

Smaller areas contain fewer species for several reasons [463, ch. 13; 470; 827, ch. 7]. They have fewer sorts of habitat, other things being equal, and fewer habitats translate into fewer species [463, ch. 13; 470; 827, ch. 7]. Even within one habitat type, the population size of any one

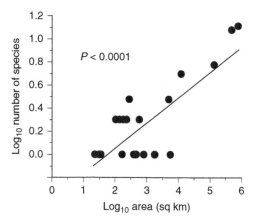

FIGURE 6.9. More primate taxa in larger areas.* Number species per Sunda Shelf island by island area. Data from [307].

*Number taxa by size Sunda island. Species, $N = 22$, r^2adj $= 0.60$, $F = 33$, $P < 0.0001$; genera, r^2adj $= 0.53$, $F = 22$, $P < 0.0001$. Data from [307].

species is likely to be smaller in the smaller area, and hence the species is more at risk of extinction. Small islands are less likely than are large islands to be found by dispersing organisms and so less likely to be stocked or restocked. Adverse effects from the surrounds of the islands, such as drying winds and invasive species (including humans) can penetrate farther into smaller areas [319; 438]. And smaller areas are often surrounded by more adverse environment than are larger areas [319].

Higher Diversity of Nonhuman Primates in Larger Areas. As usual, primates do what so many other species do. In this case, more of them live in larger areas. The effect is true in the long term and large spatial scale— across countries, for example [157, ch. 4; 636]. It is true of the Sunda Shelf islands, such as Sumatra, Java, and Bali, created by the late Pleistocene– Holocene global rise in sea levels (Fig. 6.9) [307; 335; 345; 561]. It is true of original forest areas across Africa [156]. And it is also true in the short term and small spatial scale of recent forest fragmentation [317; 491].

Among primates, as among other taxa, the slope of the species–area line is significantly less than 1. The consequences are that smaller areas have a higher density of species than do larger areas, even if the smaller areas have a lower number of species. With respect to loss of species as area available drops, species are lost at a slower rate than the decrease in area. A rough rule of thumb is a 50% loss of species with a 90% loss of area. Primates are quite close to that rule. From the 760,000 sq. km of Borneo to the 131,000 sq. km of Java, an 80% loss of area, the number of species drops by about 50%, from 13 to 6, and the number of genera from eight to four [307].

Finally, for many species–area relationships, real islands have less diversity than do neighboring mainland regions of the same area. An example is the terrestrial mammals of southeastern Asia [345; 346]. This phenomenon is due to the fact that real islands are usually far more isolated than are areas of the same size on the mainland. Therefore, fewer species manage to reach islands for a start, and if one goes extinct on an island, it is less likely to be replaced by another than would be the case for the same size of area on the mainland. The same effect is seen for habitat islands too, even if it is not quite so clear-cut [447].

Higher Diversity of Human Cultures in Larger Areas. Human diversity does the same as biodiversity. In general, the number of languages increases as area increases (Fig. 6.10) [277]. However, while the association is true of mainland countries, it is not true of island countries (Fig. 6.10) [277]. For island countries, latitude is a significant correlate. Add latitude to the model, and both latitude and area correlate significantly with number of languages (Fig. 6.10) [277].

Within a subset of the world's islands, those of Oceania (excluding New Zealand), the number of languages clearly increases as island area increases, including when latitude is statistically accounted for (Fig. 6.11A) [277]. Moreover, within a subset of islands in Oceania, the Solomon Islands, again we see that number of languages increase as area increases (Fig. 6.11B) [277; 755]. In the Solomon Islands, latitude has no effect, in large part because the Solomon Islands extend over only 4° of latitude.

When I discussed the Forster effect in chapter 5, one of the measures of diversity that I used was density of taxa and cultures. I did so to account for the effect of area available on diversity (i.e., numbers of taxa or cultures), as discussed in this chapter. I pointed out that island countries contain higher densities of languages than do mainland countries at the same latitude, a fact evident in Fig. 5.2B. The phenomenon is simply explained by the fact that the slope of the relationship between species/cultures and area is less than 1 [666, ch. 2].

The consequence of a species/culture slope of less than 1 is easily seen in the general phenomenon of a 90% decline in area being associated with only a 50% decline in numbers of taxa or languages. To get a usefully representative measure to compare equal areas, wherever they are, we need the number of taxa/cultures with area accounted for, i.e., the residual of numbers from area. This is the measure I used in Fig. 5.2A to show a greater diversity of languages in tropical countries than in temperate ones. With residuals as the measure, island countries have

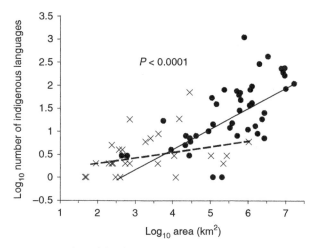

FIGURE 6.10. In mainland countries, but not island countries (unless latitude is accounted for), number of languages increases as area of country increases. Number languages by area of country.* Mainlands, •, ——. Islands, **x**, - - - . Data from [274].

*Number languages by area. Mainlands, globe: $N = 40$, r^2adj $= 0.45$, $F = 32.6$, $P < 0.0001$; Africa, Asia, South America, $N = 6$–14, $r_s = 5$–30, $P < 0.05$; Europe, $N = 6$, $r_s = 3$, $P > 0.1$. Islands, globe: $N = 26$, r^2adj $= 0.06$, $P > 0.1$. Number of languages in island countries by area and latitude. Whole model, $N = 26$, $\chi^2 = 14.7$, r^2adj $= 0.4$, $P < 0.001$, $AIC_c = 26$; area, $\chi^2 = 12$, $P < 0.001$; latitude, $\chi^2 = 11$, $P < 0.001$. All calculations exclude countries with centers $< 10°$ apart, and single-factor analyses exclude outliers. Data from [277].

the number of languages that we expect for their area and do not differ from mainland countries.[j]

One of the reasons why small areas usually contain fewer species is that populations are smaller and therefore more liable to extinction. Joseph Henrich makes essentially the same argument for loss of one aspect of culture in Tasmania, namely, the variety of tools and skills [352]. Tasmanian indigenous peoples arrived from Australia perhaps 34 kya. Then from about 8 kya, after isolation from Australia by rising sea level, they gradually lost a variety of apparently useful implements (and presumably the skill to use them). Examples include nets, barbed spears,

j. Number languages expected, given area of island, mainland. Globe, $N = 25, 45$, $F = 0.1$, $P > 0.5$. Only Africa, S. America with both samples $N \geq 5$, $P \geq 0.1$. Data from [277].

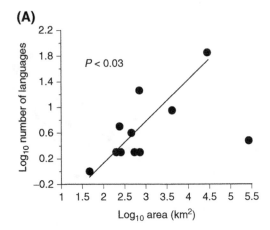

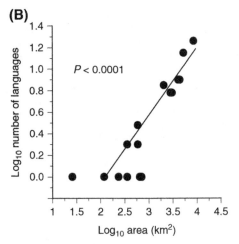

FIGURE 6.11. On Pacific islands, number of languages increases as area of country increases

(A) Oceania.* Data from [277].

(B) Solomon Islands.** Regression lines omit outliers. Data from [755, ch. 3].

*Number of languages by area, Oceania. $N = 10$, Spearman $r_s = 0.7$, $P = 0.03$; $r_{latitude} = 0.82$, $P < 0.02$. Data from [277]. Number of languages by latitude, Oceania. $r_s = 0.35$, $P > 0.1$.

** Number of languages by area, Solomon Islands. $N = 16$, $r_s = 0.87$, $P < 0.0001$. Data from [755, ch. 3].

spear-throwers, and boomerangs, all of which continued to be made and used on mainland Australia.

Henrich suggests that because of the small population, even before the genocidal British arrived, too few skilled people were present to serve as models for the unskilled, and so the skills died [352]. In a quantitative analysis of the Tasmania effect elsewhere, Michelle Kline and Robert Boyd showed that across 10 western Pacific islands, the number of tools and tool units was lower on islands with smaller populations[k] [426]. The effect could be general, extending to, for example, loss of languages on small islands [651].

The Tasmania effect is a corollary of Paul Mellars' hypothesis about the sudden rise of modern human culture [524]. He argues for Europe around 45 kya that a major increase in human population density and the resultant increased intensity of interaction had a synergistic effect that led to larger social groups, specialized roles within those groups, and other indications of greater social, cognitive, and technological complexity and abilities.

The difference is that in the Tasmanian case, an argument can be made for the direction of cause and effect. Demographic isolation began the decline in complexity. In the European case, presumably the increased complexity could have driven the demographic change, which is what Mellars argues could have happened in Africa earlier [526].

I cannot resist closing this section on the diversity–area relationship without the example of the booklouse *Psocoptera*. Those of us with large or old houses have more species of booklice than do those of us with small or new houses [46]. Old money benefits biodiversity?

Isolation and Diversity

Quadrupeds [mammals] found on islands situated near the continents generally form a part of the stock of animals belonging to the adjacent main land; but small islands remote from continents are in general altogether destitute of land quadrupeds, except such as appear to have been conveyed to them by men.

[468, Lyell, 1832, ch. 6, p. 90]

The topic of the effect of isolation on number and type of species in an area is effectively that of barriers to movement and hence was treated in chapter 4. I raise it here again, though, because degree of isolation is a

---

k. Island size by tool variety. Tools, $r_s = 0.73$, $P < 0.02$; tool units, $r_s = 0.68$, $P < 0.04$. Data from [426]. Island area by tools, tool units: not significant.

fundamental aspect of island biogeography [470]. The farther an island is from a source, the less likely are species to get to it, and thus the fewer species it contains [463, ch. 13; 470; 827]. The concept is old, as indicated by the quote above from Charles Lyell in his 1832 *Principles of Geology*, and it has been amply documented since [463, ch. 13]. An important effect of destruction of habitat is not just its fragmentation into smaller areas, but the fact that the remaining "islands" of habitat are ever more widely separated [208].

In addition to distance as a factor isolating islands, the nature of the environment between islands influences whether species can cross and which species do cross [446; 463, ch. 14; 614]. Mammals, for example, are generally not good at crossing water, as Lyell noticed (quote above). For amphibians, seawater is particularly unfavorable. Consequently, oceanic islands, those never connected to a mainland, have relatively few native amphibians [165, ch. 12]. For habitat islands, the density of human use of the land between the fragments of remaining habitat can strongly influence whether species persist in the fragments [319; 581].

Possible parallels exist in humans. For instance, people did not start to colonize the eastern Pacific until about 3,000 years ago, after the development of ocean-going canoes (chapter 2). Nevertheless, movement across water is so easy for humans now that the number of languages spoken in island countries bears no relation to their distance from the nearest mainland: area of island and latitude are the main correlates[1] [277].

Area Used (Size of Geographic Range) Matches within-Taxon Diversity

The previous section concerned the relationship of taxonomic and cultural diversity to area available to be used by a taxon or by a human culture. In this section, I turn to the area in fact used—the geographic range size—and its relation to diversity within the geographic range.

A little-remarked but seemingly widespread phenomenon of the relationship between space used and diversity is that the larger the geographic range of a taxon, the more subtaxa the taxon has. In other words, species with larger geographic ranges have more subspecies. Darwin appears to have been the first to notice the phenomenon [165, ch. 2]. Eduardo Rapo-

---

1. Number of languages on islands by isolation (distance), area, latitude. $N = 19$; whole model, $\chi^2 = 8$, r^2adj $= 0.25$, $P < 0.05$, $AIC_c = 20.7$; latitude, area, $\chi^2 = 6$, $P < 0.02$ for each, with neither alone having a significant effect; isolation, $\chi^2 = 0.3$, $P > 0.5$. Data from [277].

port substantiated it for a variety of mammalian orders, including primates [629, ch. 3]. However, he did not say what continent(s) his sample came from, and from the families shown I infer it was only South America.

At the level of species within genera, this Darwin–Rapoport effect has been documented for birds [55, ch. 13] and for pines of the genus *Pinus* [843]. Primates show the phenomenon in all continents, at both the taxonomic level of species within genera (Fig. 6.12) [309; 311; 313] and subspecies within species.[m]

Perhaps because this phenomenon of taxa with larger geographic ranges having more subtaxa is little remarked in biogeography, it is not remarked at all in anthropology as far as I have discovered. However, my analysis of the North American Indian language map of Goddard indicates that it might exist (Fig. 6.13) [268].

A proviso might be necessary concerning the relation between area covered and diversity. The number of studies or number of people studying can influence records of biodiversity [556; 632]. Widespread species, especially if they also occur at relatively high local population density, as often happens (next section), might be more studied [151; 155; 185]. If more studies provide more chance to see variation [155], or perceive it, then sampling bias might lead to the finding that widespread species or cultures are more taxonomically variable.

AREA, ENVIRONMENT, AND DENSITY

Climate and Geographic Range Size

I showed in chapter 5, when discussing why diversity of cultures might be greater in the tropics than at higher latitudes, that area occupied by a culture correlated in some analyses with productivity of the area and with the density of peoples in the area. Productivity is only one possible environmental correlate of range size. In a multivariate model of 10 variables (such as mean annual temperature, temperature of the coldest month, and so on; see chapter 5), productivity was a variable in the best model for Australian and North American hunter-gatherers, but not globally (the

m. Number subspecies per species by range size. Accounting for latitude, and for phylogeny with family–subfamily membership in generalized linear model, Globe, df = 17, $\chi^2 = 60$, $P < 0.0001$; $\log_{10}$ area, $\chi^2 = 39$, $P < 0.0001$; mid-latitude, $\chi^2 = 3$, $P < 0.1$; phylogeny, $\chi^2 = 22$, $P < 0.02$. Africa, df = 6, $\chi^2 = 15$, $P < 0.0001$; America, df = 7, $\chi^2 = 7$, $P < 0.01$; Asia, df = 7, $\chi^2 = 17$, $P < 0.0001$; Madagascar, df = 4, $\chi^2 = 4$, $P < 0.1$; for none of the continents separately was either phylogeny or latitude significant. Data from [313].

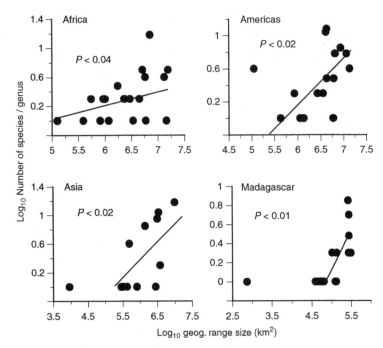

FIGURE 6.12. Primate genera with small geographic ranges have fewer species per genus than do genera with large geographic ranges.* Number of species per genus by range size of genus, per continent. Regression lines exclude outliers (one per continent). Data from [314].

*Number of primate species per genus by geographic range size, latitude, phylogeny (family membership) in generalized linear model. Globe (exc. Madagacar): $N = 48$; whole model, $\chi^2 = 23.7$, r^2adj = 0.29, $P < 0.03$, $AIC_c = 58.4$; range size, $\chi^2 = 12.8$, r^2adj = 0.18, $P < 0.0005$, $AIC_c = 37.5$; phylogeny, $\chi^2 = 11.9$, r^2adj = 0.13, $P > 0.3$; latitude, $\chi^2 = 0.8$, $P > 0.3$. Data from [309; 314].

only regions with a sufficient sample for analysis).[n] Mean annual rain was the only variable that appeared in the best model for all three regions.

In the case of languages of the Old World (excluding Australia), only length of growing season and productivity correlated with geographic area, but the variance explained was less than 5% [162].

n. Range size by climate. Globe, mean annual temperature, rain, $N = 61$, r^2adj = 0.4, $F = 23$, AICc = 109; Australia, mean annual rain, productivity, $N = 24$, r^2adj = 0.6, $F = 23$, AICc = 34; N. America, mean annual rain, mean temperature in coldest month, productivity, $N = 72$, r^2adj = 0.4, $F = 17$, AICc = 137. Data from [62].

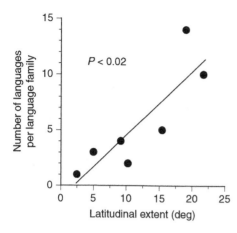

FIGURE 6.13. Darwin–Rapoport effect in human cultures: native Great Plains North American languages.* Number of languages per language family by latitudinal extent of language family. Data from map in [268].

*Darwin–Rapoport effect, native Great Plains American languages (for details see text description for Fig. 5.16E). $N = 7$, $r_s = 0.86$, $P < 0.02$. Data from map in [268].

Mainland Compared to Island Population Densities

A biogeographical characteristic of species on islands is that individuals live at higher population densities than do their mainland sister populations or species [463, ch. 14]. Two main explanations are offered for the contrast.

One is that islands have fewer predators for a given area. That is because predators with their relatively large home ranges are harder hit by loss of area than are prey [753]. Fewer predators mean fewer prey killed and, perhaps more importantly, freer foraging for resources.

The other suggested explanation for denser populations of species on islands is fewer competing species. With fewer competing species, the density of any one species can increase. The lemurs of Madagascar are the classic example for primates. They occur at higher densities for their body mass than do the primates of the other three, far larger, continents[o] [311; 315].

The high population density of lemurs correlates with the fact that Madagascar lacks many of the taxa that elsewhere could compete with lemurs [200]. Additionally, the recent arrival of humans on Madagascar coincided with the extinction of over 30% of lemur species, all of them

o. Island vs. mainland densities accounting for body mass. Madagascar vs. Africa, Americas, and Asia: $N = 12$, 47 genera; $z = 3.2$, $P < 0.01$. Data from [312; 315]. Comparing families, all four lemur families (no density for Daubentoniidae) had (a) higher residual density for mass than did all 13 nonlemur families, (b) higher density for geographic range size than all but one (Tarsiidae) nonlemur family. Mass accounted for because statistically lower for Madagascar than for other three continents.

the larger-bodied ones (chapter 10) [269], thus potentially allowing the remaining smaller-bodied species to increase in density as a result of competitive release. Increase of density of unhunted small-bodied species is a supposition, but Carlos Peres and coworkers have demonstrated its occurrence among South American primates [593].

By contrast, humans do not live at different densities on islands than they do on mainlands. The lack of difference is true of hunter-gatherers[p] [62] and of speakers of indigenous languages[q] [277].

Presumably, the effect of predators can be discounted for humans. The effect of competitors should still exist, however. The fact that it apparently does not could mean that, if findings for humans are applicable to animals, absence of predation on islands is the stronger influence on their higher population densities of island animals. However, it seems unlikely that humans would compete less intensively than other species with either their own species or with other species. What humans can do more readily than other mammalian species is leave islands on which competition is intense, so reducing their density on islands.

Parasites and pathogens can have as much of an influence on their hosts as predators on their prey (chapter 9). Yet fewer species of parasite and pathogen on smaller islands is not a main explanation among biologists for the high density of host species on islands. The absence of the explanation exists despite the fact that the apparently little study that has been done shows the expected relationship of fewer parasitic and pathogenetic taxa on smaller islands, whether real islands or habitat islands [430; 434]. Not only that, but parasites, like predators, might be more severely affected than their hosts by a reduction in area [434; 753].

One study indicated that island hosts (birds) were more heavily infected by ectoparasites than were mainland hosts of the same species [830]. However, a main suggested reason had less to do with biogeography than with the fact that other evidence suggested that the island hosts were less resistant because they were less well nourished [830].

p. Population density hunter-gatherers on islands vs. mainlands. Statistically insignificant difference of less than 10% separates median density of island compared to mainland societies (39.5 vs. 43 per 100 sq. km) across the same range of latitudes in Asia, the continent with 10 of the 11 island hunter-gatherer societies. Data from [62].

q. Population density indigenous language speakers on islands vs. mainlands within same range of size and latitude (density varies with both variables; chapters 5 and 6). No significant differences for overall density or median language's density. Generalized linear model of mainland–island comparison of overall and median density of speakers, with area and latitude as fixed variables, produces no significant effect: Whole model, $N = 14$, $\chi^2 = 2.3$, $P > 0.5$; all three fixed variables, $\chi^2 < 0.2$, $P > 0.5$. Data from [277].

Primates with small geographic range sizes have a lower diversity of protozoa and viruses than do primates with large geographic range sizes [11; 568], as do primate populations at low density, which also have lower diversity of helminth parasites [568]. A relation between host population density and number of parasite species in the population is quite widespread [567, ch. 3]. However, I know of no data for the relationship between primate parasites and pathogens and size of the space available, as opposed to size of a taxon's range.

A study on humans indicates precisely the sort of effect that an epidemiological biogeographer or a biogeographical epidemiologist would predict. Measles was less prevalent on Pacific islands with fewer people (Fig. 6.14) [32]. The original data did not include distance to the nearest mainland (America, Asia, Australia). The farther islands were from the nearest mainland, the less disease there was, again as island biogeography would have predicted. However, the effect was statistically negligible with population size accounted for.

Two islands were particularly instructive. For the size of its population (and area), Guam had a far higher incidence of measles than did any other island. The suggested explanation is the unusually high number of immigrants resulting from the American military base on the island [32]. For its population size, French Polynesia had the lowest incidence of measles. A possible explanation is that it is the group of islands in the sample most distant from either the eastern or western Pacific mainlands and therefore the most difficult for carrier immigrants to reach (even if in total, isolation is not a significant influence). This explanation is substantiated by the fact that human malaria has not reached Polynesia from Asia; it stops in Vanuatu in southeastern Melanesia, at the Buxton Line[r] [417].

These findings about the incidence of parasites and pathogens on islands means that their potential influence on differences in population density between island and mainland populations needs to be added to the currently suggested influences of contrasts in predation intensity and competition. Islands might have higher densities of hosts because islands have fewer parasites. But if so, why do humans on islands not apparently live at higher densities than do those on mainlands? Less disease should translate into a lower mortality rate on islands, and a higher density, but then the lower mortality rate would be offset by greater competition among the larger population of survivors.

---

r. Malaria in the Pacific. Avian malaria, probably introduced with chickens, has spread throughout the Pacific (chapter 10).

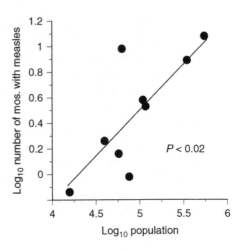

FIGURE 6.14. Less disease where fewer people.* Number months with measles cases by population size of island. Data from [32].

*Measles incidence on Pacific islands. % months with measles by population size, $N = 9$, Spearman $r_s = 0.7$, $P < 0.03$; by distance from mainland, $r_s = 0.7$, $P < 0.03$; population size but not distance significant with other accounted for ($r = 0.7$, 0.5, respectively). Data from [32].

A Relation between Local Density and Geographic Range Size

A very common biogeographic finding is that as the geographic range size of the taxon increases, so does the local population density of that taxon [77; 100; 257]. In other words, taxa that are rare or common on one measure (population density or geographic range) are also rare or common on the other measure. Consequently, the taxa at either end of the relationship are either very rare, occurring at both low density and small range, or very common, occurring at both high density and large range.

By contrast, human hunter-gatherer populations are at their least dense where geographic ranges are largest, globally and on all continents but South America (Fig. 6.15) [62].

For other primates, the taxonomic level considered affects whether we see the relationship, which is not an unusual finding in comparative analyses [336]. For species, the density–range relationship does not hold [20; 157, ch. 5; 195; 315]; it is weak across primate genera, but it is very strong across families [315].

A variety of explanations exist to explain the general relationship in nonhumans between range size and density [257]. One is straightforward sampling error: the less dense the local population, the more likely are individuals to be missed, and therefore the more likely is it that the geographic range size will be underestimated. The explanation does not work for primates. The low-density, small-geographic-range families are some of the most noticeable (apes—large bodied, loud), while one of the highest-density, largest-range taxa is one of the least noticeable (nocturnal, small-bodied bushbabies [Galagonidae]) [315].

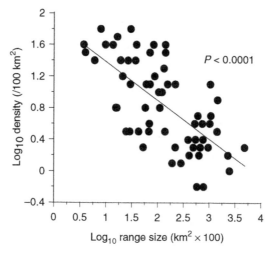

FIGURE 6.15. Lower population density of hunter-gatherer in larger geographic ranges. Population density by geographic range size.* Data from [62].

*Population density of hunter-gatherer cultures by geographic range size. Globe, $N = 63$, r^2adj $= 0.44$, $F = 50.5$, $P < 0.0001$. Density drops significantly with latitude (chapter 5), but a minimum model with latitude and range size showed only range size to influence density ($N = 61$, r^2adj. $= 0.4$, $F = 49$, $P < 0.0001$, AICc $= 69.6$); the fit was no better with latitude added (AICc $= 70.6$). Continents (N, r_s, $P <$): Asia, 15, 0.94, 0.0001; Australia, 26, 0.84, 0.0001; N. America, 75, 0.69, 0.0001; S. America, 19, 0.15, > 0.1; Africa, 15, 0.68, 0.01. (Only mainland cultures; for Globe, only one culture per State; for Continents, only cultures whose centers are farther apart than the sum of the diameters of their geographic ranges). Data from [62].

None of the biological explanations offered for other taxa work for primates, because if they work at all, they explain variation in only geographic range, not also in density [315]. Indeed, biogeographers in general are still debating explanations for the density–range relationship and concluding that probably several influences operate simultaneously to produce the relationship [77; 252; 256].

The reason for the human hunter-gatherers' contrary relationship is quite simple. Hunter-gatherer cultures with large ranges are typically those that live at high latitudes where resources are sparse (chapter 6). Therefore, not only does the culture cover a large range (Rapoport effect), but its members live at low density. Exploration of the interaction of the Rapoport effect and its explanations with the density–range size relationship and its explanations might throw some light on the causes of the density–range size relationship by, for instance, helping to control for latitudinal influences, or the variety of environmental effects that correlate with latitude.

CONCLUSION

The tiny island form of hominin, *Homo floresiensis*, might have persisted until later than 17,000 years ago. If so, it is a classic case of an island holdout, illustrating yet more similarity between nonhuman and human (or at least hominin) biogeography. I will come back to the biogeography of this phenomenon in chapters 8 and 10.

Some of the explanations offered for how the Rapoport effect (smaller geographic ranges at lower latitudes) is tied to the Forster effect (more taxonomic diversity at lower latitudes), or indeed for the Forster effect alone, do not work for humans. It remains to be seen whether the reasons they do not work can help further elucidate what is happening in other species.

Pathogens, including parasites, have a strong effect on the biology and demography of species, including humans. If I am correct concerning the relative lack of work on the biogeography of pathogens—for example, the lack of mainland–island comparisons—the biogeography of pathogens might be a subdisciplinary area of biogeography worth expanding. Anthropology, or more broadly, study of humans, might be particularly valuable here, for a large database must already exist. In other chapters, I discuss some aspects of the biogeography of disease, such as malaria (chapters 2 and 9) and diabetes (chapter 5).

In the next chapter, I continue consideration of the biogeography of human health, this time with a review of some aspects of the biogeography of diet and drugs.

7

A Biogeography of Human Diet and Drugs

Tell me what you eat, I will tell you what you are.

[91, Brillat-Savarin, 1825, p. xxxv in 1879 reprint]

SUMMARY

Humans from different parts of the world have different diets. They correspondingly have different genes and different physiologies. For instance, the milk-drinking adults of Africa and northern Europe have genes coding for the production of lactase in adulthood; others get sick if they drink milk as adults. The seaweed diet of Japanese since long ago has resulted in the Japanese, but not other peoples, being able to digest seaweed, via genes in a bacterium that feeds on seaweed transferring into one of the human stomach bacteria.

On the other hand, many Japanese peoples are extra susceptible to the break-down products of alcohol. No benefit has yet been firmly demonstrated. More generally, scattered more or less at random across the world for no obvious adaptive reason are people with variously active variants of enzymes (I discuss two) that result in considerable regional and individual differences in reactions to drugs as varied as antidepressants and analgesics.

However, not all regional differences are associated with regional differences in associated genes. Arctic peoples used to eat a diet unusually high in protein and fat. They could do so without ill effects because they were raised on that diet and therefore got used to it. Also, the lack of carbohydrate in their diet meant that for energy, they instead used the fat to metabolize the protein, and so escaped obesity.

The topics of this chapter could have been in chapters 2, 3, or 5. I have separated them from there in part because of the intrinsic interest that diet and medicine hold for many of us: half of the *New York Times'* Tuesday's "Science" section is devoted to diet and medicine. Also, the topics are an excellent example of the biogeography of the interaction of culture, environment, physiology, and genes.

Animal species that eat different diets have different digestive anatomy and physiology. Carnivores are obviously different from herbivores, but within dietary categories as well, such as the plant-eating primates, different diets are associated with different digestive anatomy and physiology [138; 139; 528; 727]. Colobine monkeys, which can have a diet heavy in leaves, have large guts, an ungulate-like digestive tract, and a ruminant-like set of bacteria and enzymes. Indeed, the stomach enzymes (lysozymes) of a foliage-eating colobus monkey are more similar to the lysozymes of a cow than they are to the lysozymes of the colobus monkey's far closer relatives, a baboon or human: 4 amino-acid differences separate a colobus' lysozyme from a cow's, but 14 separate the colobus' from a baboon's, and 18 from a human's [727].

Even within single species, the diets of separate populations can differ. A classic example in primatology is the gorilla. Populations at sea level in western Africa eat the fruit of over 100 species of plant, constituting well over 50% of the diet, whereas the eastern high-elevation mountain gorilla populations eat the fruit of less than 10 species, for less than 5% of the diet [322, ch. 4]. We do not yet know whether any physiological differences accompany the contrast in diet, but it seems likely that they would, at least during the western African fruiting season.

To Jean Brillat-Savarin's quote about humans at the head of this chapter could be added the phrase "and I will tell you where you come from." Humans, like the gorilla, have different diets in different parts of the world. For instance, potatoes (some of the most delicious I have ever tasted) were a staple of where I used to do field work, in northwestern Rwanda, whereas they are a rare food elsewhere in Africa[a] [213]. We need enzymes to break down our food (proteinases for protein, lactase for milk products, amylases for starch, alcohol dehydrogenase for our beer and wine, and so on). If what we eat varies geographically, we can expect that the associated enzymes and genes will vary geographically [303]. There is a biogeography of diet.

MILK CONSUMPTION AND LACTASE PRODUCTION

Lactase is an enzyme in the small intestine that breaks down lactose, the sugar in milk, into its simpler, absorbable sugars [193, ch. 5]. Human

---

a. Potato production and consumption. Annual consumption per head, Rwanda, 125 kg; of top 11 producer countries in Africa, data available for 9: range per head, 3–88 kg, median of 30 kg; all of Africa, median = 14 kg; of top 11 producer countries, Rwanda 46 metric tons/sq km of country, remainder, 0.5–17 t/sq km. Data from [213].

infants require the enzyme to digest their mother's milk. Most humans, about 65%, cease producing lactase after infancy [193; 389]. If they consume milk, or unprocessed milk products, they get sick: flatulence and diarrhea are common symptoms. A variety of jargon words with specific meanings describe the ability, or not, to digest lactose in adulthood. I use lactose acceptance as the general term for the ability.

The distribution of adult lactose acceptance across the world is uneven (Fig. 7.1). However, the distribution is not random. It correlates environmentally and culturally with consumption of milk in adulthood [193, ch. 5; 330; 371; 389]. This biogeographical fact is perhaps best illustrated by comparison of neighboring pastoralists and nonpastoralists. Thus in Nigeria, Rwanda, Sudan, and the Middle East, about 80% of pastoralist peoples who drink milk can digest lactose, whereas less than 30% of the nonpastoralist citizens of those countries can do so [193; 389].

Surprisingly, several pastoralist peoples who use milk cannot digest lactose well as adults [193, ch. 5; 389]. Examples are the Nuer and Dinka of Sudan, and Mongolians. Instead of drinking milk, though, they convert the milk to products in which the lactose has been broken into simpler sugars through fermentation or removed [193, ch. 5]. Yogurt is an example of a fermented milk product. Cheese is an example of a dairy product from which lactose is largely absent (the lactose remains in the whey, the liquid left behind once the cheese is formed). The fact that these dairy cultures can use milk only by denaturing or removing the lactose substantiates the hypothesis that dairying led to the evolution of the alternative, physiological means of coping with lactose as adults, namely, continued production of lactase after childhood.

Further evidence for the idea that adult lactase production is tied to consumption of milk by adults is the apparent timing of the origins of, on the one hand, ungulate domestication and, on the other hand, of adult lactose acceptance and of the genes for adult lactase production. Herding and the genes for adult lactose acceptance first appeared at (very) roughly 10,000 years ago in Eurasia[b] [330; 389; 769], but statistical analyses comparing phylogenies of peoples and lactose acceptance indicate that the herding arose first [371]. Some evidence indicates that the origins of some European lactose acceptance was in the Urals [330]. Archeological and genetic evidence indicate that lactose acceptance in northern Africa and perhaps far northern Europe might have arisen a

---

b. Lactose acceptance at 10 kya. This date has wide error, e.g., 95% confidence limits of c. 2 kya to 20kya for genes promoting adult lactase production [769].

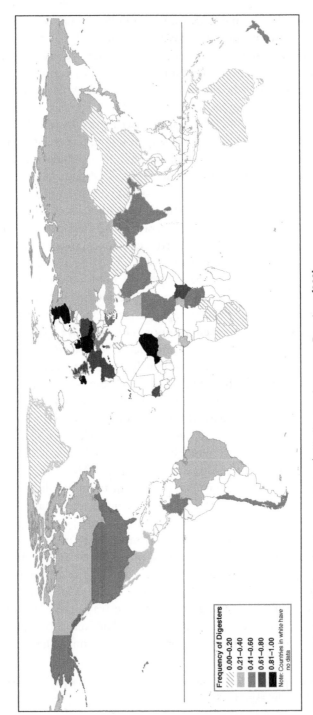

FIGURE 7.1. Global distribution of lactose acceptance/milk digestor frequency. Data from [389].

little later with the arrival of pastoralists from the Middle East and then spread down through sub-Saharan Africa perhaps about 5,000 years ago [330; 389; 769].

The fact that from such a short time ago the alleles allowing lactose acceptance are present in nearly all of any pastoralist population that shows lactose acceptance indicates a strong advantage for the ability to digest lactose and very rapid evolution and spread of that ability and genetic attribute [330]. The advantage might differ in northern Europe from Sahelian Africa, the two centers of high frequency of adult lactose acceptance. In Africa, subject to droughts, the advantage might be a source of clean liquid; in northern Europe, lacking sunlight, the advantage might be the vitamin D in milk and its influence on calcium absorption [330; 389]. However, the same phylogenetic analysis that tied lactose acceptance to dairying could not tie it statistically to either latitude or aridity [371].

The alleles for lactase production are in the same gene at the same place on the same chromosome in the European and African populations that can digest milk as adults [330; 389; 769]. However, the exact alleles responsible differ between Africans and Europeans [330; 389; 769]. The difference might be due to chance, or it could be related to the potential different advantages of lactose acceptance [330; 389]. Yet another variant in Saudi Arabians could be tied to consumption of camel milk, as opposed to cow milk [201].

A twist on the story is provided by some Somali pastoralists who drink milk as adults and yet do not produce lactase as adults [389]. These peoples would seem to negate the whole hypothesis of the connection between culture and biological evolution. It turns out that among the Somali pastoralists, yet another means of digesting lactose has evolved. Catherine Ingram suggests that among these Somali, rather than the body itself evolving to digest lactose in adulthood, the colon bacteria have changed in such a way that the people can digest the lactose [389].

These connections between pastoralism, physiology, gut ecosystem, and genes is one of the nicer examples yet of the interaction between culture (i.e., behavior), environment, physiology, and genes [193]. A change in behavior (drinking milk in addition to eating meat or blood) produced, in effect, a different environment (lactose in the diet); the different environment affected our behavior (increased association with livestock and milk ingestion into adulthood); the interaction of the two affected our gut flora and our physiology via our genes; and the changed flora and physiology in turn affected our behavior and environment (increased use of milk and closer association with more ungulates). In biology, the study of the way in which organisms alter their environment and the effect of that alteration on the organisms has become its own field of research, niche construction [573].

STARCH AND SALIVARY AMYLASE

Lactose is a disaccharide. Starch is a polysaccharide, found in grain and tubers (e.g., wheat and potatoes). Amylase is an enzyme in saliva that begins the process of digesting starch. The gene that enables the production of amylase is Amy1. It turns out that the more copies of Amy1 a person carries, the more amylase they have in their saliva, and therefore the better they can digest starch.

The biogeographic connection is the fact that the number of copies of Amy1 that a person has varies regionally with the amount of starch in the diet [596]. Japanese peoples, the Tanzanian Hadza, and European Americans all have diets high in starch (over 100 g a day). The median individual in these three regions has 6 Amy1 copies (of a maximum of 10). By contrast, the median individual of four low-starch populations has about 4.5 copies (two rainforest hunter-gather peoples, the Biaka and the Mbuti; a pastoralist peoples, the Datog of Tanzania; and a pastoralist fishing peoples, the Yakut of Eurasia). Chimpanzees have even less starch in their diet and have only 2 Amy1 copies [596].

Starch is quite difficult to digest without cooking, and so it is probably a late addition to the hominin diet [846]. Its advantage as a source of energy is that it is a resource used by hardly any other large animals, both because it is difficult to digest raw and because tubers are difficult to extract from the ground. Humans often have to use a lot of effort and digging sticks to get at tubers [341]. Richard Wrangham suggests that hominin's ability to cook (i.e., their use of fire) massively increased the amount of energy available to individuals and could have been responsible for the explosive increase in hominin brain size starting between 2 and 1 mya [846; 847].

A problem with tubers as a main dietary item, however, is that they are deficient in folates, the crucial role of which in human physiology I discussed in the section on skin color (chapter 5). But it turns out that populations that rely on tubers are more likely than others to have a gene particularly active in folate metabolism [303]. At the opposite extreme, roots and tubers can have deleteriously high concentrations of toxic glycosides. Correspondingly, genes active in production of enzymes that denature glycosides are particularly frequent where people rely on a diet of root and tubers [303].

ONLY THE JAPANESE CAN DIGEST SEAWEED

In the case of the consumption of milk and starch and our associated genetic and physiological traits, it looks as if the consumption came first and its advantages (extra nutrition) then led to genetic evolution in those populations that increased individuals' ability to use the beneficial resource [330; 389; 769].

The ability of the Japanese, but not Americans of European origin, to digest seaweed is also a story of ingestion first and evolution of increased benefit later [349]. However, the route from diet to ability is different than in the case of lactose or starch.

Much of our, or any animal's, ability to digest food comes from the huge bacterial community of the gut, which is why humans and animals can have digestive problems when they take antibiotics. The bacteria can break down complex sugars in plant parts and so make them available to the animal or human. The phenomenon is particularly vital for herbivores, a category that includes humans because we have a fair amount of plant food in our diet.

It turns out that the Japanese peoples' ability to digest seaweed, particularly the seaweed used to wrap sushi, *Porphyra* (nori), did not arise via mutations of the peoples' genes or of our gut bacterial genes [349]. Instead, the genes responsible are those of a bacterium, *Zobellia galactanivorans,* that lives on the seaweed. The Japanese can digest seaweed because the genes that allow the *Zobellia* bacterium to digest seaweed somehow transferred themselves directly into the human gut bacterium *Bacteroides thetaiotaomicron* [349]. The *Zobellia* itself cannot survive in the human gut.

Such "horizontal" transfer of genes between bacteria is quite common. It seems possible, therefore, that a more or less large proportion of our ability to digest our food could have come from our diet itself. If so, we can expect more discoveries of regional differences in digestive ability, given the many regional differences in diet.

ALCOHOL ENZYMES AND ALCOHOLISM

The biogeography of lactase and amylase production is intimately associated with the environment, as mediated by culture, and is clearly influenced by adaptive evolution. The topic is, in effect, an extension of chapter 5. In the case of the global distribution of the enzymes associated with the breakdown of alcohol and the consequent global distribution of alcoholism, we do not currently know what the balance of benefits and costs are. Adaptive evolution could be involved, but so could the theme of chapter 2, namely, the biogeographic history of the peoples' origins, as well as part of this chapter's theme, the influence of culture.

Peoples of southeastern Asian origin are on average less able than peoples from much of the rest of the world to denature one of the breakdown products of alcohol, acetaldehyde. The reason is that they have an ineffective variant of the enzyme acetaldehyde dehydrogenase [274; 529; 590; 742]. People with the ineffective form of the enzyme suffer unpleasant effects from acetaldehyde, such as facial flushing or, more seriously, essentially the effects of a hangover [742]. The inability to denature acetaldehyde, and the consequent unpleasant side-effects of alcohol consumption, appear to be associated with less alcohol consumption and less drunkenness in Asians than in Europeans [529; 742]. Within Asian populations too, individuals with the ineffective variant are less likely to be alcoholics. For instance, whereas only 2% of Japanese alcoholics had the ineffective variant, 50% of Japanese nonalcoholics had it [529].

Yi Peng et al. argued that the inability to break down acetaldehyde, and the consequent dislike of alcohol, was and is advantageous because the inability is associated with less drunkenness [590]. Their evidence is that a map of the spread of rice cultivation and inferred production of rice wine from coastal southeastern Asia matches a map of the frequency of the gene morph responsible for the ineffective variant of acetaldehyde dehydrogenase: the earlier rice was cultivated, the greater the proportion of the population that has the ineffective variant [590].

Their argument might be supported by the fact that the very peoples that have this apparently beneficial inability to break down acetaldehyde also have the ability to break down alcohol into acetaldehyde faster than do those who can denature acetaldehyde [693; 742]. In other words, they miss the more direct, perhaps pleasant, effects of alcohol, such as dizziness and sleepiness, and suffer the indirect unpleasant effects of faster build-up of acetaldehyde.

However, if indeed the drinking of alcohol is costly to the ability to survive, mate, or rear children, what has stopped other populations from evolving mechanisms to prevent individuals liking alcohol? A simple answer is that advantageous mutations are far from inevitable. Additionally, it is not only genetic and physiological factors that influence alcoholism. So also do cultural and socioeconomic ones. Ecuadorian Andean Indian populations have an extremely high incidence of ineffective acetaldehyde dehydrogenase [274], and yet they also have a high incidence of alcoholism [453]. Both cultural norms and poverty correlate with the Indians drinking to get drunk [453].

TWO ENZYMES, CODEINE, AND ANTIDEPRESSANTS

I have chosen the enzymes CYP2D6 and NAT2[c] to illustrate geographic variation in our drug-related physiology, along with the very incomplete state of our understanding of the variation. That incomplete understanding translates into ignorance in medical practice, and hence the danger of wrong treatment.

The enzymes CYP2D6 and NAT2 both exist in a variety of forms that vary in the rate at which they metabolize a variety of drugs [115; 529; 817]. Slow-acting variants can make the drugs ineffective; fast-acting variants can produce adverse side effects.

For instance, CYP2D6 variants affect the metabolism by the body of antidepressants and codeine (among many others) to the extent that in some individuals these drugs are ineffective at usual doses [529; 817]. The variants might also be associated with the incidence of various cancers [529].

NAT2 affects metabolism of drugs as widely different as the vasodilator hydralazine (used to treat high blood pressure), isoniazid (used to treat tuberculosis), and sulfonamide (an antibacterial compound) [115; 529; 817]. Its variants too seem to be associated with the incidence of various cancers, such as bladder cancer and colorectal cancer [529; 574].

The geographical distribution of some variants of CYP2D6 and NAT2 show some evidence of pattern, but the patterns are weak, and most of the variants show little to no pattern—for example, no latitudinal or longitudinal gradients [248; 419; 529; 574; 817]. With so little apparent geographic logic to the distribution of most of the variants, founder effects and genetic drift are perhaps the best explanation for their distribution [529]. If a latitudinal gradient of NAT2 exists, it is in only one region, eastern Asia [419; 529; 574]. There, slow variants are

c. CYP2D6; NAT2. Cytochrome P; N-acetyltransferase

more common near the equator, and fast variants more common at high latitudes[d] [419].

This latitudinal gradient matches the finding of a higher incidence of the slow variant of NAT2 among agriculturalists than among pastoralists [478], which correlates in turn with, on the one hand, less meat in the diet at lower latitudes (Fig. 3.3C, Fig. 7.2) and, on the other hand, a finding that meat eating might exacerbate connections between fast NAT2 variants and colorectal cancer [574]. However, the relationship of this variant of NAT2 with latitude is stronger than is the relationship of meat eating with latitude or of pastoralism with latitude. Therefore, any functional hypothesis or a connection via diet with cancer is far from confirmed.

Some have suggested that the connection between genes and their associated enzymes with metabolism of drugs might have originated with the enzymes' role in the metabolism of the huge variety of drug-like compounds in plants, such as alkaloids [529; 817]. Again, though, substantiated connections are few and far between.

THE HIGH-PROTEIN, HIGH-FAT ESKIMO/INUIT[E] DIET

Hunter-gatherers from all latitudes often ate a diet of about one-half to two-thirds animals (including fish)—a diet high in animal protein and fat [152; 488; 773]. The more large mammals in the diet, the more fat, because large mammals contain proportionately more fat than do small mammals [152]. People within the Arctic ate more animal food than anyone else did (over 90%), although within the tropics the range can go up to 60% [773] or even 80% [152].

The main difference between arctic peoples and others is the arctic peoples' access to and consumption of sea mammals and their immense amounts of fat (Fig. 7.2) [152; 188]. It seems to be a popular belief that the arctic peoples have a specialized digestive or other physiology that allowed them to subsist on such an apparently abnormal diet.

However, the hunter-gatherers' and especially arctic peoples' ability to survive, even flourish, with low cholesterol levels and low incidence of cardiovascular disease on a diet that sometimes is close to pure fat

---

d. NAT2 frequency by latitude. $N = 7$, $r_s = 0.96$, $P < 0.001$ (only one datum per population). All others ($N = 3$ with >5 data) $P > 0.1$. Data from [419].

e. Eskimo/Inuit. "Eskimo" used to be the term applied to arctic peoples, but while it is acceptable by some Alaskan arctic peoples, it is not by, for example, Canadian arctic peoples, who prefer "Inuit."

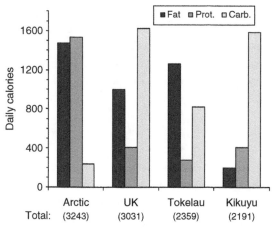

FIGURE 7.2. Diets of different peoples. UK (United Kingdom); Tokelau (Polynesia); Kikuyu (Kenya). Data from [33].

and protein is not due to a specialized physiology. Instead, the reasons include the fact that the hunter-gatherer and arctic diet is low in carbohydrates, a variety of differences between wild and domestic animals as food, and differences between hunter-gatherers and modern methods of preparation of food [152; 188].

The high protein and fat content of the diet means a low carbohydrate content, which results in the body using the fat to break down the protein as an energy source, with the result that the high-fat diet is not associated with obesity or arteriosclerosis or the other problems of high-fat/high-carbohydrate diets. A main difference between wild and domestic mammals as food is that wild mammal fat is mostly unsaturated, with a healthily low omega-6/omega-3 fatty acid ratio. Finally, arctic hunter-gatherers tend to eat their meat raw or barely cooked, meaning that a number of vitamins that are often destroyed by cooking, such as vitamin C, are available to them.

In sum, an interaction of an abnormal diet in an environment short of fuel results in peoples with a normal physiology surviving just fine in what in other circumstances could be an unhealthy diet. The large home ranges covered by high-latitude peoples (chapter 6), and especially the long distances sometimes travelled,[f] [62; 488] also probably helped [152].

f. Long distances travelled, high latitude. Female, male foraging distances by latitude, N = 6, 8 societies; $r_s \geq 0.9$, $P < 0.01$; distances per sex of 12–26 km in Arctic, 7–10 km in

This is not to say that arctic peoples do not have any adaptive genetic and hence physiological differences from people that live at lower latitudes. As discussed in chapter 5, arctic peoples seem to be able in various ways to withstand cold better than can low-latitude peoples, and these differences correlate with a variety of relevant genetic and enzymatic differences [303].

CONCLUSION

Diet and the associated enzymes and genes vary geographically. Whatever adaptive benefits the relevant enzymes had or have, the geographic variation in their incidence means that medical practice must take account of biogeography [115]. People from equatorial regions could be more prone to hypertension, especially in a modern salt-laden environment, because of the formerly beneficial ability to retain salt and to vasoconstrict under conditions of water shortage (chapter 5) [856; 857]. Yet the same people might be less reactive to some hypertension drugs (hydralazine, for instance) than are people from temperate regions. That diminished reactivity could be lethal in a medical environment that ignores biogeography. The high degree of variation among individuals within regions should not be ignored either [248].

---

tropics. Data from F. Marlowe, unpublished, [see 488]. Distance moved between camps. Average distance by latitude, $N = 36$, r^2adj $= 0.66$, $F = 68$, $P < 0.0001$; Tropical vs. temperate + arctic, $N = 15$, 21 (only 1 Arctic), medians $= 14$, 35 km, $F = 33$, $P < 0.0001$ (one hunter-gatherer peoples per state, only mainland, only mobile groups, outlier excluded). Data from [62].

PART THREE

Interaction among Cultures and Species

8

We Affect Our Biogeography

Now these are the kings of the land, which the children of
Israel smote, and possessed their land on the other side
Jordan . . . And these are the kings of the country which
Joshua and the children of Israel smote on this side
Jordan . . . and the Lord said unto him . . . there remaineth
yet very much land to be possessed. This is the land that yet
remaineth. [The Holy Bible, Joshua, 12, 1–24; 13, 1–6]

Wherever the European has trod, death seems to pursue the
aboriginal. . . . Nor is it the white man alone, that thus acts
the destroyer . . . The varieties of man seem to act on each
other; in the same way as different species of animals—the
stronger always extirpating the weaker. [167, Darwin, p. 520]

SUMMARY

Humans worldwide have competed with one another probably ever since we
were humans. When one culture is successful, the biogeography of human cultures changes, one culture expanding at the expense of another. Not all extinctions of cultures can be blamed on competition. Sometimes cultures with few
individuals die out simply because they are too small to survive the vagaries of
fate—and most cultures consist of few individuals. But cooperation can allow
cultures to expand their geographic range, as when countries such as the United
States welcome so many peoples of other nations.

So far, I have written almost as if humans live in an environment empty
of other species, as if humans were and are affected by only the physical
environment. But, of course, species and populations affect one another
and therefore affect each other's distribution. The ways in which species

and populations affect the biogeography of other species and populations is the topic of Part 3. I start with the effects of human populations on each other, then the effect of other species on human biogeography, and finally our influence on other species. Often the effects are deleterious, but some species and populations have changed or expanded their ranges as a result of interaction with other species and populations.

DELIBERATE AND INADVERTENT EXTERMINATION

Humans of one society have deliberately exterminated those of other societies for millennia. A Neanderthal male from about 40 kya has rib lesions that could indicate wounding from a projectile [140]. A 9,000-year-old skeleton from western United States had a projectile point embedded in his pelvis [134]. Ötzi, the Tyrolean "iceman" from about 5,000 years ago, was probably killed by the arrow found embedded in his shoulder [553].

These individual cases indicate for how long hominins have been competing. The extent of competition is another matter. In *The Third Chimpanzee*, Jared Diamond lists 17 genocides[a] of over 10,000 dead, and 6 of over 1 million [176]. More generally, the *Times Atlas of World History* can be read as a compendium of nonstop aggressive competition for space between whole societies [42].

Competition for land was the root of much of the atrocity. Obvious examples are the hunting down and slaughter by the British of Tasmanian aborigines, and the slaughter of Native Americans by mostly European immigrants [141; 176, ch. 16; 560, ch. 5]. It is not just the evil white male who has tried to exterminate other societies to obtain their land. Accounts of early explorers in Africa and America indicate almost perpetual territorial battle [354; 644]. The 1994 slaughter in Rwanda of Tutsi peoples by Hutu peoples in a country that for decades had been the most densely populated in sub-Saharan Africa can validly be interpreted as a direct outcome of competition for space, as can the movement of the Tutsi peoples before and after 1994 [266; 362; 790].

Competition for space and resources is not confined to humans. Groups of all sorts of animals compete against other groups, and large groups usually defeat small groups [305; 316]. Intercommunity raids

---

a. Genocide. Definitions of genocide are varied, detailed, and debated [173]. I use it to mean the killing of peoples considered as others in order to exterminate their sort, whether or not only the difference is the cause, or also competition for resources.

and killings by male chimpanzees have become famous, at least in primatology [809]. The raids do not occur unless a three-to-one advantage in numbers exists [836], a ratio true too of female lions' willingness to challenge [507]. A potential statistically verified example for humans is the demonstration that politically complex societies have larger geographic ranges than do simpler societies, which are mostly smaller [162]. However, politically complex societies also occur at higher latitudes [162]. Given that among animals and hunter-gatherer societies ranges are also larger at higher latitudes (chapter 6), the results do not necessarily support hegemony, even if they fit the hypothesis.

One way that societies outcompete others is with superior arms, as with the European invasion of so many continents from the 16th century on [178]. But another way is inadvertent and sometimes deliberate introduction of disease [32; 178; 515]. The influence of disease on human biogeography is a topic of the next chapter.

RANGE CHANGE IN THE FACE OF COMPETITION

If a population is exterminated, its geographic range is clearly affected. It has gone to zero from whatever it was before. Before the last Tasmanian died, their population of perhaps 5,000 in the 62,000 sq. km or so of the Tasmanian mainland was reduced to a couple of hundred people on the outlying 1330 sq. km of Flinders Island, and then to an even smaller remnant in a smaller area a few tens of km from the capital [176, ch. 16]. The last native Tasmanian died in 1876 [176, ch. 16], so removing Tasmania from the range of the Australian aboriginal peoples. Before the last Yana Native American died, their population was reduced from an unknown number to just one family hiding from genocidal prospectors and ranchers in what is now the Lassen National Forest, and then to one man living in and around the University of California Berkeley Museum of Anthropology, far outside his peoples' previous range [176, ch. 16].

Range reduction is exemplified by the system of Indian Reservations in the United States. These cover slightly more than 2% of the land surface of the United States, with huge areas of especially the eastern parts of the country with no reservations [107; 781]. All of the 15 largest reservations are west of the Great Plains—in other words, in some of the driest parts of the country. This exclusion of the Native American peoples to marginal areas is chillingly similar to what happened to native animal species as human immigrants spread across the continents (chapter 10) [128].

Humans can also aggressively cause other humans to greatly expand their range. The effect is exemplified by the movements of hundreds of thousands of refugees, even millions, in the face of internecine conflict. Calculations are that 1.5–2 million Rwandans dispersed into neighboring Burundi, D.R. Congo, and Tanzania at the height of the 1994 slaughter in Rwanda. Two years later, most returned, in part because of competition for land in the regions to which they had moved [266]. In the period of the Rwandan invasion and genocide, 1990–1994, nearly 7 million people worldwide were displaced from their home state to other states [772].

COMPETITION AND THE MOVE OUT OF AFRICA

Such displacement might have been happening ever since the rise of modern humans. "Migration [out of Africa] must have been a response to population growth and local overcrowding," wrote Luigi Cavalli-Sforza [125, ch. 4, p. 92]. Cavalli-Sforza goes on to suggest that the population might have numbered about 50,000 people just before our movement out of Africa 55 kya (chapter 2). Competition as the cause of humans' diaspora from Africa is a common suggestion. How do we test the hypothesis, though?

One way might be to estimate the threshold density that might cause sufficient overcrowding to force emigration. Current thinking is that modern humans probably expanded out of eastern Africa (chapter 2). But 50,000 people in the 3 million or so sq. km of eastern Africa is hardly overcrowding. If the area from which modern humans expanded was just Ethiopia (chapter 2), and all 50,000 people were there, the source area's human density is about one person per 20 sq. km. That figure is the median density of hunter-gatherer peoples in Africa in Binford's compendium [62]; i.e., it is not indicative of obvious overcrowding. However, estimates of the population size of humans 55 kya are close to guesswork, and we have no reliable estimate at all of the extent and distribution of suitable habitat then.

Studies of animals might allow a test of the idea that competition forces dispersal. Among mammals as subjects, as many studies found no effect of competition on dispersal as found an effect, while for birds twice as many studies found no effect as found as an effect [504]. Human cultures too respond differently to competition, some staying resident and intensifying use of the land, and others migrating and expanding (chapter 5).

But if competition did not cause humans to disperse out of Africa, or we cannot test whether it did or not, are there alternative explanations? We know that in several mammalian species, individuals and groups will not infrequently make long sorties outside their usual home range, sorties of several times the diameter of their home range [787]. Do we need to imagine any stimulus for a movement out of Africa beyond exploration by a large-brained, curious, inventive species ready to seize opportunities whenever encountered? Moving into areas empty of their own species, and therefore of the most effective competitors, they will presumably have done better than if they had stayed.

Seizing an opportunity obtained by exploration is not the same argument as being forced out by competition. But exploration as the cause of the movement out of Africa 55 kya is just as untestable as competition as the cause. Whatever explanation is offered, maybe it is time to insist that it comes with suggestions of how it can be tested.

DISAPPEARING CULTURES

Among them [the Siberians], I found one old woman who still remembered her native language . . . At first, only with difficulty, she recalled eleven words . . . now she had completely lost her sense of nationality, even her own mother-tongue.

[23, Arseniev, 1928 (translation), p. 143]

The dire fate of many of the world's languages is well known [164; 325; 326; 333; 476; 560]. Around 80% of languages are endemic and spoken by few people [324]. The fact that most languages are spoken by a few people in a small area is not special to languages. As I have described before, most things are rare (chapters 5 and 6). Nevertheless, rare things are more likely to disappear than are common things, and countries that are hotspots for endangered biodiversity are also hotspots for endangered languages [325].

Of the world's recorded 6,909 languages, 1,510 are spoken by less than 1,000 people, 472 by less than 100, 133 by less than 10 (Table 8.1) [455]. In 1995, although 209 indigenous languages were still spoken in North America, only 46 of them were spoken by children; i.e., 78% of the languages were likely to disappear within a generation [268, ch. 1]. Among individual countries, Australia lost the most languages in the 4 years between the last two editions of *Ethnologue*, 70 of 231 [277; 455]. Several countries gained languages, but that was because linguists discovered previously unrecorded communities, not because in 4 years new

TABLE 8.1
"NEARLY EXTINCT" LANGUAGES (I.E., ONLY A FEW ELDERLY
SPEAKERS ARE STILL LIVING)

Region	Total languages	Nearly extinct	Percent
Globe	6,909	473	6.8
Africa	2,110	46	2.2
Americas	993	182	18.3
Asia	2,322	84	3.6
Europe	234	9	3.8
Pacific	1250	152	12.2

SOURCE: Data from [455].

languages evolved. China gained the most languages, adding 57 to its previous 235 [277; 455].

For this chapter, the question is whether any of the disappearances of languages are due to competition. In the past, they certainly were. Even if ruling peoples/governments did not deliberately kill people of a culture, often they banned the speaking of the minority languages [560, ch. 6]. I doubt that any dominant colonial power was innocent in this respect. Nevertheless, not all disappearances can be laid at the door of the powerful. As I discussed in chapter 6 with reference to the depauperate tool culture of the Tasmanian aboriginal peoples, a small population means fewer people to learn from and greater chance that skills, including linguistic skills, will die out. Small populations also mean a greater chance that the peoples will die out: it takes fewer slings and arrows of outrageous fortune—fewer epidemics, fewer droughts, fewer extra-cold winters—to kill off a small population than it does a large one.

COOPERATION AND ITS BIOGEOGRAPHICAL CONSEQUENCES

I am British-born but now a U.S. citizen living in California, 8,000 km from my native country. A Brit, I have worked professionally in several countries in Africa, and in Japan. In all cases, the cooperation and generosity of the residents of those countries are what made my presence and work possible. With the story repeated worldwide, the geographic range of many peoples is far greater than it would have been without cooperation.

Millions of members of the former British Empire now live in Britain. Even more millions from other parts of the world now live in the United

States. While some competition is perceived, the national variety of the cities of Britain and the United States exists for the most part because of mutual cooperation between residents and immigrants.

Mass movements from one country to another, even if initiated by competition at both ends of the journey, are possible only with a fair amount of cooperation at the destination. Cooperation during the dispersal was probably necessary too. Certainly, anthropologists have suggested that social cooperation helped the expansion of modern humans across the globe (chapter 3).

With regard to disappearing languages, and therefore cultures, several have come back from the brink, often because of medical help from outsiders and the pride engendered in the speakers by the interest in their language shown by outsiders [113].

CONCLUSION

History can to some extent be seen as part of human biogeography, and vice versa. Human cultures expand across the globe and contract as do the geographic ranges of nonhuman species. And the ranges of both respond to the same influences—environmental change, competition, disease, and stochastic demographic effects falling more frequently on small populations in small geographic ranges. I have so far presented the book as a combination of anthropology and biogeography. Perhaps history should be added to the mix.

9

Other Species Affect
Our Biogeography

Without trypanosomiasis the whole of the sub-Saharan
continent would, like Latin America, have been readily
conquered by European forces long before the 19th Century.
[505, Maudlin, 2006, p. 679]

SUMMARY

The organisms that cause disease vary in their geographic distribution, and therefore, so also do our physiological traits and genes. Thus, people native to areas
where malaria is present have a different genetic complement than do those who
live in areas free of malaria.

Some diseases have prevented humans from occupying certain regions of the
world. For instance, sleeping sickness might have slowed European hegemony
over Africa and so preserved African cultural diversity.

People who live in cities face a differently ranked set of pathogens than do
those who live rurally, and therefore their physiologies are different. Thus, it was
not only regional differences in susceptibility that led to the devastation of indigenous populations as Europeans expanded across the globe but the fact that
the exploring Europeans were mainly city dwellers. Even now, the prevalence of
measles corresponds to the size of the infected populations.

And before we had developed advanced weapons, it is not impossible that
predators and scavengers limited our numbers and our spread across the world.

The organisms that cause disease vary in their geographic distribution.
Malaria is more prevalent in tropical countries than in temperate ones and
is absent from the Arctic. Measles is a disease of cities. And so on. Many
aspects of our genetic makeup and physiology are related to evolved responses to disease organisms. If the diseases vary geographically, so does

our physiology and our genetic constitution, our genotype. I present here some examples of this biogeographical coincidence of pathogens, disease, and human physiology. The aim is to provide examples of the biogeography of interspecific interaction and also to illustrate to biogeographers of nonhuman species the extent of information on human biogeography.

DISEASE AND HUMAN BIOGEOGRAPHY

Malaria and Sleeping Sickness

Sickle cell anemia is the classic example of how regional variation in the incidence of a disease causes regional variation in the nature of humans, in other words of how disease affects human biogeography [332, ch. 10; 530, ch. 6; 640, ch. 7].

Malaria is induced by the protozoan *Plasmodium*, the most lethal form of which is *P. falciparum*. *P. falciparum* is particularly widespread and prevalent in tropical Africa, although it now occurs throughout the tropics [714]. Sickle cell anemia, a lethal condition, is also prevalent in tropical Africa. It is characterized by individuals with it having two copies of the Hb^S gene. The gene persists in falciparum malarial regions, and people homozygous for it are born, because a person heterozygous for Hb^S is resistant to falciparum malaria and hence survives better than the homozygous Hb^A person. Outside of malarial areas, the homozygous Hb^A person survives better than the heterozygote individual, which is why sickle cell anemia is effectively nonexistent where there is no falciparum malaria.

Sickle cell anemia, or the gene that causes it when homozygous, does not occur everywhere where *P. falciparum* occurs. Rather, as with the ability to digest lactose (chapter 7), different genes, and hence chemicals, seem to have evolved to do the same job. For instance, in some areas of western Africa, another form of hemoglobin gene, Hb^C, confers resistance to malaria, indeed better resistance against severe malaria [332, ch. 10; 530, ch. 6]. It occurs together with Hb^S, but it is rare where Hb^S is common, which itself is rare where Hb^C is common, perhaps because Hb^C is more recently evolved than Hb^S and was in the process of replacing Hb^S in regions of severe falciparum malaria before malarial control campaigns [332, ch. 10; 530, ch. 6].

P. falciparum infects humans and gorillas. It used to be though that its closest relative was *P.reichenowi*, which infects chimpanzees, but it is now known that the closest relative is another *P. falciparum* of gorillas [460].

Elsewhere than Africa, malaria appears to be associated with variants of precursors of hemoglobin. While they might offer some protection against malaria, they have adverse effects in its absence, such as thalassemias, which were first recognized in and around the Mediterranean [332, ch. 10]. The argument for protection against malaria of the blood variants that produce the thalassemias is based, as with sickle cell anemia, on the fact that the otherwise costly thalassemias (and other protein variants) are at high frequencies in populations where malaria is or was prevalent.

One example comes from Sardinia, where the average incidence of thalassemia was over four times as high in the malarial lowlands as in the cold central highlands where malaria was and is rare [332, ch. 10]. A similar altitudinal variation occurs in New Guinea [540, ch. 5]. So now, in addition to altitudinal variation among humans in physiological ability to cope with low oxygen concentrations (chapter 5), we have altitudinal variation in physiological ability to cope with diseases whose prevalence depends on vectors whose prevalence in turn is determined by the environment.

Similar regional variation is seen with sleeping sickness (trypanosomiasis). Those with the disease are particularly likely to have one or two variants of a particular gene, APOL1, and people homozygous for either variant were likely to have kidney problems[a] [606]. Sleeping sickness is an African disease. Correspondingly, the APOL1 variants are far more common among people of western African ancestry (30% incidence) than among people whose families came from China, Europe, or Japan (0%); in Nigeria, 40% of people tested had one of the variants [606]. Additionally, African Americans are up to four times more likely than European Americans to suffer from long-term kidney disease [606].

As with sickle cell anemia and malaria, so with APOL1 and kidney disease; those heterozygous for the variant are almost certainly protected against sleeping sickness, because the blood of people with the APOL1 variants kills the trypanosome that causes sleeping sickness [606]. It looks as if original variants of APOL1 might have evolved to protect against the trypanosome that causes sleeping sickness, *Trypanosoma brucei*. The trypanosome then evolved a new form, *T. brucei rhodesiense*, which evaded the protein that APOL1 codes. But then some time in the last 10,000 years, the two new variants of APOL1 evolved, and the resultant proteins once again attack the trypanosome [606].

a. Kidney disease. Focal segmental glomerulosclerosis; end-stage kidney disease.

Biogeography of the Histocompatability (HLA) System

The major histocompatablity complex (MHC) is in humans termed the human leucocyte antigen (HLA) system. It is involved in recognition of alien cells or products in the body and is a vital part of our immune system, a finding arising from research into skin grafts [332, ch. 11; 530, ch. 7].

The system is highly varied (i.e., genetically polymorphic), presumably as an adaptation to the great variety of pathogens and other toxic substances that can and do enter our bodies [332, ch. 11; 530, ch. 7]. Certainly, several of the HLA forms are associated with prevalence of a variety of diseases, many of which might be autoimmune diseases [332, ch. 11; 530, ch. 7]. Most populations are heterozygous for most of the HLA alleles, implying that, as with the sickle cell alleles, the physiological effects produced by one form of the allele have costs, even if the allele might protect against one or more pathogens [530, ch. 7].

The biogeographical relevance of the system is that the variety of forms occur at different frequencies in different areas (Fig. 9.1) [125, ch. 2; 332, ch. 11; 530, ch. 7]. The reason for the variation, if it is not a result of history, of founder effects, and of drift (chapter 2), must be that pathogens vary geographically, even if medical anthropology still does not understand the details.

Additionally, a signature of the history of human dispersal out of Africa (chapter 2) might be present also, as indicated by the greater similarity between the European and African frequencies of many of the HLA variants as compared to the European and Japanese frequencies, despite the greater similarity of latitude between Europe and Japan (Fig. 9.1) [125, ch. 2; 332, ch. 11].

DISEASE AND CONQUEST

In chapter 4, I discussed the idea that disease could restrict the movement of humans and hence maintain cultural diversity by preventing homogenization [219]. The problem with the studies that I described there was that the data to support the idea were correlational (diversity of disease and cultures covaried) and could easily be used to suggest not that disease prevented movement but that a lack of human movement for other reasons maintained the diversity of disease in a region.

However, as the quotation from Ian Maudlin at the head of the chapter indicates, tsetse flies and sleeping sickness (trypanosomiasis) could

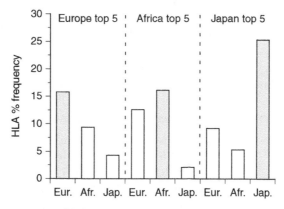

FIGURE 9.1. Median frequency of the most frequent five forms of HLA in each of Europe, Africa, and Japan (shaded columns), and the frequencies of each region's top five in the other two regions (open columns). Data from [332, ch. 11].

well have prevented Europeans from moving into sub-Saharan Africa and so protected African cultural diversity from European hegemony for a few centuries [236, ch. 4; 505]. Tsetse flies (*Glossina*) can carry the organism (*Trypanosoma*) that causes sleeping sickness in domestic livestock, including horses, and humans. Huge areas of tropical Africa are home to tsetse flies [236, ch. 3]. Depending on the species of trypanosome, those areas are effectively uninhabitable for horses and other nonnative livestock. Without horses in particular, human movement is far slower, and a cavalry is impossible [505]. Although Europeans conquered in the 16th to 18th centuries much of easily accessible South America, where no disease like sleeping sickness occurs, it is arguable that tsetse flies and sleeping sickness delayed European occupation of Africa until the 19th century [236; 505].

Indeed, it seems that in the New World, no widely lethal insect-born diseases might have existed before Europeans arrived in large numbers [515, ch. 6]. When they did, they brought with them two new deadly diseases and their vectors [515, ch. 6]. The diseases were yellow fever and falciparum malaria, and the vectors were, of course, slaves, particularly those from West Africa, along with two mosquitos new to the region.[b] The new immigrants also inadvertently altered the

---

b. Yellow fever; malaria. *Flavivirus* carried by *Aedes aegypti*; *Plasmodium* carried by *Anopheles* sp.

environment to suit the mosquitoes, for example, by supplying concentrations of the sort of containers favored by the yellow fever mosquito [515, ch. 6].

John McNeill recounts this story and persuasively argues that because the original Spanish invaders of the Caribbean had been resident there long enough to develop resistance to especially yellow fever by the time that other European nations attempted to oust them, they resisted for far longer than they otherwise might have done [515, ch. 6]. British attempts, for example, lost up to 10 times as many soldiers and sailors to disease as they did to Spanish arms [515, ch. 6]. Therefore, the global distribution of the Spanish was maintained and that of the British curtailed.

Disease, not battle losses, were a major factor in the British surrender of Yorktown in 1781 to the American revolutionary forces, McNeill argues [515, ch. 6]. That is, disease affected the global distribution of the British. Ten percent of the British forces had died from fighting by the surrender, but maybe 50% from disease, mainly malaria. The Americans fared better, because they had been in the country long enough to have developed resistance to the malaria.

Britain, Spain, and then France were arguably ousted from Haiti (then St. Domingue) by yellow fever, more than by the slave revolts eventually led by Toussaint Louverture and Jean-Jacques Dessalines [515, ch. 7]. Disease might have been responsible for as much as three-quarters of French deaths, whereas the African-born slaves were more resistant. Toussaint and Dessalines knew well the strength of disease as their ally, avoiding battle until the rainy season and yellow fever depleted the French forces, for example. The story is the same in Cuba, with the resistant revolutionaries able to let disease do their job against the Spanish [515, ch. 7].

The U.S. invasion succeeded, because by then the mosquito had been identified as the vector of yellow fever, and it was eradicated in Cuba [515, ch. 8]. America's aggressive extension of its territory was similarly successful in Panama, with the United States' building of the canal where before the French had been rebuffed by yellow fever and malaria [515, ch. 8].

If conquering a disease allowed American expansionism, escaping a disease could have caused humans' final expansions across the Pacific. Teina Rongo and coworkers argue that the long canoe voyages across the Pacific were stimulated by large-scale dinoflagellate outbreaks (algal blooms) that poisoned fish and hence the humans that relied on them [662].

DISEASE AND HUMAN POPULATION DENSITY

Disease affects human density. But human density also affects disease [14]. Smallpox, measles, and flu[c] are diseases of cities, easily transmitted by people in close contact, and quickly dying out when transmission rates are low or most of the population is infected [14; 32]. Survivors carry the pathogens but are immune to them.

As Europeans were killed in "the white man's grave" of West Africa by tropical diseases to which the Africans were more immune [163, ch. 1, 3, 4], so native peoples were killed by the new city diseases carried by exploring and acquisitive Europeans spreading across the globe from the 17th century on. Estimates vary, but from 50% to 90% of Native Americans in both continents could have been exterminated by the diseases brought by the first European invaders [32; 178, ch. 11]. The data are better for various Polynesian islands, where rates of over 50% mortality on some islands are confirmed [32].

We can see the importance of transmission and population size in the incidence of measles on Pacific islands. Measles is prevalent in more months on islands with large populations of people than ones with small populations (Fig. 6.14) [32]. The same was true of the bubonic plague. In the 1720–1722 plague pandemic in France, all cities of over 10,000 people were affected, but hundreds of villages of fewer than 100 people were not affected [734]. Isolation was presumably also an influence in addition to population size and density, small villages being more isolated than large cities.

In sum, a biogeography of human density (chapters 5 and 6) is also a biogeography of human disease.

COMPETITORS, PREDATORS, PREY, AND HUMAN BIOGEOGRAPHY

Robert Foley lists for late-Pliocene–early-Pleistocene hominins in Africa 8 genera and 19 species of carnivores that either preyed on hominins or might have competed with them [228, ch. 9]. Ten genera and 30 species could have competed for roots [228, ch. 9], which in the form of tubers could have been especially important to hominins [846; 847]. And Foley lists 10 genera and 28 species of primate, of which 18 were savannah woodland omnivores (i.e., in same habitat and with the same diet

c. Smallpox, measles, flu: *Variola, Morbillivirus, Orthomyxoviridae.*

as the hominins) and therefore potential competitors with hominins [228, ch. 9]. Similarly large numbers of large-bodied predators, prey, and competitors existed in the Middle East and Europe at the same time and into the Middle Pleistocene [250; 777]. The large ungulates were an even more important prey resource as hominins moved into high latitudes (chapter 3) [488; 773], which might have meant that any scavenging or hunting carnivores in the region were then even more important competitors than in Africa.

If resources, predation, and competitors influence the distribution of nonhuman species, we have to ask whether large mammals as competitors, predators, or prey influenced the distribution of humans. Would humans have left Africa had there not been large-bodied prey in Eurasia, Australia, Siberia, and the Americas? Would oysters and other seashore mollusks have been sufficient to fuel the diaspora (chapter 2)?

Alternatively, did large mammalian predators so deplete the prey that they prevented both the expansion of hominin populations within Africa and the movement of humans out of Africa until the later Middle Pleistocene [777]? Alan Turner suggests that up to the late Pleistocene, hominins were primarily scavengers and that hominin population numbers will have been determined by the ratio of hunting carnivores that did not consume bone with scavengers that did [777].

This ratio changed in Europe from the late Pliocene to mid-Pleistocene, with consequent effects on hominins' ability to move into Europe, Turner suggests [777]. From about 1.5 mya, the ratio was roughly equal in eastern Africa, having dropped from a greater proportion of carcass-producing species. Large-bodied carcass producers remained in Europe until much later, but so also did large-bodied scavengers. It might have been from only about 0.5 mya that Europe became more friendly to scavenging hominins, with the disappearance then of all but a few remaining scavengers, essentially those in eastern Africa nowadays [777].

Folklore accounts indicate that even with their advanced weapons, modern Native American societies feared and respected large predators [613]. Of course, predators and scavengers were not just providers of carcasses and passive competitors for those carcasses. Both could have preyed on hominins or actively fought hominins for access to carcasses. But the extent of such potential predation and contest competition is completely unknown [777], and therefore so also is its effect on human distribution.

CONCLUSION

Disease certainly affects human biogeography. But humans also affect the biogeography of disease. The two-way interaction then produces a story like that of dairying and lactose tolerance in adulthood: we affect the environment, which affects us, and we respond to the environment's effect by evolving the ability to use the altered environment. Where we grow rice, we provide a perfect environment for mosquitoes and hence for the malarial parasite; the increase in incidence of malaria increases the proportion of people resistant to the parasite, which increases the number of people available to grow rice, and so increases the extent of the environment altered for rice-growing, and yet more mosquitoes [640, ch. 7].

On a grander scale, we are warming the global climate. There can be no doubt that we are therefore altering the distribution of pathogens [203]. Preventative measures allowing, we are consequently altering the distribution of evolved disease-resistance in human populations. An obvious prediction is an increase in the incidence of genes for malarial resistance outside the traditional hotspots of malaria.

10

We Affect Other Species' Biogeography

It is clear, therefore, that we are now in an altogether
exceptional period of the earth's history. We live in a zoologi-
cally impoverished world, from which all the hugest, and
fiercest, and strangest forms have recently disappeared.
[803, Wallace, 1876, p. 150]

An adequate knowledge of the fauna and flora of the whole
world . . . can never now be obtained, owing to the reckless
destruction of forests, and with them of countless species
of plants and animals. [805, Wallace, 1880, p. 7]

SUMMARY

Ever since modern humans first spread out of Africa, and probably before then, we
have changed the range of other species, irretrievably altering the globe's biogeographical patterns. In a variety of ways, we have driven thousands of species to extinction, or in other words, reduced their geographic range to zero. Our movement
across the globe can often be mapped by the accompanying wave of extinctions.
We entered northern Europe and the Americas at a time of rapid, massive change
of climate, and probably hastened extinctions initiated by the climate change
there. Elsewhere, we were the original and main cause of extinctions. But several
species have benefited from our presence, greatly extending their range as a result
of their association with us. Our parasites and crops are the obvious examples.

An often-assumed characteristic of successful dispersing species—weed
species—is that they are not necessarily good competitors [766]. *H.
sapiens* is clearly an exception to that generalization. Humans are both
good dispersers and good competitors. As we have spread across the
world, we have driven, and continue to drive, many species to extinction,

possibly including species of our own genus *Homo*. If we want to stop that extermination, it will be helpful to understand the biogeography of the extermination and its causes, as well as of the occasional extension of species' geographic ranges that we have allowed or encouraged [828].

HUMANS VS. OTHER *HOMO* SPECIES

As modern humans spread out of Africa from roughly 55 kya, they moved into regions occupied by *Homo erectus,* which was still inhabiting Java up to maybe 30 kya [748]. The humans certainly also met Neanderthals, extant in the refugium of southern Spain up to 28 kya, maybe even 24 kya, if gone from the rest of Europe by about 30 kya [220, ch. 7; 223]. We know that the two species met because Neanderthal genes are in the human genome [288], as is the case for the Denisova *Homo*, extant in southern Siberia up to at least 40 kya [431; 637].

Erectus had disappeared from the Asian mainland so long before the arrival of modern humans, over 200,000 years beforehand [748], that we cannot be blamed for the majority of the decline of Erectus. Nevertheless, the latest dates for Erectus, Neanderthal, and Denisova coincide with dates for the spread of modern humans outside of Africa. So, did modern humans cause or hasten the extinction of any of these other *Homo* species? Or was the rapidly cooling climate of the northern hemisphere as humans dispersed out of Africa responsible for the disappearance of the final populations of Erectus, Denisova, and Neanderthal? The short answer is that we do not know.

Evidence against direct competition between modern humans and Neanderthals used to be the fact that nobody had found a site containing contemporaneous Neanderthal and modern human remains [220, ch. 7; 699; 700]. However, if evidence and argument that modern humans contain Neanderthal genes [288] is upheld, then clearly the two species must have met, and equally clearly, the possibility of competitive exclusion must be retained.

If our ancestors 30 kya were anything like us in the last few centuries, any competition could well have been active contest, with the losers being eaten [485]. Alternatively or as well, if the modern humans were more efficient hunters and gatherers, they could have depleted the game populations sufficiently or made them so fearful that Neanderthals could no longer find sufficient food for survival [250].

The supposition that the humans were more efficient food gatherers than Neanderthals, because of better tools, is speculation. But it seems likely, given that we survived and Neanderthals did not.[a] Also, a Neanderthal could well have needed more energy in hunting than a similar-sized modern human, because of the Neanderthal's inefficient locomotion (chapter 5). [425, ch. 6; 814]

Maybe humans had a better diet than did Neanderthals, because the humans used plant foods more than did Neanderthals [364], even though Nenderthals used a variety [649]. If we had a better diet, our survival and reproductive rates would have been higher. If adults survived better, maybe they had more time and energy to cooperate in raising relatives' offspring, which would have raised reproductive rate of kin and improved childhood survival [379].

But if modern humans could outcompete Neanderthals in Eurasia, by whatever means they did so, what happened in the Levant around 70 kya as the climate cooled and Neanderthals moved in as modern humans moved out (chapter 3) [220, ch. 7; 699; 700]? Competitive exclusion of modern humans by Neanderthals seems unlikely.

Rather, the expansions and contractions of each species could have been more or less independent of each other and perhaps driven by environmental fluctuations. In this scenario, modern humans retreated from the Levant before the Neanderthals moved back in. Subsequently, the Neanderthals went extinct throughout their range in Eurasia because the approach of the last glacial maximum reduced them to such small populations in such small refuges, in southern Europe, that their numbers never recovered [220, ch. 7].

In support of the idea that climate more than direct competition determined the demise of the Neanderthals, Clive Finlayson points out that until very recently, the arctic peoples of northern America survived better there than did others, Europeans, who had what we might otherwise consider a more advanced culture. The reason is that the arctic peoples' culture was better suited to the region [220, ch. 7]. No competition was involved. The northern peoples simply survived better. Similarly with Amundsen's and Scott's race to the South Pole. Neither attacked the other, neither got first to the other's supplies, and neither depleted any

---

a. Neanderthal, modern human tools. An apparent sudden advance in Neanderthal technology (Châtelperronian) just before the Neanderthals died out, about 30 kya, after a long period of little change in technology (Mousterian) indicates that by 30 kya Neanderthals were adopting modern human methods of making tools [525]. The problem with this scenario is that the apparent sudden advance is based mostly on a single site, in which it is not impossible that modern human tool types were mixed with Neanderthal remains [425, ch. 6].

resources. Instead, Amundsen, using northern peoples' clothing and technology, had a relatively easy trip to the South Pole and back, whereas Scott failed lethally with British Navy clothing and technology [791].

In sum, we do not yet know if the Neanderthals disappeared because of competition with humans. Nor do we know whether any competition was active contest competition or passive "scramble" competition. However, the extent and intensity of competition among modern humans makes a lack of contest competition between Neanderthals and modern humans unlikely. By the final stages of the last age, when Neanderthals were reduced to life in a few refuges, the presence of humans throughout habitable Eurasia would almost certainly have prevented a Neanderthal recovery.

LATE QUATERNARY AND RECENT EXTINCTIONS

Deliberately and unintentionally, directly and indirectly, we have reduced to zero the geographic range of thousands of species [392; 450; 562]. We have known about this extermination for a long time, as the two quotes from Alfred Wallace at the start indicate.

I ended the first quote before the original sentence ends, as others using the quote to comment on our eradication of other species have so ended it. However, Wallace continued, "and it is, no doubt, a much better world for us now they have gone" [803, p. 150]. Wallace was presumably glad to be able to roam the streets of London without being chased by the "hugest, and fiercest" species, the wolves and bears that used to inhabit Britain. The point is that much as we might admire our fellow carnivores, we also fear them, both as competitors and as predators [613]. We therefore exterminate them.

If all that the proverbial visiting Martians had available were time-lapse maps of the reduction to zero of the geographic range of Late Quaternary species, the movement of modern humans around the globe could be plotted and timed quite accurately. We arrive in an area empty of us, and very soon afterward, many of the larger animal species in the area disappear, whether reptiles, mammals, or birds [110; 123; 209; 264; 285; 428; 474; 493; 822]. For instance, North America lost about two-thirds of its terrestrial mammalian genera of over 44 kg (100 lb) [123, ch. 2], and Australia lost roughly 85% of its terrestrial fauna of over 44 kg [535]. The threshold happens to be about the weight of an adult male chimpanzee.

The coincidence of extinction of large-bodied species and arrival of humans is particularly evident on islands [123, ch. 2; 421; 494; 720;

721]. A classic case is Mauritius and the dodo (*Raphus cucullatus*). Mauritius, an island of 2,000 sq. km in the Indian Ocean east of Madagascar, was settled in the mid-1600s; the dodo (a 20 kg flightless pigeon) was gone less than a century later, maybe even less than half a century later [620, ch. 4]. Its fate was the same as that of the six genera and 11 species of moa (a flightless bird) of New Zealand, the largest of which weighed over 200 kg [123, ch. 2; 150; 493, ch. 32, 33, 34]. Similarly, the two genera and perhaps a dozen species of elephant birds of Madagascar (one of which weighed over 350 kg [174]), the ostrich-sized genyornis of Australia [264; 535], and the giant ducks of Hawaii [493, ch. 35] all disappeared shortly after the arrival of humans on the islands. These examples are all of birds because mammals rarely make it to oceanic islands. But not just large birds went; so did large plants [611].

For primates, Madagascar is a prime example of human-caused extinction. Just two millennia ago, the largest primate then extant, *Archaeoindris*, lived on Madagascar, along with the largest bird in the world, the elephant bird (*Aepyornis maximus*). *Archaeoindris* weighed perhaps 200 kg, larger than a gorilla.[b] But about 2,000 years ago humans arrived, and within a few centuries *Archaeoindris* was gone. So was every other lemur species of over 10 kg (Fig. 10.1) [109; 175; 269; 276; 411; 594; 645]. Madagascar was not unique in its loss of primates. Humans began to arrive in the Caribbean about 4,500 years ago [546], and now none of the at least three genera of Caribbean island primates remains [226, ch. 14].

It might or might not be relevant that Erectus and Neanderthals were also large-bodied primates, weighing well over 44 kg [264].

The extra susceptibility of larger species, including primates, is still true today [306; 618]. In the first place, hunters tend to go after the larger species [71; 285; 391; 579; 592; 593]. Also, large animals, including primates, are likely to reproduce slowly, to live in small populations at low densities, and to use large areas [177; 285; 391; 579], all traits that increase susceptibility to extinction [98; 119; 306; 310].

Chris Johnson argued that slow reproduction, not large body size, was the susceptible trait in the Late Quaternary extinctions caused by humans [406]. However, my analysis of his data indicates that either both influences were at play, or that body mass was the more important

---

b. Largest lemur. A recent study states 161 kg [269], but several others state c. 200 kg. 161 kg would make *Archaeoindris* the same size as a male gorilla, and therefore female *Archaeoindris* twice the size of female gorillas.

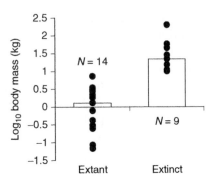

FIGURE 10.1. Heaviest lemur genera went extinct after humans' arrival in Madagascar, c. 2,000 years ago. Circles, individual genera; columns, median (1.3 kg, 22 kg). Data from [270; 411; 645].

influence[c]. That both mass and reproductive rate might have been influential is not surprising, given how closely correlated across species body size and reproductive rate are[d].

CLIMATE OR HUMANS?

At the same time as humans were dispersing through North America, the climate was warming rapidly as the last ice age ended. So was it humans or the changing climate and environment that were the main cause of the extinctions?[e] Since Paul Martin and Richard Klein's *Quaternary Extinctions* of 25 years ago, which for the most part strongly argued that humans caused the megafaunal extinctions, the human vs. climate debate has intensified [280; 281; 285–287; 428; 851]. The same discussion occurs over the extinctions in Europe and Australia that accompanied humans' arrival tens of thousands of years earlier than in the Americas [96; 815; 850], as well over the burst of extinctions soon after humans' arrival in Madagascar just 2,000 years ago [108; 269; 270; 793].

The other conversation concerns how humans might have caused the extinctions, if they did. Did they do it by hunting of populations of animals unused to humans as predators? This hypothesis is often described as the blitzkrieg or overkill hypothesis. Alternatively, or in addition, did

c. Extinction via body size as well as reproductive rate. Extinct vs. extant species, $N = 177$: log mass, $\chi^2 = 9.6$, $P < 0.003$, log reprod. rate, $\chi^2 = 42$, $P < 0.0001$, binomial logit generalized linear model. Add family as variable: mass, $\chi^2 = 50$, $P < 0.0001$, reprod. rate, $\chi^2 = 0.4$, ns; family, $\chi^2 = 53$, $P < 0.0001$. All data from [406].

d. Reproductive rate (birth interval) by mass. $N = 17$ primate families, r^2adj.$= 0.6$, $F = 29$, $P < 0.0001$. Data from [313].

e. An extraterrestrial impact causing the cooling of the Younger Dryas period at the time has been suggested, but several studies negate the hypothesis [745].

humans' domestic animals carry diseases to which the native fauna were not resistant? Did humans alter the habitat, particularly with fire, to such an extent that they exterminated native large mammals? Did humans and their livestock simply outcompete the native fauna?

I will consider the climate-or-humans debate first and then briefly examine the variety of ways in which humans could have been so destructive.

The issue, especially for the Americas, but also northern Europe, is that as humans moved into the new territory (back into old territory in the case of Europe), the globe was coming out of a major ice age, and therefore the environment was changing rapidly. The conservation literature, and even the popular media now, is full of accounts of species' inability to cope with global warming. Similarly, in the lead-up to the Holocene, species might have been equally incapable of coping.

For instance, the giant deer (*Megaloceros giganteus,* the "Irish elk") disappeared in Ireland at a time when the climate and environment were changing rapidly but before humans arrived in Ireland [39; 739]. The giant deer moved into Ireland about 12,000 years ago as the climate warmed at the end of the ice age, but they disappeared 1,500 years later, a full 1,500 years before any sign of humans' arrival in Ireland. It was the cold of the Younger Dryas and disappearance of winter browse that doomed the deer [39]. It did not, though, disappear because of lack of nutrients to grow its giant antlers. Any food shortage would have resulted simply in smaller antlers. Much more importantly, in polygynous species such as deer, it is the reproductive output of the females, not that of the males, that determines the fate of populations.

More generally, it looks as if many species in Eurasia and North America declined, sometimes precipitously, as the world warmed after the last ice age, with many of the declines starting before humans' arrival [41; 263; 270; 428; 564; 739]. Differences between species in when and where they declined and eventually disappeared can be expected given the different requirements of different species [739].

The differences can also be expected, however, if species differ in how well represented they are in the record. It turns out that at least in North America, the late-Pleistocene mammalian genera supposedly extinct before the arrival of humans are mostly ones poorly represented by fossil or subfossil remains [209]. If one mathematically accounts for representation in the record and for geographic distribution of the mammals and of fossil finds, almost all genera probably went extinct within 2,500 years of humans' arrival [209].

For Australia and Madagascar, the argument is that increasing aridity starting before humans arrived was the precipitating cause of the extinctions, even if humans might have hastened the final event [108; 269; 270; 850; 851]. A more emphatic denial of the role of humans in Australia is that dating of Australian extinctions is so imprecise that less than 20% of them can be placed within a reasonable period of humans' arrival [850; 851].

Nevertheless, as we can see happening now, environmental zones and the species in them shift with climate, rather than disappear completely [463, ch. 9; 583; 738]. Only where species could not move with the shifting environmental zones would a changing climate then be a cause of extinction. The giant deer in Ireland is an example, stuck as it was on effectively a peninsula of western Europe. In Russia, it survived far later [739]. Similarly, mountain tops limit upward movement in response to warming [99; 547]. Long winters and short growing seasons at the highest latitudes could also be a barrier to further movement north [39].

Despite the undoubted impact of climate change on Late Quaternary extinctions, especially in North America and Eurasia, a variety of lines of evidence indicate that humans were often a major final cause.

One corroborating fact is that it was the largest animals that were the ones most prone to extinction soon after humans' arrival, because we know that nowadays hunting affects the larger species in a community more than the smaller [71; 285; 391; 579; 592; 593].

Also, extinctions in North America over the 60 million years before the arrival of humans indicate little difference between large-bodied and small-bodied species in the rate of extinction from one million-year interval to the next, whereas at the end-Pleistocene extinction pulse, extinct species were on average twice the mass of all previously extinct species and twice the mass of survivors [8]. Although problems attend comparisons of extinction rates over time, including the lack of equivalence of taxonomic levels and the decreasing probability of detecting extinctions the further back in time the period being investigated [55, ch. 14], these contrasts seem too great to be explained by sampling error.

Furthermore, many of the American, Australian, Eurasian, and Malagasy species that disappeared soon after humans arrived had survived plenty of previous equally severe climatic changes as those that occurred around the time of human arrival [40; 108; 110; 269; 535; 739]. For instance, most of the large-bodied taxa that disappeared with humans' arrival in North America about 15 kya had less than 200,000 years previously survived an equally intense, rapid change of climate [494]. To match the North American extinction rates in the late Pleistocene, we have to go back millions of years to the boundaries between the Tertiary eras, from 60 to 3 mya [8].

The same argument applies to the mass extinctions of other regions, Australia, Madagascar, and the Pacific Islands. The species survived previous climatic change and then died out as humans arrived, which happened to be during a period when the climate was relatively stable, even unchanging according to some, such as less than 1,000 years ago in New Zealand [110; 269; 535; 657].

By contrast, in Africa the end of the last northern ice age was accompanied by the climate becoming far wetter, and tropical forest expanded from remnants on mountains and along rivers to the whole of the Congo Basin [148; 412]. Nevertheless, because hunting humans had lived in the regions for tens of thousands of years, and so a naïve or otherwise susceptible prey population no longer existed, extinctions of large mammals stayed close to background levels [123, ch. 2; 492]. The same difference in extinction rates between long occupied and recently occupied regions is seen with the extinction of birds on Pacific islands [603].

IF HUMANS, HOW?

I have so far concentrated on hunting as the main way in which humans might have driven species to extinction, because I have concentrated on the obvious extinctions of megafauna. Hunting is far from the only means by which we might have exterminated species in the late Pleistocene, and continue to exterminate them now. Imported disease, imported livestock and vermin, fire, deforestation, and competition for resources are all possible and debated ways in which we could have driven other species to extinction, either independently of the effects of climate change or by intensifying or speeding the effects of climate change [10; 41; 97; 110; 111; 270; 285; 287; 428; 474; 657; 850; 851].

Hunting

I have already discussed some of the evidence for hunting as the means by which humans exterminated species. I will mention here two further sorts of evidence.

First, some quantitative models that take into account, for example, our population size and the reproductive rate of large mammals, indicate that hunting alone could have caused extinctions [9; 837]. But other models indicate that hunting could not have done so [859], and as so often, the models are susceptible enough to the assumptions that as yet they cannot in and of themselves be considered reliable means by which to judge the likelihood of extermination by hunting alone [96].

Second, humans are principally diurnal hunters, and thousands of years ago, before the invention of the bow, were open-country hunters, taking terrestrial prey. If our hunting were a cause of the Late Quaternary extinctions, then surviving species should have been nocturnal, forest dwelling, and arboreal. They were. The surviving species with reproductive rates as low as the putative reproductive rates of extinct species, which might therefore have been expected to have been exterminated (29 of them), were indeed likely to be nocturnal ($N = 13$), forest dwelling ($N = 20$), and arboreal ($N = 14$) [406].

Disease

The domestic animals that humans brought with them to new regions of the world would likely have carried pathogens to which native animal populations would not have been resistant. We have seen in historic times the devastation caused by introduced diseases, such as the rinderpest epidemic that swept across Africa in the late 1800s, when probably millions of wild animals died [605]. It would be surprising if introduced diseases of mammalian livestock had not adversely affected susceptible native species [285; 475], but evidence is hard to obtain.

Competition

Humans must have directly and indirectly competed for resources with both herbivores and carnivores. If we were as deadly hunters as some suggest, we must have considerably reduced the prey populations of other carnivores, including Neanderthals [250].

Particular contested resources might have been caves—hence perhaps the extinction of the cave bear [285]. And in arid areas, water resources were probably contested, especially by pastoralists. Such might have been the cause of the final demise of some of the larger-bodied terrestrial lemurs in Madagascar, some of which survived for some time after humans arrived, as if something with less acute effects than hunting was the cause of their extinction [108; 109].

Habitat Alteration, Including Species-Area Effects

Humans have used fire for thousands of years. For instance, a considerable increase in charcoal deposits occurred about 10 kya in eastern North America, concomitant with the arrival of humans and the disappearance of large-bodied herbivores [263]. In Australia, a possible cause of the extinction of the genyornis was a fire-induced change from grassland to arid shrubland [534]. However, the fact that grazing mammals in Australia were no more likely to go extinct than were browsing mammals at this time indicates that the effect, if present, might not have been general [405].

Before all natural habitat is removed, whether it is removed by fire or not, it is often reduced to scattered fragments of its former extent. As discussed in chapter 6, both the reduction in total area of the natural habitat and the isolation of the fragments in a sea of unsuitable habitat will have led to the extinction of many species [463, ch. 16].

REDUCTION OF GEOGRAPHIC RANGE

Direct Influence

We have considerably reduced the geographic range of hundreds of mammalian species [472], and thousands of species in total [392], even if we have not yet exterminated them.

In an analysis of how in historic times we have reduced the geographic range of other species, Rob Channell and Mark Lomolino showed how in four continents with sufficient data (Africa, Australia, Eurasia, North America), the geographic ranges of both animals and plants had obviously (i.e., statistically significantly) contracted toward the edge rather than the center of their historic range in all but Africa[f] [128; 129; 463, ch. 16]. The study also showed that the initial reduction began at the periphery. The finding was a surprise because the biogeographical assumption up to then had been that with worsening conditions (such as the arrival of humans) species retreated to the center of their ranges, because the center was where conditions were best.

---

f. Reduction to edge of range. Statistically, Africa showed the effect of reduction to edge of historic range, but two outliers largely determined the association; otherwise, reduction to center and edge were equally likely [129].

Channel and Lomolino's interpretation was that as we humans expanded across the continents, so we drove species ahead of us to peripheral refuges. That did not happen in Africa, because our long presence in the continent had left it with only species already resistant to us. This is the same explanation as that given for the lack of late Pleistocene extinctions in Africa (this chapter). Even now, Africa has the lowest percentage of threatened primate species[g] [157, ch. 10; 318], if not of mammals [472].

Contraction to the edge was not the only surprise from the study. We have long known that island taxa have suffered a far greater proportion of extinctions at our hands than have mainland taxa [123, ch. 2; 494; 720] and that about twice as great a proportion of island as of mainland taxa are now threatened with extinction [472]. Nevertheless, where species occurred on both the mainland and islands, an unexpectedly high proportion of them survive now on only the islands [128]. The Tasmanian tiger (*Thylacinus cynocephalus*) might be one of the best-known examples [407]. Extinct in Australia by the late 1800s, it persisted in Tasmania until deliberately exterminated in 1930.

A reasonable interpretation of the finding that some species had or have their last refuge on small islands is that humans arrived later on the islands than on the mainlands [128]. An example is the woolly mammoth (*Mammuthus primigenius*), reduced to a single population of miniaturized individuals (see chapter 6) on Wrangel Island off the north coast of northeastern Siberia, where it survived until perhaps 4,000 years ago, presumably in part because humans did not get to the island until then [788]. Another is the giant kangaroo (*Protemnodon*), which lasted 5,000 longer on Tasmania than in Australia, concomitant with the later arrival dates of humans in Tasmania than in Australia [780].

The Flores hominin is another example of an island hold-out (chapter 6), as is Erectus on Java, where it persisted for scores of thousands of years after its demise on the Asian mainland [748].

Climate Warming

Added to the past direct change of habitat of other species, we are now indirectly affecting their (and our) environment through the change that we are causing to the heat-retaining capacity of the atmosphere [463, ch. 16]. As we warm the atmosphere, so animals' and plants'

g. Africa fewest Threatened primates. Africa 17% of 58; next lowest, Americas 30% of 66; Africa vs. all others together, $\chi^2 = 26$, $P < 0.0001$. Data from [318].

ranges are shifting poleward and upward. Species in the northern hemisphere are retreating from the southern edge of their range and extending the northern edge [463, ch. 16; 582; 583], as they did in the late Pleistocene and early Holocene from about 15 kya as the climate warmed [72; 564]. Species, including gelada baboons, are also moving up mountains [191; 547; 631], again as they did at the end of the last ice age [99].

The altitudinal move is currently more alarming perhaps than the latitudinal move. Animals moving up mountains can run out of mountain to move up [191; 547]. By contrast, as long as connecting habitat exists, animals can continue to move north. The proviso is an important one. The huge difference between now and past major changes of climate is that we have converted so much of the land to our ends that often connecting habitat does not exist [463, ch. 16].

Global warming is going to cause a greater rise of temperature at high latitudes than at low latitudes [624]. Nevertheless, for at least three interrelated reasons tropical species are by no means safe.

First, because of the relatively even temperatures of the tropics, the tropical species have not adapted to cope with large changes in temperature [402; 761]. I discussed this topic in chapter 4 with reference to mountains as barriers to distribution.

Second, temperatures change with latitude less at lower latitudes than they do at higher latitudes, and consequently for a given temperature increase, tropical species will need to shift their ranges farther to stay within a temperature range to which they are adapted [848]. Exacerbating this problem of the long distance that tropical species will need to move to stay within the temperate ranges to which they are adapted is that two of the most biodiverse regions of the world, the Amazon and Congo Basin, are very flat, and temperatures change less in a given distance on flat terrain than on terrain with topography.

Lastly, we return to the Rapoport effect, the fact that many species, including primates, have smaller geographic ranges in the tropics than at higher latitudes (chapter 6). The consequence is that for a given area of change of environment in the tropics, more species are likely to be adversely affected than is the case at higher latitudes.

Biodiversity and Overall Human Population Density

In chapter 5, I discussed influences on the biogeography of human density. Understanding the biogeography of human density is a crucial component of understanding our influence on other species and hence perhaps managing that influence. Generally, where many humans live, not only are there more of us to kill other species, but we leave less space for them [318; 319; 466; 581]. Correspondingly, for primates anyway, in the geographic ranges of species classified as Threatened, human population density is on average significantly higher than in the ranges of species classified as Low Risk [318].

Nevertheless, in general but with extremely wide scatter in the relationship, many species live where many people live [35; 143; 466]. Thus, Richard Cincotta and coauthors found that 20% of the human population lived in the 12% of the world's land surface covered by the identified 25 hottest spots of biodiversity [143]. Furthermore, the rate of increase of the human population is higher in the world's terrestrial biodiversity hotspots than it is elsewhere [143]. The alarming correlation of biodiversity and human density is surprising, given that humans tend to outcompete so many other taxa (this chapter).

The correlation is explained in part by the fact that similar geographic and biogeographic influences operate on humans as they do on other species [355]. Humans and many other species proliferate where productivity is relatively high, and do not do well where productivity is very low (chapter 5) [35; 481].

However, correlations do not necessarily indicate causes; the finding of many species where there are many humans could arise in part from the fact that where there are many humans, there are many human eyes, binoculars, and magnifying glasses to record species [192; 556; 632]. For instance, in a sample of over 75,000 locality records of specimens of over 1,000 species of birds in Africa, observations were significantly concentrated around the fringes of cities and along roads and rivers, i.e., where ornithologists might go to look for birds [632].

The studies that find correlations between biodiversity and human population density all find a large amount of scatter in the data. Thus, none of the hotspots of human density already mentioned (central Europe, the Himalayan foothills, or central China) is highly biodiverse [143]. Specifically with regard to nonhuman primates, on none of the four continents on which primates exist is there a match between hotspots of primate genera and hotspots of human population density. Indeed, in Africa, the Americas, and Asia, the primate generic hotspots are in areas that have some of the very lowest human population densities—for instance, the western Amazon in South America [308]. Much of the disparity is explained by the fact that primates are at their most diverse in tropical forest, an environment in which humans nowadays occur at low densities and in which hunter-gathers are at no higher densities than elsewhere (chapter 3) [62, table 5.10].

A biogeographic correlation of many humans with many wild species means that we will see in the near future a massive extinction pulse, as we exterminate actively and passively the remaining wildlife. We will not affect all species equally. If we can combine an understanding of the principles of human distribution with biogeography and with understanding of the biology of extinction risk, especially the distribution of extinction-prone species [118], we should be in a far better position than we are to predict future extinctions, and to mitigate them.

CHANGE IN COMMUNITY STRUCTURE

A community lacking "the hugest, and fiercest, and strangest forms" is very different from one that still has those forms. As I previously described, the sorts of species susceptible to human-caused extinction differ biologically from the less susceptible species [306; 311; 314]. Also, which species are most susceptible changes with the nature of the threat [391]. Thus, we change not simply the number of species in a community but the very nature of the community [403; 849].

The changes that we cause can have knock-on effects on the community. Species compete with one another. Large-bodied species often outcompete small-bodied ones in direct competition [239]. Consequently, when large-bodied primate species are hunted to extinction, populations of the smaller-bodied ones can increase [593]. Such is one explanation for the relatively high population density of species on islands, which have fewer species for their area than do mainlands and hence face less interspecific competition (chapter 6).

The changes that we cause can be quite complex. Often the first species to go under the adverse influence of humans are the large predators [753]. Prey species can then increase in density [753; 754]. And in at least one case, the prey has not simply increased in density but itself become a main predator: with elimination of predators from the land remaining as small islands in the waters of a hydroelectric dam in Venezuela, the capuchin monkey (*Cebus*) increased in density and became a main predator of bird nests [754].

It is not only animals that we affect. For instance, the current community of old trees in the supposedly natural tropical forest of Cameroon has a large proportion of trees that regenerate well after shifting cultivation. In other words, the forest has grown back on land heavily used by humans [786]. By comparison, the younger trees are ones that prefer natural gaps in the forest. In other words, humans from centuries previously are determining the composition of today's tropical forest, including the Amazonian forest [114].

HUMANS BENEFIT MANY SPECIES

When I first visited Hawaii in 2001, I was shocked to see its dense population of Indian mynah birds (*Acridotheres tristis*). The bird's native range is southern Asia. It did not get to Hawaii naturally. Humans took it there, as well as to Madagascar, southern Africa, Australia, and New Zealand, in all of which places it is a pest [75].

We have probably been expanding the geographic range of other species ever since we ourselves left Africa to spread across the earth, starting perhaps 50 kya (chapter 2). Two worldwide human parasites, for example, are most closely related to parasites of African animals, indicating that they evolved with us in Africa. The human tapeworm (*Taenia*) is related to African carnivore tapeworms [363]. And the two genera of human lice are related to the chimpanzee and gorilla lice [633]. HIV is just a recent example of an out-of-Africa pathogenic organism carried by humans worldwide [378].

The movement around the world of one member of our highly populous and diverse gut bacteria, *Helicobacter,* traced through DNA analysis, closely matches the movement of humans, first out of Africa, and then spreading from Asia (chapter 2) [212]. Another co-traveler from Asia is the black rat (*R. rattus*). It carried the fleas that carried the bacterium *Yersinia pestis* that caused bubonic plague, or the black death of 14th-century Europe [246]. And I have already mentioned smallpox carried by European explorers and travelers worldwide in previous centuries (chapter 9) but now eradicated [350].

While our tapeworms, lice, gut flora, and blood pathogens affect only us, many of the other species whose geographic range we have expanded can affect the environment in general, sometimes to the environment's and our detriment, sometimes not. Several animal species, such as the Indian mynah, have done so well where we have helped them expand their range that they are major pests [199]. They then tend to be termed "invasives." Invasives cause billions of dollars worth of damage annually in just the United States [602; 800].

One such invasive is a fellow primate, the vervet monkey (*Chlorocebus*). Native to Africa, it was introduced to the Caribbean island of Barbados where it is now a crop raider [78]. Our interference in its distribution increased by two-thirds its longitudinal extent and gave it by far the greatest longitudinal extent of any primate.

Plants too can be pestilential invasives. One example is kudzu (*Pueraria*), a vine from Japan that in the southern United States is smothering thousands of hectares of native plants [800]. However, most plant species do not turn into pests, at least on islands [682]. That is in stark contrast to animals on islands, which are a major cause of insular extinctions [67; 463, ch. 16]. Indeed, there is little evidence of any widespread extinction of native plant species caused by the introduced plant species, with the result that island plant diversity has effectively doubled as a result of our plant introductions [682].

Star thistle (*Centaurea solstitialis*) is an exception. Introduced from Mediterranean Europe, it is a scourge to ranch land in the Mediterranean environment of California [451]. Nevertheless, I suspect that it could be a major summer resource for honey bees, judging from how much they use it [43] and the fact that it can be the only flowering plant over many square kilometers at the height of the parched summer here in California's Central Valley and surrounding foothills. Honey bees, too, are an introduced species to the United States. Yet they are far from being pests, as the current intense worry over "colony collapse disorder" demonstrates [408].

If a plant might indirectly benefit us through benefiting honey bees, many plant and animal species are of direct benefit to us, and hence we have spread them widely throughout the globe. We deliberately introduced over 200 European species of bird, fish, and mammal into New Zealand to make the country feel more like the UK, the birthplace of most of New Zealand's early immigrants [717]. We have taken crops from South America and Asia to Africa and Europe, from Africa to the Americas, from Europe to the Americas and Africa, and so on [178, ch. 7, 8; 323]. Similarly, we have massively increased the global geographic range of our domestic animals. Domestic horses, cows, dogs, sheep, pigs, and cats, all now with near-global geographic ranges, originated around the fertile crescent of the Middle East [178, ch. 9; 189].

We have also spread ornamental plants throughout the world. A classic example is the ginkgo tree (*Ginkgo biloba*). The genus was once distributed throughout Laurasia but survives in the wild now in only a small area of central southern China [702]. We must have increased its chances of survival by planting it widely in Europe, North America, and South Africa, even though outside botanical gardens most specimens are males because the fruits smell unpleasant.

I started this section on how we have benefitted many species by markedly extending their geographic distribution with examples of a few species that live in and on us. In fact, hundreds of species do so, almost all bacteria. For them, the different parts of our body provide very different environments, and so we have different species on different parts of our body. Our three louse species, the body louse, the head louse (both *Pediculus*), and the pubic louse (*Pthirus*), are obvious examples. A more impressive example, because it includes so many more taxa and so many more environments, is that of the distribution of bacteria in and on us. The bacterial communities in the gut, mouth, ear, nostrils, head hair, and on over 10 sites on the skin all differ from one another [154]. A biogeography of bacteria in and on the human body could provide the means to study in microcosm general biogeographical principles, including many having to do with historical biogeography, given the high evolutionary rate of bacteria.

CONCLUSION

An adequate knowledge of the fauna and flora of the whole world . . . can never now be obtained, owing to the reckless destruction of forests, and with them of countless species of plants and animals.
[805, Wallace, 1880, p. 7]

The human species is . . . an environmental hazard. . . . Perhaps a law of evolution is that intelligence usually extinguishes itself.
[835, Wilson, 1993, p. 25]

Mark Lomolino and coauthors present in their *Biogeography* an impassioned plea for not just increased biogeographical understanding to improve our ability to conserve nature but also increased biogeographical knowledge. They write about the lack of information on past and present distributions of species and of the geography of humans [463, ch. 16, 17].

A problem with current biogeographic analysis is that, as Alfred Wallace pointed out, we had irrevocably altered the biogeography of many species, including the human species, long before we got around to studying biogeography [805, ch. 1]. Others have subsequently made the same point [285]. Although we have probably not altered any

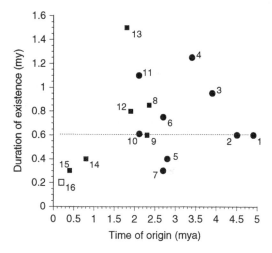

FIGURE 10.2. Duration of existence of hominin species (million years) by time of origin.* Line = median. 1–7, 10, 11 [•] = *Ardipithecus, Australopithecus, Paranthropus*; 8, 9, 12–16 [■] = *Homo*; 16 [□] = modern humans. Data from [232; 840].

*Duration of existence. 1 = *Ardipithecus ramidus*, 2 = *Australopithecus anamensis*, 3 = *A. afarensis*, 4 = *A africanus*, 5 = *A. bahrelghazali*, 6 = *A. gahri*, 7 = *Paranthropus aethiopicus*, 8 = *P. boisei*, 9 = *P. robustus*, 10 = *Homo habilis*, 11 = *H. rudolfensis*, 12 = *H. ergaster*, 13 = *H. erectus*, 14 = *H. heidelbergensis*, 15 = *H. neanderthalensis*, 16 = *H. sapiens*.

fundamental pattern, we must have affected the quantitative expression of many patterns [255]. How much have we altered, for example, the species-area effect on islands? We have reduced the number of animal species on islands but increased the number of plant species [682]. If we have altered the quantitative expression of the patterns, how much more difficult have we made discovery of the biogeographical processes behind the natural patterns [255; 805, ch. 1]?

A final question on the topic of human biogeography has to be how much more difficult we have made our own survival.

Our species, humans (*Homo sapiens*), has been on earth for 200,000 year so far (Fig. 10.2). We have another 400,000 years to go to be just an average hominin (Fig. 10.2) [232; 840]. The hominin species that lasted the longest, as far as current studies indicate, is *Homo erectus*. Erectus was on earth for over 1 million years (Fig. 10.2). Its most advanced tool was a stone hand axe [153].

How soon are the massive biogeographical changes that we have wrought going to reduce the human species' geographic range size to zero? As our brain size increased from Erectus' 900 cc to humans' 1,350 cc, our ever more efficient tools of extraction have massively increased the damage that we cause and the amount of resources that we consume. The efficiency of our competitive tools (our weapons) has increased perhaps even more. It seems unlikely that the combination of disappearing resources and increasingly efficient weapons will allow us to survive even the next 10,000 years, let alone another 400,000. Our salvation will come only if our evolved cooperative tendencies and abilities supersede our competitive ones [360; 361].

References

1. Agetsuma, N., and Nakagawa, N. 1998. Effects of habitat differences on feeding behaviors of Japanese monkeys: comparison between Yakushima and Kinkazan. *Primates in Medicine*, 39, 275–289.

2. Agresti, A. 1990. *Categorical Data Analysis*. John Wiley & Sons, New York.

3. Agustí, J., Blain, H.-A., Cuenca-Bescós, G., and Bailon, S. 2009. Climate forcing of first hominid dispersal in Western Europe. *Journal of Human Evolution*, 57, 815–821.

4. Aiello, L.C. 2010. Five years of *Homo floresiensis*. *American Journal of Physical Anthropology*, 142, 167–179.

5. Aldenderfer, M., and Yinong, Z. 2004. The prehistory of the Tibetan plateau to the seventh century A.D.: perspectives and research from China and the West since 1950. *Journal of World Prehistory*, 18, 1–55.

6. Aldenderfer, M.S. 2003. Moving up in the world. *American Scientist*, 91, 542–549.

7. Allen, J.A. 1877. The influence of physical conditions in the genesis of species. *Radical Review*, 1, 108–140.

8. Alroy, J. 1999. Putting North America's end-Pleistocene megafaunal extinction in context. Large-scale analyses of spatial patterns, extinction rates, and size distributions. In R.D.E. MacPhee, editor. *Extinction in Near Time: Causes, Contexts, and Consequences*. Kluwer Academic/Plenum Publishers, New York.

9. Alroy, J. 2001. A multispecies overkill simulation of the end-Pleistocene megafaunal mass extinction. *Science*, 292, 1893–1896.

10. Alroy, J., Grayson, D.K., Slaughter, R., and Skulan, J. 2001. Did human hunting cause mass extinction? *Science*, 294, 1459–1462.

11. Altizer, S., Nunn, C.L., and Lindenfors, P. 2007. Do threatened hosts have fewer parasites? A comparative study in primates. *Journal of Animal Ecology,* 76, 304–314.

12. Ambrose, S.H. 2003. Did the super-eruption of Toba cause a human population bottleneck? Reply to Gathorne-Hardy and Harcourt-Smith. *Journal of Human Evolution,* 45, 231–237.

13. Amos, W., and Hoffman, J.I. 2010. Evidence that two main bottleneck events shaped modern human genetic diversity. *Proceedings of the Royal Society B: Biological Sciences,* 277, 131–137.

14. Anderson, R.M., and May, R.M. 1991. *Infectious Diseases of Humans. Dynamics and Control.* Oxford University Press, Oxford.

15. Anikovich, M.V., Sinitsyn, A.A., Hoffecker, J.F., Holliday, V.T., Popov, V.V., Lisitsyn, S.N., et al. 2007. Early Upper Paleolithic in Eastern Europe and implications for the dispersal of modern humans. *Science,* 315, 223–226.

16. Archibald, J.D., and Deutschman, D.H. 2001. Quantitative analysis of the timing of the origin and diversification of extant placental orders. *Journal of Mammalian Evolution,* 8, 107–124.

17. Archibold, O.W. 1995. *Ecology of World Vegetation.* Chapman & Hall, London.

18. Argue, D., Donlon, D., Groves, C., and Wright, R. 2006. *Homo floresiensis*: microcephalic, pygmoid, *Australopithecus,* or *Homo*? *Journal of Human Evolution,* 51, 360–374.

19. Argue, D., Morwood, M.J., Sutikna, T., Jatmiko, and Saptomo, E.W. 2009. *Homo floresiensis*: a cladistic analysis. *Journal of Human Evolution,* 57, 623–639.

20. Arita, H.T., Robinson, J.G., and Redford, K.H. 1990. Rarity in Neotropical forest mammals and its ecological correlates. *Conservation Biology,* 4, 181–192.

21. Armitage, S.J., Drake, N.A., Stokes, S., El-Hawat, A., Salem, M.J., White, K., et al. 2007. Multiple phases of North African humidity recorded in lacustrine sediments from the Fazzan Basin, Libyan Sahara. *Quaternary Geochronology,* 2, 181–186.

22. Armitage, S.J., Jasim, S.A., Marks, A.E., Parker, A.G., Usik, V.I., and Uerpmann, H.-P. 2011. The southern route "Out of Africa": Evidence for an early expansion of modern humans into Arabia. *Science,* 331, 453–456.

23. Arseniev, V.K. 1996. *Dersu the Trapper.* McPherson & Co., New York.

24. Atkinson, Q.D. 2011. Phonemic diversity supports a serial founder effect model of language expansion from Africa. *Science,* 332, 346–349.

25. Austin, C.C. 1999. Lizards took express train to Polynesia. *Nature,* 397, 113–114.

26. Avise, J.C. 2000. *Phylogeography. The History and Formation of Species.* Harvard University Press, Cambridge, MA.

27. Avise, J.C. 2009. Phylogeography: retrospect and prospect. *Journal of Biogeography,* 36, 3–15.

28. Ayres, J.M., and Clutton-Brock, T.H. 1992. River boundaries and species range size in Amazonian primates. *American Naturalist,* 140, 531–537.

29. Baab, K.L., and McNulty, K.P. 2009. Size, shape, and asymmetry in fossil hominins: the status of the LB1 cranium based on 3D morphometric analyses. *Journal of Human Evolution,* 57, 608–622.

30. Bailey, G.N., Reynolds, S.C., and King, G.C.P. 2011. Landscapes of human evolution: models and methods of tectonic geomorphology and the reconstruction of hominin landscapes. *Journal of Human Evolution,* 60, 257–280.

31. Bailey, R.C., and Headland, T.N. 1991. The tropical rain-forest: is it a productive environment for human foragers? *Human Ecology,* 19, 261–285.

32. Baker, P.T. 1988. Infectious disease. In G.A. Harrison et al., editors. *Human Biology. An Introduction to Human Evolution, Variation, Growth, and Adaptability.* Oxford University Press, Oxford.

33. Baker, P.T. 1988. Nutritional stress. In G.A. Harrison et al., editors. *Human Biology. An Introduction to Human Evolution, Variation, Growth, and Adaptability.* Oxford University Press, Oxford.

34. Baker, P.T. 1988. The physical environment. In G.A. Harrison et al., editors. *Human Biology. An Introduction to Human Evolution, Variation, Growth, and Adaptability.* Oxford University Press, Oxford.

35. Balmford, A., Moore, J.L., Brooks, T., Burgess, N., Hansen, L.A., Williams, P., et al. 2001. Conservation conflicts across Africa. *Science,* 291, 2616–2619.

36. Banning, C. 1946. The Netherlands during German occupation. Food shortage and public health, first half of 1945. *Annals of the American Academy of Political and Social Science,* 245, 93–110.

37. Barbujani, G., and Sokal, R.R. 1990. Zones of sharp genetic change in Europe are also linguistic boundaries. *Proceedings of the National Academy of Sciences* (USA), 87, 1816–1819.

38. Barlow, L.K., Sadler, J.P., Ogilvie, A.E.J., Buckland, P.C., Amorosi, T., Ingimundarson, J.H., et al. 1997. Interdisciplinary investigations of the end of the Norse Western Settlement in Greenland. *Holocene,* 7, 489–499.

39. Barnosky, A.D. 1986. "Big game" extinction caused by late Pleistocene climatic change: Irish elk (*Megaloceros giganteus*) in Ireland. *Quaternary Research,* 25, 128–135.

40. Barnosky, A.D., Bell, C.J., Emslie, S.D., Goodwin, H.T., Mead, J.I., Repenning, C.A., et al. 2004. Exceptional record of mid-Pleistocene vertebrates helps differentiate climatic from anthropogenic ecosystem perturbations. *Proceedings of the National Academy of Sciences* (USA), 101, 9297–9302.

41. Barnosky, A.D., Koch, P.L., Feranec, R.S., Wing, S.L., and Shabel, A.B. 2004. Assessing the causes of late Pleistocene extinctions on the continents. *Science,* 306, 70–75.

42. Barraclough, G. 1984. *The Times Atlas of World History.* Times Books, London.

43. Barthell, J.F., Randall, J.M., Thorp, R.W., and Wenner, A.M. 2001. Promotion of seed set in yellow star-thistle by honey bees: evidence of an invasive mutualism. *Ecological Applications,* 11, 1870–1883.

44. Barton, R.N.E., Bouzouggar, A., Collcutt, S.N., Schwenninger, J.-L., and Clark-Balzan, L. 2009. OSL dating of the Aterian levels at Dar es-Soltan I (Rabat, Morocco) and implications for the dispersal of modern *Homo sapiens. Quaternary Science Reviews,* 28, 1914–1931.

45. Basell, L.S. 2008. Middle Stone Age (MSA) site distributions in eastern Africa and their relationship to Quaternary environmental change, refugia and the evolution of *Homo sapiens. Quaternary Science Reviews,* 27, 2484–2498.

46. Baz, A., and Monserrat, V.J. 1999. Distribution of domestic Psocoptera in Madrid apartments. *Medical and Veterinary Entomology,* 13, 259–264.

47. Beall, C.M. 2006. Andean, Tibetan, and Ethiopian patterns of adaptation to high-altitude hypoxia. *Integrative and Comparative Biology,* 46, 18–24.

48. Beall, C.M. 2007. Two routes to functional adaptation: Tibetan and Andean high-altitude natives. *Proceedings of the National Academy of Sciences* (USA), 104, 8655–8660.

49. Beall, C.M., and Steegmann, A.T. 2000. Human adaptation to climate: temperature, ultraviolet radiation, and altitude. In S. Stinson et al., editors. *Human Biology: An Evolutionary and Biocultural Perspective.* Wiley-Liss, New York.

50. Beals, K.L. 1972. Head form and climatic stress. *American Journal of Physical Anthropology,* 37, 85–92.

51. Beard, K.C. 2008. The oldest North American primate and mammalian biogeography during the Paleocene–Eocene Thermal Maximum. *Proceedings of the National Academy of Sciences* (USA), 105, 3815–3818.

52. Beehner, J.C., and McCann, C. 2008. Seasonal and altitudinal effects on glucocorticoid metabolites in a wild primate (*Theropithecus gelada*). *Physiology & Behavior,* 95, 508–514.

53. Bell, G. 2001. Neutral macroecology. *Science,* 293, 2413–2418.

54. Bennett, A., Sain, S.R., Vargas, E., and Moore, L.G. 2008. Evidence that parent-of-origin affects birth-weight reductions at high altitude. *American Journal of Human Biology,* 20, 592–597.

55. Bennett, P.M., and Owens, I.P.F. 2002. *Evolutionary Ecology of Birds: Life Histories, Mating Systems and Extinction.* Oxford University Press, Oxford.

56. Benson, L., Petersen, K., and Stein, J. 2007. Anasazi (pre-Columbian Native-American) migrations during the middle-12th and late-13th centuries—were they drought induced? *Climatic Change*, 83, 187–213.

57. Benton, M.J., and Donoghue, M.J. 2007. Paleontological evidence to date the tree of life. *Molecular Biology and Evolution*, 24, 26–53.

58. Berger, L.R., de Ruiter, D.J., Churchill, S.E., Schmid, P., Carlson, K.J., Dirks, P.H.G.M., et al. 2010. *Australopithecus sediba*: a new species of *Homo*-like australopith from South Africa. *Science*, 328, 195–204.

59. Bergmann, C. 1847. Ueber die Verhältnisse der Wärmeökonomie der Thiere zu ihrer Grösse. *Gottinger Studien*, 3, 595–708.

60. Betti, L., Balloux, F., Hanihara, T., and Manica, A. 2010. The relative role of drift and selection in shaping the human skull. *American Journal of Physical Anthropology*, 141, 76–82.

61. Bindon, J.R., and Baker, P.T. 1997. Bergmann's rule and the thrifty genotype. *American Journal of Physical Anthropology*, 104, 201–210.

62. Binford, L.R. 2001. *Constructing Frames of Reference: An Analytical Method for Archaeological Theory Building Using Hunter-Gatherer and Environmental Data Sets*. University of California Press, Berkeley.

63. Bininda-Emonds, O.R.P., Jones, K.E., MacPhee, R.D.E., Beck, R.M.D., Grenyer, R., et al. 2007. The delayed rise of present-day mammals. *Nature*, 446, 507–512.

64. Bird, D.W., and Bird, R.B. 2000. The ethnoarchaeology of juvenile foragers: shellfishing strategies among Meriam children. *Journal of Anthropological Archaeology*, 19, 461–476.

65. Bird, D.W., Richardson, J.L., Veth, P.M., and Barham, A.J. 2002. Explaining shellfish variability in middens on the Meriam Islands, Torres Strait, Australia. *Journal of Archaeological Science*, 29, 457–469

66. Bischof, N. 1975. Comparative ethology of incest avoidance. In R. Fox, editor. *Biosocial Anthropology*. John Wiley & Sons, New York.

67. Blackburn, T.M., Cassey, P., Duncan, R.P., Evans, K.L., and Gaston, K.J. 2004. Avian extinction and mammalian introductions on oceanic islands. *Science*, 305, 1955–1958.

68. Blackburn, T.M., and Gaston, K.J., editors. 2003. *Macroecology: Concepts and Consequences*. Blackwell Publishing, Oxford.

69. Blackburn, T.M., Gaston, K.J., and Loder, N. 1999. Geographic gradients in body size: a clarification of Bergmann's rule. *Diversity and Distributions*, 5, 165–174.

70. Blackburn, T.M., and Hawkins, B.A. 2004. Bergmann's rule and the mammal fauna of northern North America. *Ecography*, 27, 715–724.

71. Bliege Bird, R.L., and Bird, D.W. 1997. Delayed reciprocity and tolerated theft: the behavioral ecology of food-sharing strategies. *Current Anthropology*, 38, 49–78.

72. Boeskorov, G.G. 2006. Arctic Siberia: refuge of the Mammoth fauna in the Holocene. *Quaternary International,* 142, 119–123.

73. Bogin, B., and Rios, L. 2003. Rapid morphological change in living humans: implications for modern human origins. *Comparative Biochemistry and Physiology A—Molecular and Integrative Physiology,* 136, 71–84.

74. Böhm, M., and Mayhew, P.J. 2005. Historical biogeography and the evolution of the latitudinal gradient of species richness in the papionini (Primata: Cercopithecidae). *Biological Journal of the Linnean Society,* 85, 235–246.

75. Bomford, M., and Sinclair, R. 2002. Australian research on bird pests: impact, management and future directions. *Emu,* 102, 29–45.

76. Bonnefille, R. 1995. A reassessment of the Plio-Pleistocene pollen record of East Africa. In E.S. Vrba et al., editors. *Paleoclimate and Evolution with Emphasis on Human Origins.* Yale University Press, New Haven, CT.

77. Borregaard, M.K., and Rahbek, C. 2010. Causality of the relationship between geographic distribution and species abundance. *Quarterly Review of Biology,* 85, 3–25.

78. Boulton, A.M., Horrocks, J.A., and Baulu, J. 1996. The Barbados vervet monkey (*Cercopithecus aethiops sabaeus*): changes in population size and crop damage, 1980–1994. *International Journal of Primatology,* 17, 831–844.

79. Bouzouggar, A., Barton, N., Vanhaeren, M., d'Errico, F., Collcutt, S., Higham, T., et al. 2007. 82,000-year-old shell beads from North Africa and implications for the origins of modern human behavior. *Proceedings of the National Academy of Sciences* (USA), 104, 9964–9969.

80. Bowler, J.M., Johnston, H., Olley, J.M., Prescott, J.R., Roberts, R.G., Shawcross, W., et al. 2003. New ages for human occupation and climatic change at Lake Mungo, Australia. *Nature,* 421, 837–840.

81. Bradley, B.J., and Mundy, N.I. 2008. The primate palette: the evolution of primate coloration. *Evolutionary Anthropology,* 17, 97–111.

82. Bradley, S.B., and Deavers, D.R. 1980. A re-examination of the relationship between thermal conductance and body weight in mammals. *Comparative Biochemistry and Physiology,* 65A, 465–476.

83. Bradley, B., and Kamilar, J.M. 2011. Variation in primate coat colour supports Gloger's rule. *Journal of Biogeography,* 38, 2270–2277.

84. Bramanti, B., Thomas, M.G., Haak, W., Unterlaender, M., Jores, P., Tambets, K., et al. 2009. Genetic discontinuity between local hunter-gatherers and central Europe's first farmers. *Science,* 326, 137–140.

85. Bramble, D.M., and Lieberman, D.E. 2004. Endurance running and the evolution of *Homo*. *Nature,* 432, 345–352.

86. Brandon-Jones, D. 1996. The Asian Colobinae (Mammalia: Cercopithecidae) as indicators of Quaternary climatic change. *Biological Journal of the Linnean Society,* 59, 327–350.

87. Brandon-Jones, D. 1998. Pre-glacial Bornean primate impoverishment and Wallace's line. In R. Hall and D. Holloway, editors. *Biogeography and Geological Evolution.* Backhuys Publishers, Leiden, The Netherlands.

88. Brandon-Jones, D., Eudey, A.A., Geissmann, T., Groves, C.P., Melnick, D.J., Morales, J.C., et al. 2004. Asian primate classification. *International Journal of Primatology,* 25, 97–164.

89. Brès, P.L.J. 1986. A century of progress in combating yellow fever. *Bulletin of the World Health Organization,* 64, 775–786.

90. Brierley, C.M., Federov, A.V., Liu, Z., Herbert, T.D., Lawrence, K.T., and LaRiviere, J.P. 2009. Greatly expanded tropical warm pool, and weakened Hadley Circulation in the early Pliocene. *Science,* 323, 1714–1718.

91. Brillat-Savarin, J.A. 1825. *Physiologie du Goût.* Paris.

92. Broadhurst, C.L., Wang, Y.Q., Crawford, M.A., Cunnane, S.C., Parkington, J.E., and Schmidt, W.F. 2002. Brain-specific lipids from marine, lacustrine, or terrestrial food resources: potential impact on early African *Homo sapiens. Comparative Biochemistry and Physiology B—Biochemistry & Molecular Biology,* 131, 653–673.

93. Bromage, T.G., and Schrenk, F., editors. 1999. *African Biogeography, Climate Change, & Human Evolution.* Oxford University Press, Oxford.

94. Bromham, L., and Cardillo, M. 2007. Primates follow the "island rule": implications for interpreting *Homo floresiensis. Biology Letters,* 3, 398–400.

95. Bromham, L., Phillips, M.J., and Penny, D. 1999. Growing up with dinosaurs: molecular dates and the mammalian radiation. *Trends in Ecology and Evolution,* 14, 113–118.

96. Brook, B.W., and Bowman, D.M.J.S. 2002. Explaining the Pleistocene megafaunal extinctions: models, chronologies, and assumptions. *Proceedings of the National Academy of Sciences* (USA), 99, 14624–14627.

97. Brook, B.W., and Bowman, D.M.J.S. 2004. The uncertain blitzkrieg of Pleistocene megafauna. *Journal of Biogeography,* 31, 517–523.

98. Brook, B.W., and Bowman, D.M.J.S. 2005. One equation fits overkill: why allometry underpins both prehistoric and modern body size-biased extinctions *Population Ecology,* 47, 137–141.

99. Brown, J.H. 1971. Mammals on mountaintops: nonequilibrium insular biogeography. *American Naturalist,* 105, 467–478.

100. Brown, J.H. 1984. On the relationship between abundance and distribution of species. *American Naturalist,* 124, 255–279.

101. Brown, J.H. 1995. *Macroecology.* University of Chicago Press, Chicago.

102. Brown, J.H., Gillooly, J.F., Allen, A.P., Savage, V.M., and West, G.B. 2004. Toward a metabolic theory of ecology. *Ecology,* 85, 1771–1789.

103. Brown, P., Sutikna, T., Morwood, M.J., Soejono, R.P., Jatmiko, Wayhu Saptomo, E., et al. 2004. A new small-bodied hominin from the late Pleistocene of Flores, Indonesia. *Nature,* 431, 1055–1061.

104. Browne, J. 1983. *The Secular Ark. Studies in the History of Biogeography.* Yale University Press, New Haven, CT.

105. Brumm, A., Jensen, G.M., van den Bergh, G.D., Morwood, M.J., Kurniawan, I., Aziz, F., et al. 2010. Hominins on Flores, Indonesia, by one million years ago. *Nature*, 464, 748–752.

106. Brunet, M., Guy, F., Pilbeam, D., Taisso, H., Mackaye, H.T., Ahounta, D., et al. 2002. A new hominid from the Upper Miocene of Chad, Central Africa. *Nature*, 418, 145–151.

107. Bureau of Indian Affairs. 2006. *Indian Reservations in the Continental United States.* Bureau of Indian Affairs, Washington DC, 1.

108. Burney, D.A. 1997. Theories and facts regarding Holocene environmental change before and after human colonization. In S.M. Goodman and B.D. Patterson, editors. *Natural Change and Human Impact in Madagascar.* Smithsonian Institution Press, Washington, DC.

109. Burney, D.A., Burney, L.P., Godfrey, L.R., Jungers, W.L., Goodman, S.M., Wright, H.T., et al. 2004. A chronology for late prehistoric Madagascar. *Journal of Human Evolution*, 47, 25–63.

110. Burney, D.A., and Flannery, T.F. 2005. Fifty millennia of catastrophic extinctions after human contact. *Trends in Ecology and Evolution*, 20, 395–401.

111. Burney, D.A., and Flannery, T.F. 2006. Response to Wroe *et al.*: island extinctions versus continental extinctions. *Trends in Ecology and Evolution*, 21, 63–64.

112. Burnham, K.P., and Anderson, D.R. 2001. Kullback-Leibler information as a basis for strong inference in ecological studies. *Wildlife Research*, 28, 111–119.

113. Cahill, M. 2006. From endangered to less endangered. Case histories from Brazil and Papua New Guinea. SIL Electronic Working Papers 2004-004, August 2004. Available from www.sil.org/silewp/2004/silewp2004-004.htm. [Cited 2010.]

114. Campbell, D.G. 1994. Scale and patterns of community structure in Amazonian forests. In P.J. Edwards, R.M. May, and N.R. Webb, editors. *Large-scale Ecology and Conservation Biology.* Blackwell Science Ltd., Oxford.

115. Campbell, M.C., and Tishkoff, S.A. 2008. African genetic diversity: implications for human demographic history, modern human origins, and complex disease mapping. *Annual Review of Genomics and Human Genetics*, 9, 403–433.

116. Cann, R.L., Stoneking, M., and Wilson, A.C. 1987. Mitochondrial DNA and human evolution. *Nature*, 325, 31–36.

117. Carbone, C., and Gittleman, J.L. 2002. A common rule for the scaling of carnivore density. *Science*, 295, 2273–2276.

118. Cardillo, M., Mace, G.M., Gittleman, J.L., and Purvis, A. 2006. Latent extinction risk and the future battlegrounds of mammal conservation. *Proceedings of the National Academy of Sciences* (USA), 103, 4157–4161.

119. Cardillo, M., Mace, G.M., Jones, K.E., Bielby, J., Bininda-Emonds, O.R.P., Sechrest, W., et al. 2005. Multiple causes of high extinction risk in large mammal species. *Science,* 309, 1239–1241.

120. Carey, J.W., and Steegman, A.T. 1981. Human nasal projection, latitude and climate. *American Journal of Physical Anthropology,* 56, 313–319.

121. Carto, S.L., Weaver, A.J., Hetherington, R., Lam, Y., and Wiebe, E.C. 2009. Out of Africa and into an ice age: on the role of global climate change in the late Pleistocene migration of early modern humans out of Africa. *Journal of Human Evolution,* 56, 139–151.

122. Cashdan, E. 2001. Ethnic diversity and its environmental determinants: effects of climate, pathogens, and habitat diversity. *American Anthropologist,* 103, 968–991.

123. Caughley, G., and Gunn, A. 1996. *Conservation Biology in Theory and Practice.* Blackwell Science, Cambridge, MA.

124. Cavalli-Sforza, L.L. 2000. *Genes, Peoples, and Languages.* University of California Press, Berkeley.

125. Cavalli-Sforza, L.L., Menozzi, P., and Piazza, A. 1994. *The History and Geography of Human Genes.* Princeton University Press, Princeton, NJ.

126. Cavalli-Sforza, L.L., Piazza, A., Menozzi, P., and Mountain, J. 1988. Reconstruction of human evolution: bringing together genetic, archaeological, and linguistic data. *Proceedings of the National Academy of Sciences* (USA), 85, 6002–6006.

127. Chakraborty, D., Ramakrishnan, U., Panor, J., Mishra, C., and Sinha, A. 2007. Phylogenetic relationships and morphometric affinities of the Arunachal macaque *Macaca munzala,* a newly described primate from Arunachal Pradesh, northeastern India. *Molecular Phylogenetics and Evolution,* 44, 838–849.

128. Channell, R., and Lomolino, M.V. 2000. Dynamic biogeography and conservation of endangered species. *Nature,* 403, 84–86.

129. Channell, R., and Lomolino, M.V. 2000. Trajectories to extinction: spatial dynamics of the contraction of geographical ranges. *Journal of Biogeography,* 27, 169–179.

130. Chaplin, G. 2004. Geographic distribution of environmental factors influencing human skin coloration. *American Journal of Physical Anthropology,* 125, 292–302.

131. Chaplin, G., and Jablonski, N.G. 2009. Vitamin D and the evolution of human depigmentation. *American Journal of Physical Anthropology,* 309, 451–461.

132. Chase, J.M. 2010. Stochastic community assembly causes higher biodiversity in more productive environments. *Science*, 328, 1388–1391.

133. Chase, J.M., and Leibold, M.A. 2002. Spatial scale dictates the productivity-biodiversity relationship. *Nature*, 416, 427–430.

134. Chatters, J.C. 2000. The recovery and first analysis of an early Holocene human skeleton from Kennewick, Washington. *American Antiquity*, 65, 291–316.

135. Chen, F.C., and Li, W.H. 2001. Genomic divergences between humans and other hominoids and the effective population size of the common ancestor of humans and chimpanzees. *American Journal of Human Genetics*, 68, 444–456.

136. Cherry-Garrard, A.G.B. 1922. *The Worst Journey in the World: Antarctic 1910–1913*. Doran, New York.

137. Chiaroni, J., Underhill, P.A., and Cavalli-Sforza, L.L. 2009. Y chromosome diversity, human expansion, drift, and cultural evolution. *Proceedings of the National Academy of Sciences* (USA), 106, 20174–20179.

138. Chivers, D.J. 1994. Functional anatomy of the gastrointestinal tract. In A.G. Davies and J.F. Oates, editors. *Colobine monkeys. Their ecology, behavior and evolution*. Cambridge University Press, Cambridge.

139. Chivers, D.J., and Hladik, C.M. 1980. Morphology of the gastrointestinal tract in primates: comparisons with other mammals in relation to diet. *Journal of Morphology*, 166, 337–386.

140. Churchill, S.E., Franciscus, R.G., McKean-Peraza, H.A., Daniel, J.A., and Warren, B.R. 2009. Shanidar 3 Neandertal rib puncture wound and paleolithic weaponry. *Journal of Human Evolution*, 57, 163–178.

141. Churchill, W. 1997. *A Little Matter of Genocide. Holocaust and Denial in the Americas, 1492 to the Present*. City Light Books, San Francisco.

142. CIESIN and CIAT. 2005. Gridded Population of the World (GPW) Version 3. Center for International Earth Science Information Network (CIESIN); Centro Internacional de Agricultura Tropical (CIAT). Available from http://sedac.ciesin.columbia.edu/gpw/#. [Cited 2006.]

143. Cincotta, R.P., Wisnewski, J., and Engelman, R. 2000. Human population in the biodiversity hotspots. *Nature*, 404, 990–992.

144. Ciochon, R.L., and Bettis, E.A. 2009. Asian *Homo erectus* converges in time. *Nature Genetics*, 458, 153–154.

145. Colinvaux, P.A., de Oliveira, P.E., Moreno, J.E., Miller, M.C., and Bush, M.B. 1996. A long pollen record from lowland Amazonia: forest and cooling in glacial times. *Science*, 274, 85–88.

146. Collard, I.F., and Foley, R.A. 2002. Latitudinal patterns and environmental determinants of recent human cultural diversity: do humans follow biogeographical rules? *Evolutionary Ecology Research*, 4, 371–383.

147. Coller, M. 2007. Sahul time. Monash University, Melbourne, Australia. Available from http://sahultime.monash.edu.au/explore.html. [Cited 2011.]

148. Colyn, M., Gautier-Hion, A., and Verheyen, W. 1991. A re-appraisal of palaeoenvironmental history in Central Africa: evidence for a major fluvial refuge in the Zaire Basin. *Journal of Biogeography,* 18, 403–407.

149. Committee on the Earth System Context for Hominin Evolution. 2010. *Understanding Climate's Influence on Human Evolution.* National Research Council, Washington DC.

150. Cooper, A., Atkinson, I.A.E., Lee, W.G., and Worthy, T.H. 1993. Evolution of the moa and their effect on the New Zealand flora. *Trends in Ecology and Evolution,* 8, 433–437.

151. Coppeto, S.A., and Harcourt, A.H. 2005. Is a biology of rarity in primates yet possible? *Biodiversity and Conservation,* 14, 1017–1022.

152. Cordain, L., Eaton, S.B., Miller, J.B., Mann, N., and Hill, K. 2002. The paradoxical nature of hunter-gatherer diets: meat-based, yet non-atherogenic. *European Journal of Clinical Nutrition,* 56, S42–S52.

153. Corvinus, G. 2004. *Homo erectus* in East and Southeast Asia, and the questions of the age of the species and its association with stone artifacts, with special attention to handaxe-like tools. *Quaternary International,* 117, 141–151.

154. Costello, E.K., Lauber, C.L., Hamady, M., Fierer, N., Gordon, J.I., and Knight, R. 2009. Bacterial community variation in human body habitats across space and time. *Science,* 326, 1694–1697.

155. Cotgreave, P., and Pagel, M. 1997. Predicting and understanding rarity: the comparative approach. In W.E. Kunin and K.J. Gaston, editors. *The Biology of Rarity.* Chapman & Hall, London.

156. Cowlishaw, G. 1999. Predicting the pattern of decline of African primate diversity: an extinction debt from historical deforestation. *Conservation Biology,* 13, 1183–1193.

157. Cowlishaw, G., and Dunbar, R. 2000. *Primate Conservation Biology.* Chicago University Press, Chicago.

158. Cowlishaw, G., and Hacker, J.E. 1997. Distribution, diversity, and latitude in African primates. *American Naturalist,* 150, 505–512.

159. Cox, C.B., and Moore, P.D. 2005. *Biogeography: An Ecological and Evolutionary Approach.* 7th. ed. Blackwell Publishing, Oxford.

160. Cox, M.P., Redd, A.J., Karafet, T.M., Ponder, C.A., Lansing, J.S., Sudoyo, H., et al. 2007. A Polynesian motif on the Y chromosome: population structure in remote Oceania. *Human Biology,* 79, 525–535.

161. Crawford, M.H. 1998. *The Origins of Native Americans.* Cambridge University Press, Cambridge.

162. Currie, T.E., and Mace, R. 2009. Political complexity predicts the spread of ethnolinguistic groups. *Proceedings of the National Academy of Sciences* (USA), 106, 7339–7344.

163. Curtin, P.D. 1998. *Disease and Empire: The Health of European Troops in the Conquest of Africa.* Cambridge University Press, Cambridge.

164. Dalby, A. 2003. *Language in Danger.* Columbia University Press, New York.

165. Darwin, C. 1859. *On the Origin of Species by Means of Natural Selection: Or The Preservation of Favoured Races in the Struggle for Life.* John Murray, London.

166. Darwin, C. 1871. *The Descent of Man, and Selection in Relation to Sex.* John Murray, London.

167. Darwin, C.R. 1839. *Voyages of the Adventure and Beagle, Vol. III.* Henry Colburn, London.

168. de Filippo, C., Heyn, P., Barham, L., Stoneking, M., and Pakendorf, B. 2010. Genetic perspectives on forager-farmer interaction in the Luangwa Valley of Zambia. *American Journal of Physical Anthropology,* 141, 382–394.

169. de Lumley, M.-A. and Lordkipanidze, D. 2006. L'homme de Dmanissi (*Homo georgicus*), il y a 1 810 000 ans. *Comptes Rendus Palevol,* 5, 273–281.

170. de Vos, J. 2009. Neanderthal Man, *Homo erectus* and *Homo floresiensis*: l'histoire se répète. *Journal of the History of Biology,* 42, 361–379.

171. deMenocal, P.B. 2004. African climate change and faunal evolution during the Pliocene-Pleistocene. *Earth and Planetary Science Letters,* 220, 3–24.

172. Dennell, R., and Roebroeks, W. 2005. An Asian perspective on early human dispersal from Africa. *Nature,* 438, 1099–1104.

173. Destexhe, A. 1995. *Rwanda and Genocide in the Twentieth Century.* New York University Press, New York.

174. Dewar, R.E. 1984. Extinctions in Madagascar. In P.S. Martin and R.G. Klein, editors. *Quaternary Extinctions: A Prehistoric Revolution.* University of Arizona Press, Tucson.

175. Dewar, R.E. 1997. Were people responsible for the extinction of Madagascar's subfossils, and how will we ever know? In S.M. Goodman and B.D. Patterson, editors. *Natural Change and Human Impact in Madagascar.* Smithsonian Institution Press, Washington, DC.

176. Diamond, J. 1992. *The Third Chimpanzee. The Evolution and Future of the Human Animal.* HarperPerennial, New York.

177. Diamond, J.M. 1984. Historic extinctions: a Rosetta Stone for understanding prehistoric extinctions. In P.S. Martin and R.G. Klein, editors. *Quaternary Extinctions. A Prehistoric Revolution.* The University of Arizona Press, Tucson.

178. Diamond, J.M. 1997. *Guns, Germs, and Steel: The Fates of Human Societies.* W.W. Norton, New York.

179. Dillehay, T.D., Ramírez, C., Pino, M., Collins, M.B., Rossen, J., and Pino-Navarro, J.D. 2008. Monte Verde: seaweed, food, medicine, and the peopling of South America. *Science,* 320, 784–786.

180. Disotell, T.R. 1999. Human evolution: origins of modern humans still look recent. *Current Biology*, 9, R647–R650.

181. Disotell, T.R. 1999. Human evolution: sex-specific contributions to genome variation. *Current Biology*, 9, R29–R31.

182. Disotell, T.R. 1999. Human evolution: the southern route to Asia. *Current Biology*, 9, R925–R928.

183. Ditlevsen, P.D., Svensmark, H., and Johnsen, S. 1996. Contrasting atmospheric and climate dynamics of the last-glacial and Holocene periods. *Nature*, 379, 810–812.

184. Dodo, Y., Kondo, O., Muhesen, S., and Akazawa, T. 1998. Anatomy of the Neanderthal infant skeleton from Dederiyeh Cave, Syria. In T. Akazawa, K. Aoki, and O. Bar-Yosef, editors. *Neandertals and Modern Humans in Western Asia*. Plenum Press, New York.

185. Doherty, D.A., and Harcourt, A.H. 2004. Are rare primate taxa specialists or simply less studied? *Journal of Biogeography*, 31, 57–61.

186. Donoghue, P.C.J., and Benton, M.J. 2007. Rocks and clocks: calibrating the Tree of Life using fossils and molecules. *Trends in Ecology and Evolution*, 22, 424–431.

187. Dow, M.M., Cheverud, J.M., and Friedlaender, J.S. 1987. Partial correlation of distance matrices in studies of population structure. *American Journal of Physical Anthropology*, 72, 343–352.

188. Draper, H.H. 1977. The aboriginal Eskimo diet in modern perspective. *American Anthropologist*, 79, 309–316.

189. Driscoll, C.A., Macdonald, D.W., and O'Brien, S.J. 2009. From wild animals to domestic pets, an evolutionary view of domestication. *Proceedings of the National Academy of Sciences* (USA), Suppl. 1, 106, 9971–9978.

190. Dunbar, R.I.M. 1990. Environmental determinants of intraspecific variation in body weight in baboons (*Papio* spp.). *Journal of Zoology*, 220, 157–169.

191. Dunbar, R.I.M. 1998. Impact of global warming on the distribution and survival of the gelada baboon: a modelling approach. *Global Change Biology*, 4, 293–304.

192. Duncan, J.R., and Lockwood, J.L. 2001. Extinction in a field of bullets: a search for causes in the decline of the world's freshwater fishes. *Biological Conservation*, 102, 97–105.

193. Durham, W.H. 1991 *Coevolution: Genes, Culture, and Human Diversity*. Stanford University Press, Stanford, CA.

194. Eckhardt, R.B., and Henneberg, M. 2010. LB1 from Liang Bua, Flores: craniofacial asymmetry confirmed, plagiocephaly diagnosis dubious. *American Journal of Physical Anthropology*, 143, 331–334.

195. Eeley, H.A.C., and Foley, R.A. 1999. Species richness, species range size and ecological specialisation among African primates: geographical

patterns and conservation implications. *Biodiversity and Conservation,* 8, 1033–1056.

196. Eeley, H.A.C., and Lawes, M.J. 1999. Large scale patterns of species richness and species range size in anthropoid primates. In J.G. Fleagle, C.H. Janson, and K.E. Reed, editors. *Primate Communities.* Cambridge University Press, Cambridge.

197. Efron, B., and Thisted, R. 1976. Estimating the number of unseen species: how many words did Shakespeare know? *Biometrika,* 63, 435–447.

198. Egan, T. 2006. *The Worst Hard Time. The Untold Story of Those Who Survived the Great American Dust Bowl.* Houghton Mifflin Co., Boston.

199. Elton, C.S. 1958. *The Ecology of Invasions by Animals and Plants.* Methuen & Co., London.

200. Emmons, L.H. 1999. Of mice and monkeys: primates as predictors of mammal community richness. In J.G. Fleagle, C.H. Janson, and K.E. Reed, editors. *Primate Communities.* Cambridge University Press, Cambridge.

201. Enattah, N.S., Jensen, T.G.K., Nielsen, M., Lewinski, R., Kuokkanen, M., Rasinpera, H., et al. 2008. Independent introduction of two lactase-persistence alleles into human populations reflects different history of adaptation to milk culture. *American Journal of Human Genetics,* 82, 57–72.

202. Endicott, P., Ho, S.Y.W., Metspalu, M., and Stringer, C. 2009. Evaluating the mitochondrial timescale of human evolution. *Trends in Ecology and Evolution,* 24, 515–521.

203. Enserink, M. 2010. Yellow fever mosquito shows up in Northern Europe. *Science,* 329, 736.

204. Erlandson, J.M., Rick, T.C., Braje, T.J., Casperson, M., Culleton, B., Fulfrost, B., et al. 2011. Paleoindian seafaring, maritime technologies, and coastal foraging on California's Channel Islands. *Science,* 331, 1181–1185.

205. Eshleman, J.A., Malhi, R.S., and Smith, D.G. 2003. Mitochondrial DNA studies of native Americans: conceptions and misconceptions of the population prehistory of the Americas. *Evolutionary Anthropology,* 12, 7–18.

206. Estrada-Mena, B., Estrada, F.J., Ulloa-Arvizu, R., Guido, M., Méndez, R., Coral, R., et al. 2010. Blood group O alleles in Native Americans: implications in the peopling of the Americas. *American Journal of Physical Anthropology,* 142, 85–94.

207. Fagan, B. 2000. *The Little Ice Age: How Climate Made History, 1300–1850.* Basic Books, New York.

208. Fahrig, L. 2003. Effects of habitat fragmentation on biodiversity. *Annual Review of Ecology, Evolution, and Systematics,* 34, 487–515.

209. Faith, J.T., and Surovell, T.A. 2009. Synchronous extinction of North America's Pleistocene mammals. *Proceedings of the National Academy of Sciences,* (USA), 106, 20641–20645.

210. Falk, D., Hildebolt, C., Smith, K., Jungers, W., Larson, S., Morwood, M., et al. 2009. The type specimen (LB1) of *Homo floresiensis* did not have Laron syndrome. *American Journal of Physical Anthropology,* 140, 52–63.

211. Falk, D., Hildebolt, C., Smith, K., Morwood, M.J., Sutikna, T., Jatmiko, et al. 2007. Brain shape in human microcephalics and *Homo floresiensis. Proceedings of the National Academy of Sciences* (USA), 104, 2513–2518.

212. Falush, D., Wirth, T., Linz, B., Pritchard, J.K., Stephens, M., Kidd, M., et al. 2003. Traces of human migrations in *Helicobacter pylori* populations. *Science,* 299, 1582–1585.

213. FAO. 2008. International Year of the Potato. 2008. Potato World. Food and Agriculture Organization, Rome. Available from www.potato2008 .org/en/world/index.html. [Cited 2009.]

214. Fernandez, M.H., and Vrba, E.S. 2005. Body size, biomic specialization and range size of African large mammals. *Journal of Biogeography,* 32, 1243–1256.

215. Field, J.S., and Lahr, M.M. 2005. Assessment of the southern dispersal: GIS-based analyses of potential routes at oxygen isotopic stage 4. *Journal of World Prehistory,* 19, 1–45.

216. Finarelli, J.A., and Clyde, W.C. 2004. Reassessing hominoid phylogeny: evaluating congruence in the morphological and temporal data. *Paleobiology,* 30, 614–651.

217. Finch, V.A., and Western, D. 1977. Cattle colors in pastoral herds: natural selection or social preference. *Ecology,* 58, 1384–1392.

218. Fincher, C.L., and Thornhill, R. 2008. Assortative sociality, limited dispersal, infectious disease and the genesis of the global pattern of religion diversity. *Proceedings of the Royal Society B: Biological Sciences,* 275, 2587–2594.

219. Fincher, C.L., and Thornhill, R. 2008. A parasite-driven wedge: infectious diseases may explain language and other biodiversity. *Oikos,* 117, 1289–1297.

220. Finlayson, C. 2004. *Neanderthals and Modern Humans: An Ecological and Evolutionary Perspective.* Cambridge University Press, Cambridge.

221. Finlayson, C. 2005. Biogeography and evolution of the genus *Homo. Trends in Ecology and Evolution,* 20, 457–463.

222. Finlayson, C., and Carrión, J.S. 2007. Rapid ecological turnover and its impact on Neanderthal and other human populations. *Trends in Ecology and Evolution,* 22, 213–222.

223. Finlayson, C., Giles Pacheco, F., Rodríguez-Vidal, J., Fa, D., Gutiérrez, J.M., Santiago, A., et al. 2006. Late survival of Neanderthals at the southernmost extreme of Europe. *Nature,* 443, 850–853.

224. Fitzhugh, W.W. 1997. Biogeographical archaeology in the Eastern North American Arctic. *Human Ecology,* 25, 385–418.

225. Flannery, T. 1991. The mystery of the meganesian meat-eaters. *Australian Natural History*, 23, 722–729.

226. Fleagle, J.G. 1999. *Primate Adaptation and Evolution*, 2nd. ed. Academic Press, Inc., New York.

227. Fleagle, J.G., and Gilbert, C.C., editors. 2008. Modern human origins in Africa. *Evolutionary Anthropology*, 17(1), 1–80.

228. Foley, R. 1987. *Another Unique Species: Patterns in Human Evolutionary Ecology*. Longman Scientific & Technical, Harlow, UK.

229. Foley, R.A. 1993. African terrestrial primates: the comparative evolutionary biology of *Theropithecus* and the Hominidae. In N. Jablonski, editor. *Theropithecus. The Rise and Fall of a Primate Genus*. Cambridge University Press, Cambridge.

230. Foley, R.A. 1994. Speciation, extinction and climatic change in hominid evolution. *Journal of Human Evolution*, 26, 275–289.

231. Foley, R.A. 1999. Evolutionary geography of Pliocene African hominids. In T.G. Bromage and F. Schrenk, editors. *African Biogeography, Climate Change, & Human Evolution*. Oxford University Press, Oxford.

232. Foley, R.A. 2002. Adaptive radiations and dispersals in hominin evolutionary ecology. *Evolutionary Anthropology*, Suppl. 1, 32–37.

233. Fooden, J. 1971. Female genitalia and taxonomic relationships on *Macaca assamensis. Primates*, 12, 63–73.

234. Fooden, J. 1971. Male external genitalia and systematic relationships of the Japanese macaque (*Macaca fuscata* Blyth 1875). *Primates*, 12, 305–311.

235. Fooden, J., and Albrecht, G.H. 1999. Tail-length evolution in *fascicularis*-group macaques (Cercopithecidae: *Macaca*). *International Journal of Primatology*, 20, 431–440.

236. Ford, J. 1971. *The Role of Trypanosomiases in African Ecology: A Study of the Tsetse Fly Problem*. Clarendon Press, Oxford.

237. Forster, J.R. 1778. Excerpts from Remarks on the organic bodies, Ch. 5 in Observations made during a voyage round the world. In M.V. Lomolino, D.F. Sax, and J.H. Brown, editors. *Foundations of Biogeography*. University of Chicago Press, Chicago.

238. Forster, P. 2004. Ice Ages and the mitochondrial DNA chronology of human dispersals: a review. *Philosophical Transactions of the Royal Society B: Biological Sciences, 359*, 255–264.

239. French, A.R., and Smith, T.B. 2005. Importance of body size in determining dominance hierarchies among diverse tropical frugivores. *Biotropica*, 37, 96–101.

240. Frey, B.S., Savage, D.A., and Torgler, B. 2010. Interaction of natural survival instincts and internalized social norms exploring the *Titanic* and *Lusitania* disasters. *Proceedings of the National Academy of Sciences* (USA), 107, 4862–4865.

241. Frisancho, A.R. 1993. *Human Adaptation and Accommodation.* University of Michigan Press, Ann Arbor.

242. Frisancho, A.R., Borkan, G.A., and Klayman, J.E. 1975. Pattern of growth of lowland and highland Peruvian Quechua of similar genetic composition. *Human Biology,* 47, 233–243.

243. Frisancho, A.R., and Greksa, L.P. 1989. Developmental responses in the acquisition of functional adaptation to high altitude. In M.A. Little and J.D. Haas, editors. *Human Population Biology.* Oxford University Press, New York.

244. Froehle, A.W. 2008. Climate variables as predictors of basal metabolic rate: new equations. *American Journal of Human Biology,* 20, 510–529.

245. Frumkin, A., Bar-Yosef, O., and H.P., S. 2011. Possible paleohydrologic and paleoclimatic effects on hominin migration and occupation of the Levantine Middle Paleolithic. *Journal of Human Evolution,* 60, 437–451.

246. Gage, K.L., and Kosoy, M.Y. 2005. Natural history of plague: perspectives from more than a century of research. *Annual Review of Entomology,* 50, 505–528.

247. Gagneux, P., Wills, C., Gerloff, U., Tautz, D., Morin, P.A., Boesch, C., et al. 1999. Mitochondrial sequences show diverse evolutionary histories of African hominoids. *Proceedings of the National Academy of Sciences (USA),* 96, 5077–5082.

248. Garcia-Martin, E. 2008. Interethnic and intraethnic variability of NAT2 single nucleotide polymorphisms. *Current Drug Metabolism,* 9, 487–497.

249. Garn, S.M., Clark, L.C., and Harper, R.V. 1953. The sex difference in the basal metabolic rate. *Child Development,* 24, 215–224.

250. Garrard, A.N. 1984. Community ecology and Pleistocene extinctions in the Levant. In R. Foley, editor. *Hominid Evolution and Community Ecology.* Academic Press, London.

251. Gaston, K.J. 2000. Global patterns in biodiversity. *Nature,* 405, 220–227.

252. Gaston, K.J. 2003. *The Structure and Dynamics of Geographic Ranges.* Oxford University Press, Oxford.

253. Gaston, K.J., and Blackburn, T.M. 1999. A critique for macroecology. *Oikos,* 84, 353–368.

254. Gaston, K.J., and Blackburn, T.M. 2000. *Pattern and Process in Macroecology.* Blackwell Science, Oxford.

255. Gaston, K.J., and Blackburn, T.M. 2003. Macroecology and conservation biology. In T.M. Blackburn and K.J. Gaston, editors. *Macroecology: Concepts and Consequences.* Blackwell Publishing, Oxford.

256. Gaston, K.J., Blackburn, T.M., Greenwood, J.J.D., Gregory, R.D., Quinn, R.M., and Lawton, J.H. 2000. Abundance-occupancy relationships. *Journal of Applied Ecology,* 37, Suppl. 1, 39–59.

257. Gaston, K.J., Blackburn, T.M., and Lawton, J.H. 1997. Interspecific abundance-range size relationships: an appraisal of mechanisms. *Journal of Animal Ecology*, 66, 579–601.

258. Gaston, K.J., Blackburn, T.M., and Spicer, J.I. 1998. Rapoport's rule: time for an epitaph? *Trends in Ecology and Evolution*, 13, 70–74.

259. Gaston, K.J., and Williams, P.H. 1996. Spatial patterns in taxonomic diversity. In K.J. Gaston, editor. *Biodiversity: A Biology of Numbers and Differences*. Blackwell Science, Oxford.

260. George, D., Southon, J., and Taylor, R.E. 2005. Resolving an anomalous radiocarbon determination on mastodon bone from Monte Verde, Chile. *American Antiquity*, 70, 766–772.

261. Ghalambor, C.K., Huey, R.B., Martin, P.R., Tewksbury, J.J., and Wang, G. 2006. Are mountain passes higher in the tropics? Janzen's hypothesis revisited. *Integrative and Comparative Biology*, 46, 5–17.

262. Gilbert, M.T.P., Jenkins, D.L., Götherstrom, A., Naveran, N., Sanchez, J.J., Hofreiter, M., et al. 2008. DNA from Pre-Clovis human coprolites in Oregon, North America. *Science*, 320, 786–789.

263. Gill, J.L., Williams, J.W., Jackson, S.T., Lininger, K.B., and Robinson, G.S. 2009. Pleistocene megafaunal collapse, novel plant communities, and enhanced fire regimes in North America. *Science*, 326, 1100–1103.

264. Gillespie, R. 2008. Updating Martin's global extinction model. *Quaternary Science Reviews*, 27, 2522–2529.

265. Gillooly, J.F., Allen, A.P., West, G.B., and Brown, J.H. 2005. The rate of DNA evolution: effects of body size and temperature on the molecular clock. *Proceedings of the National Academy of Sciences* (USA), 102, 140–145.

266. GlobalSecurity.org. 2005. Rwanda civil war. GlobalSecurity.org. Available from www.globalsecurity.org/military/world/war/rwanda.htm. [Cited 2009.]

267. Gloger, C.W.L. 1833. *Das Abändern der Vögel durch Einfluss des Klima's*. August Schulz, Breslau, Germany.

268. Goddard, I., editor. 1996. *Handbook of the North American Indians. Vol 17. Languages*. Smithsonian Institution, Washington, DC.

269. Godfrey, L.R., and Irwin, M.T. 2007. The evolution of extinction risk: past and present anthropogenic impacts on the primate communities of Madagascar. *Folia Primatologica*, 78, 405–419.

270. Godfrey, L.R., and Jungers, W.J. 2003. The extinct sloth lemurs of Madagascar. *Evolutionary Anthropology*, 12, 252–263.

271. Godfrey, L.R., Jungers, W.L., Reed, K.E., Simons, E.L., and Chatrath, P.S. 1997. Subfossil lemurs: inferences about past and present primate communities in Madagascar. In S.M. Goodman and B.D. Patterson, editors. *Natural Change and Human Impact in Madagascar*. Smithsonian Institution Press, Washington, DC.

272. Goebel, T. 1999. Pleistocene human colonization of Siberia and peopling of the Americas: an ecological approach. *Evolutionary Anthropology,* 8, 208–227.

273. Goebel, T., Waters, M.R., and O'Rourke, D.H. 2008. The late Pleistocene dispersal of modern humans in the Americas. *Science,* 319, 1497–1502.

274. Goedde, H.W., Agarwal, D.P., Harada, S., Meier-Tackmann, D., Du Ruofu, Bienzle, U., et al. 1983. Population genetic studies on aldehyde dehydrogenase isozyme deficiency and alcohol sensitivity. *American Journal of Human Genetics,* 35, 769–772.

275. Goodman, S.M., and Ganzhorn, J.U. 2004. Biogeography of lemurs in the humid forests of Madagascar: the role of elevational distribution and rivers. *Journal of Biogeography,* 31, 47–55.

276. Goodman, S.M., and Patterson, B.D., editors. 1997. *Natural Change and Human Impact in Madagascar.* Smithsonian Institution Press, Washington, DC.

277. Gordon, R.G., editor. 2005. *Ethnologue: Languages of the World,* 15th ed. Available from www.ethnologue.com/. SIL International, Dallas, TX.

278. Gravlee, C.C., Bernard, H.R., and Leonard, W.R. 2003. Heredity, environment, and cranial form: a reanalysis of Boas's immigrant data. *American Anthropologist,* 105, 125–138.

279. Gray, R.D., Drummond, A.J., and Greenhill, S.J. 2009. Language phylogenies reveal expansion pulses and pauses in Pacific settlement. *Science,* 323, 479–483.

280. Grayson, D.K. 1984. Explaining pleistocene extinctions. Thoughts on the structure of a debate. In P.S. Martin and R.G. Klein, editors. *Quaternary Extinctions: A Prehistoric Revolution.* The University of Arizona Press, Tucson.

281. Grayson, D.K. 1989. The chronology of North American late Pleistocene extinctions. *Journal of Archaeological Science,* 16, 153–165.

282. Grayson, D.K. 1990. Donner Party deaths: a demographic assessment. *Journal of Anthropological Research,* 46, 223–242.

283. Grayson, D.K. 1993. Differential mortality and the Donner Party disaster. *Evolutionary Anthropology,* 2, 151–159.

284. Grayson, D.K. 1996. Human mortality in a natural disaster: the Willie Handcart Company. *Journal of Anthropological Research,* 52, 185–205.

285. Grayson, D.K. 2001. The archaeological record of human impacts on animal populations. *Journal of World Prehistory,* 15, 1–68.

286. Grayson, D.K. 2007. Deciphering North American Pleistocene extinctions. *Journal of Anthropological Research,* 63, 185–213.

287. Grayson, D.K., Alroy, J., Slaughter, R., Skulan, J., and Alroy, J. 2001. Did human hunting cause mass extinction? *Science,* 294, 1459–1462.

288. Green, R.E., Krause, J., Briggs, A.W., Maricic, T., Stenzel, U., Kircher, M., et al. 2010. A draft sequence of the Neandertal genome. *Science*, 328, 710–722.

289. Greenberg, J.H. 1987. *Language in the Americas*. Stanford University Press, Stanford, CA.

290. Greksa, L.P., and Beall, C.M. 1989. Development of chest size and lung function at high altitude. In M.A. Little and J.D. Haas, editors. *Human Population Biology*. Oxford University Press, New York.

291. Groube, L., Chappell, J., Muke, J., and Price, D. 1986. A 40,000 year-old human occupation site at Huon Peninsula, Papua New Guinea. *Nature*, 324, 453–455.

292. Groves, C.P. 2001. *Primate Taxonomy*. Smithsonian Institution Press, Washington, DC.

293. Grubb, P. 1990. Primate geography in the Afro-tropical rain forest biome. In G. Peters and R. Hutterer, editors. *Vertebrates in the Tropics*. Alexander Koenig Zoological Research Institute and Zoological Museum, Bonn.

294. Grubb, P., Sandrock, O., Kullmer, O., Kaiser, T.M., and Schrenk, F. 1999. Relationships between Eastern and Southern African mammal faunas. In T.G. Bromage and F. Schrenk, editors. *African Biogeography, Climate Change, & Human Evolution*. Oxford University Press, Oxford.

295. Guernier, V., Hochberg, M.E., and Guégan, J.-F. 2004. Ecology drives the worldwide distribution of human diseases. *PLOS Biology*, 2, 0740–0746; e141 doi:10.1371/journal.pbio.0020141.

296. Gunz, P., Bookstein, F.L., Mitteroecker, P., Stadlmayr, A., Seidler, H., and Weber, G.W. 2009. Early modern human diversity suggests subdivided population structure and a complex out-of-Africa scenario. *Proceedings of the National Academy of Sciences* (USA), 106, 6094–6098.

297. Haak, W., Balanovsky, O., Sanchez, J.J., Koshel, S., Zaporozhchenko, V., Adler, C.J., et al. 2010. Ancient DNA from European Early Neolithic farmers reveals their Near Eastern affinities. *PLOS Biology*, 8, 16.

298. Haffer, J. 1969. Speciation in Amazonian forest birds. *Science*, 165, 131–137.

299. Hage, P., and Marck, J. 2003. Matrilineality and the Melanesian origin of Polynesian Y chromosomes. *Current Anthropology*, 44(S5), 121–127.

300. Hall, R., Roy, D., and Boling, D. 2004. Pleistocene migration routes into the Americas: human biological adaptations and environmental constraints. *Evolutionary Anthropology*, 13, 132–144.

301. Hammer, M.F., Spurdle, A.B., Karafet, T., Bonner, M.R., Wood, E.T., Novelletto, A., et al. 1997. The geographic distribution of human Y chromosome variation. *Genetics*, 145, 787–805.

302. Hancock, A.M., and Di Rienzo, A. 2008. Detecting the genetic signature of natural selection in human populations: models, methods, and data. *Annual Review of Anthropology*, 37, 197–217.

303. Hancock, A.M., Witonsky, D.B., Ehler, E., Alkorta-Aranburu, G., Beall, C., Gebremedhin, A., et al. 2010. Human adaptations to diet, subsistence, and ecoregion are due to subtle shifts in allele frequency. *Proceedings of the National Academy of Sciences* (USA), 107, 8924–8930.

304. Hancock, A.M., Witonsky, D.B., Gordon, A.S., Eshel, G., Pritchard, J.K., Graham Coop, G., et al. 2008. Adaptations to climate in candidate genes for common metabolic disorders. *PLOS Genetics*, 4, e32, 13 pp.

305. Harcourt, A.H. 1992. Coalitions and alliances: are primates more complex than non-primates? In A.H. Harcourt and F.B.M. de Waal, editors. *Coalitions and Alliances in Humans and Other Animals*. Oxford University Press, Oxford.

306. Harcourt, A.H. 1998. Ecological indicators of risk for primates, as judged by susceptibility to logging. In T.M. Caro, editor. *Behavioral Ecology and Conservation Biology*. Oxford University Press, New York.

307. Harcourt, A.H. 1999. Biogeographic relationships of primates on south-east Asian islands. *Global Ecology and Biogeography*, 8, 55–61.

308. Harcourt, A.H. 2000. Coincidence and mismatch in hotspots of primate biodiversity: a worldwide survey. *Biological Conservation*, 93, 163–175.

309. Harcourt, A.H. 2000. Latitude and latitudinal extent: a global analysis of the Rapoport effect in a tropical mammalian taxon: primates. *Journal of Biogeography*, 27, 1169–1182.

310. Harcourt, A.H. 2002. Empirical estimates of minimum viable population sizes for primates: tens to tens of thousands? *Animal Conservation*, 5, 237–244.

311. Harcourt, A.H. 2006. Rarity in the tropics: biogeography and macroecology of the primates. *Journal of Biogeography*, 33, 2077–2087.

312. Harcourt, A.H. unpublished. Unpublished analysis.

313. Harcourt, A.H. unpublished. Unpublished data.

314. Harcourt, A.H., Coppeto, S.A., and Parks, S.A. 2002. Rarity, specialization and extinction in primates. *Journal of Biogeography*, 29, 445–456.

315. Harcourt, A.H., Coppeto, S.A., and Parks, S.A. 2005. The distribution-abundance (i.e. density) relationship: its form and causes in a tropical mammal order, Primates. *Journal of Biogeography*, 32, 565–579.

316. Harcourt, A.H., and de Waal, F.B.M., editors. 1992. *Coalitions and Alliances in Humans and Other Animals*. Oxford University Press, Oxford.

317. Harcourt, A.H., and Doherty, D.A. 2005. Species-area relationships of primates in tropical forest fragments: a global analysis. *Journal of Applied Ecology*, 42, 630–637.

318. Harcourt, A.H., and Parks, S.A. 2003. Threatened primate taxa experience high human densities: adding an index of threat to the IUCN Red List criteria. *Biological Conservation*, 109, 137–149.

319. Harcourt, A.H., Parks, S.A., and Woodroffe, R. 2001. Human density as an influence on species/area relationships: double jeopardy for small African reserves? *Biodiversity and Conservation,* 10, 1011–1026.

320. Harcourt, A.H., and Schreier, B.M. 2009. Diversity, body mass, and latitudinal gradients in primates. *International Journal of Primatology,* 30, 283–300.

321. Harcourt, A.H., and Schwartz, M.W. 2001. Primate evolution: a biology of Holocene extinction and survival on the south-east Asian Sunda Shelf islands. *American Journal of Physical Anthropology,* 114, 4–17.

322. Harcourt, A.H., and Stewart, K.J. 2007. *Gorilla Society: Conflict, Compromise and Cooperation Between the Sexes.* University of Chicago Press, Chicago.

323. Harlan, J.R. 1995. *The Living Fields: Our Agricultural Heritage.* Cambridge University Press, Cambridge.

324. Harmon, D. 1995. The status of the world's languages as reported in *Ethnologue. Southwest Journal of Linguistics,* 14, 1–33.

325. Harmon, D. 1996. Losing species, losing languages: connections between biological and linguistic diversity. *Southwest Journal of Linguistics,* 15, 89–108.

326. Harmon, D. 2001. On the meaning and moral imperative of diversity. In L. Maffi, editor. *On Biocultural Diversity. Linking Language, Knowledge, and the Environment.* Smithsonian Institute Press, Washington, DC.

327. Harpending, H. 2000. Genetic perspectives on human origins and differentiation. *Annual Review of Genomics and Human Genetics,* 1, 361–385.

328. Harpending, H.C., Batzer, M.A., Gurven, M., Jorde, L.B., Rogers, A.R., and Sherry, S.T. 1998. Genetic traces of ancient demography. *Proceedings of the National Academy of Sciences* (USA), 95, 1961–1967.

329. Harris, E.E., 2010. Nonadaptive processes in primate and human evolution. *Yearbook of Physical Anthropology,* 143, Suppl. 51, 13–45.

330. Harris, E.E., and Meyer, D. 2006. The molecular signature of selection underlying human adaptations. *Yearbook of Physical Anthropology,* 49, 89–130.

331. Harris, L.D. 1984. *The Fragmented Forest.* University of Chicago Press, Chicago.

332. Harrison, G.A., Tanner, J.M., Pilbeam, D.R., and Baker, P.T., editors. 1988. *Human Biology: An Introduction to Human Evolution, Variation, Growth, and Adaptability.* Oxford University Press, Oxford.

333. Harrison, K.D. 2007. *When Languages Die.* Oxford University Press, Oxford.

334. Harrison, T. 2010. Apes among the tangled branches of human origins. *Science,* 327, 532–534.

335. Harrison, T., Krigbaum, J., and Manser, J. 2006. Primate biogeography and ecology on the Sunda Shelf islands: a paleontological and zooarchaeological perspective. In S.M. Lehman and J.G. Fleagle, editors. *Primate Biogeography*. Springer, New York.

336. Harvey, P.H., and Clutton-Brock, T.H. 1985. Life history variation in primates. *Evolution*, 39, 559–581.

337. Harvey, P.H., and Pagel, M.D. 1991. *The Comparative Method in Evolutionary Biology*. Oxford University Press, Oxford.

338. Harvey, P.H., Promislow, D.E.L., and Read, A.F. 1989. Causes and correlates of life history differences among mammals. In V. Standen and R.A. Foley, editors. *Comparative Socioecology*. Blackwell Scientific Publications, Oxford.

339. Haug, G.H., Günther, D., Peterson, L.C., Sigman, D.M., Hughen, K.A., and Aeschlimann, B. 2003. Climate and the collapse of Maya civilization. *Science*, 299, 1731–1735.

340. Hawkes, K., O'Connell, J.F., Blurton-Jones, N.G., Alvarez, H., and Charnov, E.L. 1998. Grandmothering, menopause, and the evolution of human life histories. *Proceedings of the National Academy of Sciences (USA)*, 95, 1336–1339.

341. Hawkes, K., O'Connell, J.F., and Blurton Jones, N.G. 1989. Hardworking Hadza grandmothers. In V. Standen and R.A. Foley, editors. *Comparative Socioecology: The Behavioral Ecology of Humans and Other Mammals*. Blackwell Scientific Publishers, Oxford.

342. Hawkins, B.A., and Diniz-Filho, J.A.F. 2004. "Latitude" and geographic patterns in species richness. *Ecography*, 27, 268–272.

343. Heads, M. 2010. Evolution and biogeography of primates: a new model based on molecular phylogenetics, vicariance and plate tectonics. *Zoologica Scripta*, 39, 107–127.

344. Heaney, L.R. 1978. Island area and body size of insular mammals: evidence from the tri-colored squirrel (*Calliosciurus prevosti*) of Southwest Africa. *Evolution*, 32, 29–44.

345. Heaney, L.R. 1984. Mammalian species richness on islands on the Sunda Shelf, southeast Asia. *Oecologia*, 61, 11–17.

346. Heaney, L.R. 1986. Biogeography of mammals in southeast Asia: estimates of rates of colonization, extinction and speciation. *Biological Journal of the Linnean Society*, 28, 127–166.

347. Heaney, L.R. 1991. A synopsis of climatic and vegetational change in Southeast Asia. *Climatic Change*, 19, 53–61.

348. Heesy, C.P., Stevens, N.J., and Samonds, K.E. 2006. Biogeographic origins of primate higher taxa. In S.M. Lehman and J.G. Fleagle, editors. *Primate Biogeography*. Springer, New York.

349. Hehemann, J.H., Correc, G., Barbeyron, T., Helbert, W., Czjkek, M., and Michel, G. 2010. Transfer of carbohydrate-active enzymes from marine bacteria to Japanese gut microbiota. *Nature,* 464, 908–914.

350. Henderson, D.A. 2009. *Smallpox. The Death of a Disease: The Inside Story of Eradicating a Worldwide Killer.* Prometheus Books, Amherst, MA.

351. Henn, B.M., Gignouz, C., Lin, A.A., Oefner, P.J., Shen, P., Scozzari, R., et al. 2008. Y-chromosomal evidence of a pastoralist migration through Tanzania to southern Africa. *Proceedings of the National Academy of Sciences* (USA), 105, 10693–10698.

352. Henrich, J. 2004. Demography and cultural evolution: how adaptive cultural processes can produce maladaptive losses—the Tasmanian case. *American Antiquity,* 69, 197–214.

353. Hershkovitz, I., Kornreich, L., and Laron, Z. 2007. Comparative skeletal features between *Homo floresiensis* and patients with primary growth hormone insensitivity (Laron syndrome). *American Journal of Physical Anthropology,* 134, 198–208.

354. Hibbert, C. 1982. *Africa Explored: Europeans in the Dark Continent, 1769–1889.* Allen Lane, London.

355. Hibbs, D.A. and Olsson, O. 2004. Geography, biogeography, and why some countries are rich, and others poor. *Proceedings of the National Academy of Sciences* (USA), 101, 3715–3720.

356. Hiernaux, J., Rudan, P., and Brambati, A. 1975. Climate and the weight/height relationship in sub-Saharan Africa. *Journal of Human Biology,* 2, 3–12.

357. High Level Committee on the Indian Diaspora. 2004. Report of the High Level Committee on the Indian Diaspora. Ministry of External Affairs, India, New Delhi. Available from http://indiandiaspora.nic.in/. [Cited 2010.]

358. Hill, K.R., Walker, R.S., Božicevic, M., Eder, J., Headland, T., Hewlett, B., et al. 2011. Co-residence patterns in hunter-gatherer societies show unique human social structure. *Science,* 331, 1286–1289.

359. Hillebrand, H. 2004. On the generality of the latitudinal diversity gradient. *American Naturalist,* 163, 192–211.

360. Hinde, R.A. 2002. *Why Good is Good: The Sources of Morality.* Routledge, London.

361. Hinde, R.A., and Groebel, J., editors. 1991. *Cooperation and Prosocial Behaviour.* Cambridge University Press, Cambridge.

362. Hintjens, H.M. 1999. Explaining the 1994 genocide in Rwanda. *The Journal of Modern African Studies,* 37, 241–286.

363. Hoberg, E.P., Alkire, N.L., de Queiroz, A., and Jones, A. 2001. Out of Africa: origins of the *Taenia* tapeworms in humans. *Proceedings of the Royal Society B: Biological Sciences,* 268, 781–787.

364. Hockett, B., and Haws, J.A. 2005. Nutritional ecology and the human demography of Neandertal extinction. *Quaternary International,* 137, 21–34.

365. Hodgkinson, A., Ladoukakis, E., and Eyre-Walker, A. 2009. Cryptic variation in the human mutation rate. *PLoS Biology,* 7, 1–7; e1000027 doi:10.1371/journal.pbio.1000027.

366. Hoffecker, J.F. 2005. *A Prehistory of the North.* Turger University Press, New Brunswick.

367. Hoffecker, J.F. 2009. The spread of modern humans in Europe. *Proceedings of the National Academy of Sciences* (USA), 106, 16040–16045.

368. Hoffecker, J.F., and Elias, S.A. 2003. Environment and archeology in Beringia. *Evolutionary Anthropology,* 12, 34–49.

369. Hoffecker, J.F., and Elias, S.A. 2007. *Human Ecology of Beringia.* Columbia University Press, New York.

370. Hofmeyr, M.D. 1985. Thermal properties of the pelages of selected African ungulates. *South African Journal of Zoology,* 20, 179–189.

371. Holden, C., and Mace, R. 1997. Phylogenetic analysis of the evolution of lactose digestion in adults. *Human Biology,* 69, 605–628.

372. Holliday, T.W. 1997. Body proportions in late Pleistocene Europe and modern human origins. *Journal of Human Evolution,* 32, 423–447.

373. Holliday, T.W. 1997. Postcranial evidence of cold adaptation in European Neanderthals. *American Journal of Physical Anthropology,* 104, 245–258.

374. Holliday, T.W. 2000. Evolution at the crossroads: modern human emergence in Western Asia. *American Anthropologist,* 102, 54–68.

375. Holliday, T.W., and Falsetti, A.B. 1995. Lower-limb length of European early-modern humans in relation to mobility and climate. *Journal of Human Evolution,* 29, 141–153.

376. Holliday, T.W., and Hilton, C.E. 2010. Body proportions of circumpolar peoples as evidenced from skeletal data: Ipiutak and Tigara (Point Hope) versus Kodiak Island Inuit. *American Journal of Physical Anthropology,* 142, 287–302.

377. Holton, N.E., and Franciscus, R.G. 2009. The paradox of a wide nasal aperture in cold-adapted Neanderthals: a causal assessment. *Journal of Human Evolution,* 55, 942–951.

378. Hooper, E. 1997. Sailors and star-bursts, and the arrival of HIV. *British Medical Journal,* 315, 1689–1691.

379. Hrdy, S.B. 2009. *Mothers and Others: The Evolution of Mutual Understanding.* Belknap Press, Cambridge, MA.

380. Hubbell, S.P. 2001. *The Unified Neutral Theory of Biodiversity and Biogeography.* Princeton University Press, Princeton, NJ.

381. Hubbell, S.P., and Lake, J.K. 2003. The neutral theory of biodiveristy and biogeography, and beyond. In T.M. Blackburn and K.J. Gaston, editors.

Macroecology: Concepts and Consequences. Blackwell Scientific Publishers, Oxford.

382. Hublin, J.J. 2009. The origin of Neanderthals. *Proceedings of the National Academy of Sciences* (USA), 106, 16022–16027.

383. Hudjashov, G., Kivisild, T., Underhill, P.A., Endicott, P., J.J., S., Lin, A.A., et al. 2007. Revealing the prehistoric settlement of Australia by Y chromosome and mtDNA analysis. *Proceedings of the National Academy of Sciences* (USA), 104, 8726–8730.

384. Huffman, O.F. 2001. Geologic context and age of the Perning/Mojokerto *Homo erectus*, East Java. *Journal of Human Evolution*, 40, 353–362.

385. Hunter, J.P., and Janis, C.M. 2006. Spiny Norman in the Garden of Eden? Dispersal and early biogeography of Placentalia. *Journal of Mammalian Evolution*, 13, 89–123.

386. Huntford, R. 1999. *The Last Place on Earth: Scott and Amundsen's Race to the South Pole.* Modern Library, New York.

387. Huxley, T.H. 1863. *Man's Place in Nature.* Modern Library Science, New York.

388. Huxley, T.H. 1903. *Man's Place in Nature and Other Anthropological Essays.* D. Appleton & Co., New York.

389. Ingram, C.J.E., Mulcare, C.A., Itan, Y., Thomas, M.G., and Swallow, D.M. 2009. Lactose digestion and evolutionary genetics of lactase persistence. *Human Genetics*, 124, 579–591.

390. Institute of Historical Research. 2006. History in Focus. 11. Migration. University of London, London. Available from www.history.ac.uk/ihr/Focus/Migration/articles/house.html. [Cited 2010.]

391. Isaac, N.J.B., and Cowlishaw, G. 2004. How species respond to multiple extinction threats. *Proceedings of the Royal Society B: Biological Sciences*, 271, 1135–1141.

392. IUCN. 2009. *IUCN Red List of Threatened Species.* The World Conservation Union, Gland.

393. Jablonski, D. 1993. The tropics as a source of evolutionary novelty through geological time. *Nature*, 364, 142–144.

394. Jablonski, D. 2005. Mass extinctions and macroevolution. *Paleobiology*, 31, 192–210.

395. Jablonski, D., Roy, K., and Valentine, J.W. 2006. Out of the tropics: evolutionary dynamics of the latitudinal diversity gradient. *Science*, 314, 102–106.

396. Jablonski, N.G., and Chaplin, G. 2000. The evolution of human skin coloration. *Journal of Human Evolution*, 39, 57–106.

397. Jablonski, N.G., Whitfort, M.J., Roberts-Smith, N., and Qinqi, X. 2000. The influence of life history and diet on the distribution of catarrhine

primates during the Pleistocene in eastern Asia. *Journal of Human Evolution,* 39, 131–157.

398. Jacob, T., Indriati, E., Soejono, R.P., Hsü, K., Frayer, D.W., Eckhardt, R.B., et al. 2006. Pygmoid Austromelanesian *Homo sapiens* skeletal remains from Liang Bua, Flores: population affinities and pathological abnormalities. *Proceedings of the National Academy of Sciences* (USA), 103, 13421–13426.

399. Jacobs, Z., Roberts, R.G., Galbraith, R.F., Deacon, H.J., Rainer, G., Mackay, A., et al. 2008. Ages for the Middle Stone Age of Southern Africa: implications for human behavior and dispersal. *Science,* 322, 733–735.

400. James, F.C. 1970. Geographic size variation in birds and its relationship to climate. *Ecology,* 51, 365–390.

401. Janson, C.H., and van Schaik, C.P., editors. 1988. Food competition in primates. *Behaviour,* 105(1, 2), 1–186,

402. Janzen, D.H. 1967. Why mountain passes are higher in the tropics. *American Naturalist,* 101, 233–249.

403. Jernvall, J., and Wright, P.C. 1998. Diversity components of impending primate extinctions. *Proceedings of the National Academy of Sciences* (USA), 95, 11279–11283.

404. Jobling, M.A., Hurles, M.E., and Tyler-Smith, C. 2004. *Human Evolutionary Genetics: Origins, Peoples & Disease.* Garland Science, New York.

405. Johnson, B.J., and Prideaux, G.J. 2004. Extinctions of herbivorous mammals in the late Pleistocene of Australia in relation to their feeding ecology: no evidence for environmental change as cause of extinction. *Austral Ecology,* 29, 553–557.

406. Johnson, C.N. 2002. Determinants of loss of mammal species during the Late Quaternary "megafauna" extinctions: life history and ecology, but not body size. *Proceedings of the Royal Society B: Biological Sciences,* 269, 2221–2227.

407. Johnson, C.N., and Wroe, S. 2003. Causes of extinction of vertebrates during the Holocene of mainland Australia: arrival of the dingo, or human impact? *Holocene,* 13, 941–948.

408. Johnson, R. 2008. *Recent Honey Bee Colony Declines.* Congressional Research Service, Washington, DC.

409. Johnson, S. 1755. *Dictionary of the English Language.* J. & P. Knapton, London.

410. Jolly, A. 1986. Lemur survival. In K. Benirschke, editor. *Primates: The Road to Self-Sustaining Populations.* Springer-Verlag, New York.

411. Jolly, A. 1986. Lemur survival. In K. Benirschke, editor. *Primates: The Road to Self-Sustaining Populations.* Springer-Verlag, New York.

412. Jolly, D., Taylor, D., Marchant, R., Hamilton, A., Bonnefille, R., Buchet, G., et al. 1997. Vegetation dynamics in central Africa since 18,000 yr BP:

pollen records from the interlacustrine highlands of Burundi, Rwanda and western Uganda. *Journal of Biogeography,* 24, 495–512.

413. Jones, C.B. 1997. Rarity in primates: implications for conservation. Mastozoología Neotropical, 4, 35–47.

414. Joordens, J.C.A., Wesselingh, F.P., de Vos, J., Vonhof, H.B., and Kroon, D. 2009. Relevance of aquatic environments for hominins: a case study from Trinil (Java, Indonesia). *Journal of Human Evolution,* 57, 656–671.

415. Jordan, F.M., Gray, R.D., Greenhill, S.J., and Mace, R. 2009. Matrilocal residence is ancestral in Austronesian societies. *Proceedings of the Royal Society B: Biological Sciences,* 276, 1957–1964.

416. Jungers, W.L., Harcourt-Smith, W.E.H., Wunderlich, R.E., Tocheri, M.W., Larson, S.G., Sutikna, T., et al. 2009. The foot of *Homo floresiensis. Nature,* 459, 81–84.

417. Kaneko, A., Taleo, G., Kalkoa, M., Yaviong, J., Reeve, P.A., Gancza-kowski, M., et al. 1998. Malaria epidemiology, glucose-6-phosphate dehydrogenase deficiency, and human settlement in the Vanuatu Archipelago. *Acta Tropica,* 70, 285–302.

418. Kari, J., and Potter, B.A. 2010. The Dene-Yeniseian connection: bridging Asia and North America. *Anthropological Papers of the University of Alaska, New Series,* 5, 1–24.

419. Karim, A.K.M.B., Elfellah, M.S., and Evans, D.A.P. 1981. Human acetylator polymorphism: estimate of allele frequency in Libya and details of global distribution. *Journal of Medical Genetics,* 18, 325–330.

420. Kayser, M., Lao, O., Saar, K., Brauer, S., Wang, X., Nürnberg, P., et al. 2008. Genome-wide analysis indicates more Asian than Melanesian ancestry of Polynesians. *American Journal of Human Genetics,* 82, 194–198.

421. Kirch, P.V., and Kahn, J.G. 2007. Advances in Polynesian prehistory: a review and assessment of the past decade. *Journal of Archaeological Research,* 15, 191–238.

422. Kirkpatrick, R.C. 2007. The Asian colobines. Diversity among leaf-eating monkeys. In C.J. Campbell et al., editors. *Primates in Perspective.* Oxford University Press, New York.

423. Klein, R.G. 2000. Archeology and the evolution of human behavior. *Evolutionary Anthropology,* 9, 17–36.

424. Klein, R.G. 2008. Out of Africa and the evolution of human behavior. *Evolutionary Anthropology,* 17, 267–281.

425. Klein, R.G. 2009. *The Human Career: Human Biological and Cultural Origins,* 3rd. ed. University of Chicago, Chicago.

426. Kline, M.A., and Boyd, R. 2010. Population size predicts technological complexity in Oceania. *Proceedings of the Royal Society B: Biological Sciences,* 277, 2559–2564.

427. Knowles, L.L. 2008. Why does a method that fails continue to be used? *Evolution*, 62, 2713–2717.

428. Koch, P.L., and Barnosky, A.D. 2006. Late Quaternary extinctions: state of the debate. *Annual Review of Ecology and Systematics*, 37, 215–250.

429. Köhler, M., Moyà-Sola, S., and Wrangham, R.W. 2008. Island rules cannot be broken. *Trends in Ecology and Evolution*, 23, 6–7.

430. Korkea-Aho, T.L., Vainikka, A., Kortet, R., and Taskinen, J. 2009. Factors affecting between-lake variation in the occurrence of epidermal papillomatosis in roach, *Rutilus rutilus* (L.). *Journal of Fish Diseases*, 32, 263–270.

431. Krause, J., Fu, Q.M., Good, J.M., Viola, B., Shunkov, M.V., Derevianko, A.P., et al. 2010. The complete mitochondrial DNA genome of an unknown hominin from southern Siberia. *Nature*, 464, 894–897.

432. Krebs, J.R., and Davies, N.B., editors. 1993. *An Introduction to Behavioural Ecology*, 3rd. ed. Blackwell Scientific Publications, Oxford.

433. Kroeber, A.L. 1962. The nature of land-holding groups in aboriginal California. In *Report of the University of California Archaeological Survey*. University of California, Berkeley, 19–58.

434. Kruess, A., and Tscharntke, T. 2000. Species richness and parasitism in a fragmented landscape: experiments and field studies with insects on *Vicia sepium*. *Oecologia*, 122, 129–137.

435. Kuklick, H. 1996. Islands in the Pacific: Darwinian biogeography and British anthropology. *American Ethnologist*, 23, 611–638.

436. Kumar, S., Filipski, A., Swarna, V., Walker, A., and Hedges, S.B. 2005. Placing confidence limits on the molecular age of the human-chimpanzee divergence. *Proceedings of the National Academy of Sciences* (USA), 102, 18842–18847.

437. Kunin, W.E., and Gaston, K.J., editors. 1997. *The Biology of Rarity: Causes and Consequences of Rare-Common Differences*. Chapman & Hall, London.

438. Kupfer, J.A., Malanson, G.P., and Franklin, S.B. 2006. Not seeing the ocean for the islands: the mediating influence of matrix-based processes on forest fragmentation effects. *Global Ecology and Biogeography*, 15, 8–20.

439. Kurlansky, M. 2002. *Salt: A World History*. Walker and Co., New York.

440. Kuzmina, E.E. 2008. *The Prehistory of the Silk Road*. University of Pennsylvania Press, Philadelphia.

441. Ladle, R.J. 2008. Catching fairies and the public representation of biogeography. *Journal of Biogeography*, 35, 388–391.

442. Lahr, M.M., and Foley, R. 1994. Multiple dispersals and modern human origins. *Evolutionary Anthropology*, 3, 48–60.

443. Lahr, M.M., and Foley, R.A. 1998. Towards a theory of modern human origins: geography, demography, and diversity in recent human evolution. *Yearbook of Physical Anthropology*, 41, 137–176.

444. Lai, Y.-C., Shiroishi, T., Moriwaki, K., Motokawa, M., and Yu, H.-T. 2008. Variation of coat color in house mice throughout Asia. *Journal of Zoology*, 274, 270–276.

445. Lande, R. 1977. Statistical tests for natural selection on quantitative characters. *Evolution*, 31, 442–444.

446. Laurance, W.F. 1990. Comparative responses of five arboreal marsupials to tropical forest fragmentation. *Journal of Mammalogy*, 71, 641–653.

447. Laurance, W.F., and Bierregaard, R.O., editors. 1997. *Tropical Forest Remnants: Ecology, Management, and Conservation of Fragmented Communities*. University of Chicago Press, Chicago.

448. Lavely, W. 1991. Marriage and mobility under rural collectivism. In R. Watson and P.B. Ebrey, editors. *Marriage and Inequality in Chinese Society*. University of California Press, Berkeley.

449. Lawes, M.J., and Eeley, H.A.C. 2000. Are local patterns of anthropoid primate diversity related to patterns of diversity at a larger scale? *Journal of Biogeography*, 27, 1421–1435.

450. Lawton, J.H., and May, R.M., editors. 1995. *Extinction Rates*. Oxford University Press, Oxford.

451. LeJeune, K.D., and Seastedt, T.R. 2001. *Centaurea* species: the forb that won the west. *Conservation Biology*, 15, 1568–1574.

452. Lell, J.T., Brown, M.D., Schurr, T.G., Sukernik, R.I., Starikovskaya, Y.B., Torroni, A., et al. 1997. Y chromosome polymorphisms in Native American and Siberian populations: identification of Native American chromosome haplotypes. *Human Genetics*, 100, 536–543.

453. Lentz, C. 1999. Alcohol consumption between community ritual and political economy: case studies from Ecuador and Ghana. In C. Lentz, editor. *Changing Food Habits: Case Studies from Africa, South America and Europe*. Harwood Academic Publishers, Amsterdam.

454. Leonard, W.R., Snodgrass, J.J., and Sorensen, M.V. 2005. Metabolic adaptation in indigenous Siberian populations. *Annual Review of Anthropology*, 34, 451–471.

455. Lewis, M.P., editor. 2009. *Ethnologue: Languages of the World*, 16th ed. Available from www.ethnologue.com/. SIL International, Dallas, TX.

456. Li, J.Z., Absher, D.M., Tang, H., Southwick, A.M., Casto, A.M., Ramachandran, S., et al. 2008. Worldwide human relationships inferred from genome-wide patterns of variation. *Science*, 319, 1100–1104.

457. Lindstedt, A.L., and Boyce, M.S. 1985. Seasonality, fasting endurance, and body size in mammals. *American Naturalist*, 125, 873–878.

458. Liu, H., Prugnolle, F., Manica, A., and Balloux, F. 2006. A geographically explicit genetic model of worldwide human-settlement history. *American Journal of Human Genetics,* 79, 230–237.

459. Liu, W., Jin, C.-Z., Zhang, Y.-Q., Cai, Y.-J., Xing, S., Wu, X.-J., et al. 2010. Human remains from Zhirendong, South China, and modern human emergence in East Asia. *Proceedings of the National Academy of Sciences* (USA), 107, 19201–19206.

460. Liu, W., Li, Y., Learn, G.H., Rudicell, R.S., Robertson, J.D., Keele, B.F., et al. 2010. Origin of the human malaria parasite *Plasmodium falciparum* in gorillas. *Nature,* 467, 420–425.

461. Loh, J., and Harmon, D. 2005. A global index of biocultural diversity. *Ecological Indicators,* 5, 231–241.

462. Lomolino, M.V. 2005. Body size evolution in insular vertebrates: generality of the island rule. *Journal of Biogeography,* 32, 1683–1699.

463. Lomolino, M.V., Riddle, B.R., Whittaker, R.J., and Brown, J.H. 2010. *Biogeography,* 4th ed. Sinauer Associates, Sunderland, MA.

464. Lomolino, M.V., Sax, D.F., and Brown, J.H., editors. 2004. *Foundations of Biogeography.* University of Chicago Press, Chicago.

465. Lowen, C., and Dunbar, R.I.M. 1994. Territory size and defendability in primates. *Behavioral Ecology and Sociobiology,* 35, 347–354.

466. Luck, G.W. 2007. A review of the relationships between human population density and biodiversity. *Biological Reviews,* 82, 607–645.

467. Lum, C.L., Jeyanthi, S., Prepageran, N., Vadivelu, J., and Raman, R. 2009. Antibacterial and antifungal properties of human cerumen. *The Journal of Laryngology and Otology,* 123, 375–378.

468. Lyell, C. 1832. *Principles of Geology, Vol. 2.* John Murray, London.

469. MacArthur, R.H. 1972. *Geographical Ecology: Patterns in the Distribution of Species.* Princeton University Press, Princeton, NJ.

470. MacArthur, R.H., and Wilson, E.O. 1967. *The Theory of Island Biogeography.* Princeton University Press, Princeton, NJ.

471. Macaulay, V., Hill, C., Achilli, A., Rengo, C., Clarke, D., Meehan, W., et al. 2005. Single, rapid coastal settlement of Asia revealed by analysis of complete mitochondrial genomes. *Science,* 308, 1034–1036.

472. Mace, G.M., and Balmford, A. 2000. Patterns and processes in contemporary mammalian extinction. In A. Entwistle and N. Dunstone, editors. *Priorities for the Conservation of Mammalian Diversity: Has the Panda Had Its Day?* Cambridge University Press, Cambridge.

473. Mace, R., and Pagel, M. 1995. A latitudinal gradient in the density of human languages in North America. *Proceedings of the Royal Society B: Biological Sciences,* 261, 117–121.

474. MacPhee, R.D.E., editor. 1999. *Extinction in Near Time: Causes, Contexts, and Consequences.* Kluwer Academic/Plenum Publishers, New York.

475. MacPhee, R.D.E., and Marx, P.A. 1997. The 40,000-year plague: humans, hyperdisease, and first-contact extinctions. In S.M. Goodman and B.D. Patterson, editors. *Natural Change and Human Impact in Madagascar.* Smithsonian Institution Press, Washington, DC.

476. Maffi, L., editor. 2001. *On Biocultural Diversity: Linking Language, Knowledge, and the Environment.* Smithsonian Institute Press, Washington, DC.

477. Maffi, L. 2005. Linguistic, cultural and biological diversity. *Annual Review of Anthropology,* 34, 599–617.

478. Magalon, H., Patin, E., Austerlitz, F., Hegay, T., Aldashev, A., Quintana-Murci, L., et al. 2008. Population genetic diversity of the NAT2 gene supports a role of acetylation in human adaptation to farming in Central Asia. *European Journal of Human Genetics,* 16, 243–251.

479. Malhi, R.S., Gonzalez-Oliver, A., Schroeder, K.B., Kemp, B.M., Greenberg, J.H., Dobrowski, S.Z., et al. 2008. Distribution of Y chromosomes among Native North Americans: a study of Athapaskan population history. *American Journal of Physical Anthropology,* 137, 412–424.

480. Malhi, R.S., Mortensen, H.M., Eshleman, J.A., Kemp, B.M., Lorenz, J.G., Kaestle, F.A., et al. 2003. Native American mtDNA prehistory in the American Southwest. *American Journal of Physical Anthropology,* 120, 108–124.

481. Malthus, T.R. 1798. *An Essay on the Principle of Population.* Macmillan, London.

482. Mandryk, C.A.S., Josenhans, H., Fedje, D.W., and Mathewes, R.W. 2001. Late Quaternary paleoenvironments of Northwestern North America: implications for inland versus coastal migration routes. *Quaternary Science Reviews,* 20, 301–314.

483. Manne, L.L. 2003. Nothing has lasted forever: current and threatened levels of biological and cultural diversity. *Evolutionary Ecology Research,* 5, 517–527.

484. Marcello, A., and Thomas, K.D. 2002. Depletion of a resource? The impact of prehistoric human foraging on intertidal mollusc communities and its significance for human settlement, mobility and dispersal. *World Archaeology,* 33, 452–474.

485. Marlar, R.A., Banks, L.L., Billman, B.R., Lambert, P.M., and Marlar, J.E. 2000. Biochemical evidence of cannibalism at a prehistoric Puebloan site in southwestern Colorado. *Nature,* 407, 74–78.

486. Marler, P. 1957. Specific distinctiveness in the communication signals of birds. *Behaviour,* 11, 13–39.

487. Marlowe, F.W. 2004. Marital residence among foragers. *Current Anthropology,* 45, 277–284.

488. Marlowe, F.W. 2005. Hunter-gatherers and human evolution. *Evolutionary Anthropology,* 14, 54–67.

489. Marquet, P.A., Keymer, J.E., and Cofre, H. 2003. Breaking the stick in space: of niche models, metacommunities and patterns in the relative abundance of species. In T.M. Blackburn and K.J. Gaston, editors. *Macroecology: Concepts and Consequences.* Blackwell Scientific Publishers, Oxford.

490. Marr, K.L., Allen, G.A., and Hebda, R.J. 2008. Refugia in the Cordilleran ice sheet of western North America: chloroplast DNA diversity in the Arctic-alpine plant *Oxyria digyna. Journal of Biogeography,* 35, 1323–1334.

491. Marsh, L.K., editor. 2003. *Primates in Fragments: Ecology and Conservation.* Kluwer Academic/Plenum Publishers, New York.

492. Martin, P.S. 1984. Prehistoric overkill: the global model. In P.S. Martin and R.G. Klein, editors. *Quaternary Extinctions: A Prehistoric Revolution.* University of Arizona Press, Tucson.

493. Martin, P.S., and Klein, R.G., editors. 1984. *Quaternary Extinctions: A Prehistoric Revolution.* University of Arizona Press, Tucson.

494. Martin, P.S., and Steadman, D.W. 1999. Prehistoric extinctions on islands and continents. In R.D.E. MacPhee, editor. *Extinction in Near Time: Causes, Contexts, and Consequences.* Kluwer Academic/Plenum Publishers, New York.

495. Martin, R.D. 1990. *Primate Origins and Evolution.* Princeton University Press, Princeton, NJ.

496. Martin, R.D., MacLarnon, A.M., Phillips, J.L., and Dobyns, W.B. 2006. Flores hominid: new species or microcephalic dwarf? *The Anatomical Record, Part A,* 288A, 1123–1145.

497. Martin, R.D., Soligo, C., and Tavaré, S. 2007. Primate origins: implications of a Cretaceous ancestry. *Folia Primatologica,* 78, 277–296.

498. Martínez-Navarro, B., and Rabinovich, R. 2011. The fossil Bovidae (Artiodactyla, Mammalia) from Gesher Benot Ya'aqov, Israel: out of Africa during the Early-Middle Pleistocene transition. *Journal of Human Evolution,* 60, 375–386.

499. Martinón-Torres, M., Dennell, R., and Bermúdez de Castro, J.M. 2011. The Denisova hominin need not be an out of Africa story. *Journal of Human Evolution,* 60, 251–255.

500. Masters, J.C. 2006. When, where and how? Reconstructing a timeline for primate evolution using molecular and fossil data. In L. Sineo and R. Stanyon, editors. *Primate Cytogenetics and Comparative Genomics.* Florence University Press, Florence.

501. Masters, J.C., de Wit, M.J., and Asher, R.J. 2006. Reconciling the origins of Africa, India and Madagascar with vertebrate dispersal scenarios. *Folia Primatologica,* 77, 399–418.

502. Matisoo-Smith, E., and Robins, J. 2009. Mitochondrial DNA evidence for the spread of Pacific rats through Oceania. *Biological Invasions,* 11, 1521–1527.

503. Matisoo-Smith, E., and Robins, J.H. 2004. Origins and dispersals of Pacific peoples: evidence from mtDNA phylogenies of the Pacific rat. *Proceedings of the National Academy of Sciences* (USA), 101, 9167–9172.

504. Matthysen, E. 2005. Density-dependent dispersal in birds and mammals. *Ecography*, 28, 403–416.

505. Maudlin, I. 2006. African trypanosomiasis. *Annals of Tropical Medicine & Parasitology*, 100, 679–701.

506. McCain, C.M. 2009. Vertebrate range sizes indicate that mountains may be "higher" in the tropics. *Ecology Letters*, 12, 550–560.

507. McComb, K., Packer, C., and Pusey, A.E. 1994. Roaring and numerical assessment in contests between groups of female lions, *Panthera leo*. *Animal Behaviour*, 47, 379–387.

508. McCurdy, S.A. 1994. Epidemiology of disaster: The Donner Party (1846–1847). *Western Journal of Medicine*, 4, 338–342.

509. McDougall, I., Brown, F.H., and Fleagle, J.G. 2005. Stratigraphic placement and age of modern humans from Kibish, Ethiopia. *Nature*, 433, 733–736.

510. McHenry, H. 1994. Tempo and mode in human evolution. *Proceedings of the National Academy of Sciences* (USA), 91, 6780–6786.

511. McMahon, T.A., and Bonner, J.T. 1983. *On Size and Life*. Scientific American Library, New York.

512. McNab, B.K. 1999. On the comparative ecological and evolutionary significance of total and mass-specific rates of metabolism. *Physiological and Biochemical Zoology*, 72, 642–644.

513. McNab, B.K. 2002. Minimizing energy expenditure facilitates vertebrate persistence on oceanic islands. *Ecology Letters*, 5, 693–704.

514. McNab, B.K. 2002. *The Physiological Ecology of Vertebrates: A View from Energetics*. Comstock Publishing Associates, Ithaca, NY.

515. McNeill, J.R. 2010. *Mosquito Empires: Ecology and War in the Greater Caribbean, 1620–1914*. Cambridge University Press, Cambridge.

516. Meijaard, E. 2003. Mammals of south-east Asian islands and their Late Pleistocene environments. *Journal of Biogeography*, 30, 1245–1257.

517. Meijaard, E., and Groves, C.P. 2006. The geography of mammals and rivers in mainland southeast Asia. In S.M. Lehman and J.G. Fleagle, editors. *Primate Biogeography*. Springer, New York.

518. Meijer, H.J.M., van den Hoek Ostende, L.W., van den Berghe, G.D., and de Vos, J. 2010. The fellowship of the hobbit: the fauna surrounding *Homo floresiensis*. *Journal of Biogeography*, 37, 995–1006.

519. Meiri, S., Cooper, N., and Purvis, A. 2008. The island rule: made to be broken? *Proceedings of the Royal Society B: Biological Sciences*, 275, 141–148.

520. Meiri, S., and Dayan, T. 2003. On the validity of Bergmann's rule. *Journal of Biogeography,* 30, 331–351.

521. Meiri, S., Meijaard, E., Wich, S.A., Groves, C.P., and Helgen, K.M. 2008. Mammals of Borneo—small size on a large island. *Journal of Biogeography,* 35, 1087–1094.

522. Meiri, S., Raia, P., and Phillimore, A.B. 2011. Slaying dragons: limited evidence for unusual body size evolution on islands. *Journal of Biogeography,* 38, 89–100.

523. Meiri, S., Yom-Tov, Y., and Geffen, E. 2007. What determines conformity to Bergmann's rule? *Global Ecology and Biogeography,* 16, 788–794.

524. Mellars, P. 1996. The emergence of biologically modern populations in Europe: a social and cognitive revolution? In W.G. Runciman, J. Maynard Smith, and R.I.M. Dunbar, editors. *Evolution of Social Behaviour Patterns in Primates and Man.* Oxford University Press, Oxford.

525. Mellars, P. 2004. Neanderthals and the modern human colonization of Europe. *Nature,* 432, 461–465.

526. Mellars, P. 2006. Why did modern human populations disperse from Africa *ca.* 60,000 years ago? A new model. *Proceedings of the National Academy of Sciences* (USA), 103, 9381–9386.

527. Mercador, J. 2002. Forest people: the role of African rainforests in human evolution and dispersal. *Evolutionary Anthropology,* 11, 117–124.

528. Messier, W., and Stewart, C.-B. 1997. Episodic adaptive evolution of primate lysozymes. *Nature,* 385, 151–154.

529. Meyer, D. 1999. Medically relevant genetic variation of drug effects. In S.C. Stearns, editor. *Evolution in Health & Disease.* Oxford University Press, Oxford.

530. Mielke, J.H., Konigsberg, L.W., and Relethford, J.H. 2006. *Human Biological Variation.* Oxford University Press, New York.

531. Migliano, A.B., Vinicius, L., and Lahr, M.M. 2007. Life history trade-offs explain the evolution of human pygmies. *Proceedings of the National Academy of Sciences* (USA), 104, 20216–20219.

532. Mijares, A.S., Détroit, F., Piper, P., Grün, R., Bellwood, P., Aubert, M., et al. 2010. New evidence for a 67,000-year-old human presence at Callao Cave, Luzon, Philippines. *Journal of Human Evolution,* 59, 123–132.

533. Milanich, J.T., and Hudson, C. 1993. *Hernando de Soto and the Indians of Florida.* University Press of Florida; Florida Natural History Museum, Gainesville.

534. Miller, G.H., Fogel, M.L., Magee, J.W., Gagan, M.K., Clarke, S.J., and Johnson, B.J. 2005. Ecosystem collapse in pleistocene Australia and a human role in megafaunal extinction. *Science,* 309, 287–290.

535. Miller, G.H., Magee, J.W., Johnson, B.J., Fogel, M.L., Spooner, N.A., McCulloch, M.T., et al. 1999. Pleistocene extinction of *Genyornis*

newtoni: human impact on Australian megafauna. *Science, 283,* 205–208.

536. Milton, K. 1991. Comparative aspects of diet in Amazonian forest dwellers. *Proceedings of the Royal Society B: Biological Sciences, 334,* 253–263.

537. Mitani, J.C., and Rodman, P.S. 1979. Territoriality: the relation of ranging pattern and home range size to defendability, with an analysis of territoriality among primate species. *Behavioral Ecology and Sociobiology, 5,* 241–251.

538. Mithen, S., and Reed, M. 2002. Stepping out: a computer simulation of hominid dispersal from Africa. *Journal of Human Evolution, 43,* 433–502.

539. Mittermeier, R.A., Ganzhorn, J.U., Konstant, W.R., Glander, K., Tattersall, I., Groves, C.P., et al. 2008. Lemur diversity in Madagascar. *International Journal of Primatology, 29,* 1607–1656.

540. Molnar, S. 2006. *Human Variation: Races, Types, and Ethnic Groups,* 6th. ed. Pearson Prentice Hall, Upper Saddle River, NJ.

541. Moodley, Y., Linz, B., Yamaoka, Y., Windsor, H.M., Breurec, S., Wu, J.-Y., et al. 2009. The peopling of the Pacific from a bacterial perspective. *Science, 323,* 527–530.

542. Moore, J.L., Manne, L., Brooks, T., Burgess, N.D., Davies, R., Rahbek, C., et al. 2002. The distribution of cultural and biological diversity in Africa. *Proceedings of the Royal Society B: Biological Sciences, 269,* 1645–1653.

543. Moore, L.G. 1998. Human adaptation to high altitude: regional and life-cycle perspectives. *Yearbook of Physical Anthropology, 41,* 25–64.

544. Moore, L.G. 2001. Human genetic adaptation to high altitude. *High Altitude Medicine & Biology, 2,* 257–279.

545. Moran, E.E. 2000. *Human Adaptability: An Introduction to Ecological Anthropology,* 2nd. ed. Westview Press, Boulder, CO.

546. Morgan, G.S., and Woods, C.A. 1986. Extinction and the zoogeography of west-indian land mammals. *Biological Journal of the Linnean Society, 28,* 167–203.

547. Moritz, C., Patton, J.L., Conroy, C.J., Parra, J.L., White, G.C., and Beissinger, S.R. 2008. Impact of a century of climate change on small-mammal communities in Yosemite National Park, USA. *Science, 322,* 261–264.

548. Morwood, M.J., Brown, P., Jatmiko, Sutikna, T., Wayhu Saptomo, E., Westaway, K.E., et al. 2005. Further evidence for small-bodied hominins from the Late Pleistocene of Flores, Indonesia. *Nature, 437,* 1012–1017.

549. Morwood, M.J., and Jungers, W.L. 2009. Conclusions: implications of the Liang Bua excavations for hominin evolution and biogeography. *Journal of Human Evolution, 57,* 640–648.

550. Mourre, V., Villa, P., and Henshilwood, C.S. 2010. Early use of pressure flaking on lithic artifacts at Blombos Cave, South Africa. *Science,* 330, 659–662.

551. Munguía, M., Peterson, A.T., and Sánchez-Cordero, V. 2008. Dispersal limitation and geographical distributions of mammal species. *Journal of Biogeography,* 35, 1879–1887.

552. Munoz, S.E., Gajewski, K., and Peros, M.C. 2010. Synchronous environmental and cultural change in the prehistory of the northeastern United States. *Proceedings of the National Academy of Sciences* (USA), 107, 22008–22013.

553. Murphy, W.A., zur Nedden, D., Gostner, P., Knapp, R., Recheis, W., and Seidler, H. 2003. The Iceman: discovery and imaging. *Radiology,* 226, 614–629.

554. Nash, L.T., Bearder, S.K., and Olson, T.R. 1989. Synopsis of galago species characteristics. *International Journal of Primatology,* 10, 57–80.

555. Nee, S. 2003. The unified phenomenological theory of biodiversity. In T.M. Blackburn and K.J. Gaston, editors. *Macroecology: Concepts and Consequences.* Blackwell Scientific Publishers, Oxford.

556. Nelson, B.W., Ferreira, C.A.C., da Silva, M.F., and Kawasaki, M.L. 1990. Endemism, refugia and botanical collection density in Brazilian Amazon. *Nature,* 345, 714–716.

557. Nettle, D. 1998. Explaining global patterns of language diversity. *Journal of Anthropological Archaeology,* 17, 354–374.

558. Nettle, D. 1999. Linguistic diversity of the Americas can be reconciled with a recent colonization. *Proceedings of the National Academy of Sciences* (USA), 96, 3325–3329.

559. Nettle, D. 2009. Ecological influences on human behavioural diversity: a review of recent findings. *Trends in Ecology and Evolution,* 24, 618–624.

560. Nettle, D., and Romaine, S. 2000. *Vanishing Voices.* Oxford University Press, Oxford.

561. Nijman, V., and Meijaard, E. 2008. Zoogeography of primates in insular Southeast Asia: species-area relationships and the effects of taxonomy. *Contributions to Zoology,* 77, 117–126.

562. Nitecki, M.H., editor. 1984. *Extinctions.* University of Chicago Press, Chicago.

563. Niven, J.E. 2007. Brains, islands and evolution: breaking all the rules. *Trends in Ecology and Evolution,* 22, 57–59.

564. Nogués-Bravo, D., Rodríguez, J., Hortal, J., Batra, P., and Araújo, M.B. 2008. Climate change, humans, and the extinction of the woolly mammoth. *PLOS Biology,* 6, e79; 0685–0692.

565. Norton, H.L., Kittles, R.A., Parra, E., McKeigue, P., Mao, X., Cheng, K., et al. 2007. Genetic evidence for the convergent evolution of light skin in

Europeans and East Asians. *Molecular Biology and Evolution,* 24, 710–722.

566. Nudds, R.L., and Oswald, S.A. 2007. An interspecific test of Allen's rule: evolutionary implications for endothermic species. *Evolution,* 61, 2839–2848.

567. Nunn, C.L., and Altizer, S. 2006. *Infectious Diseases in Primates. Behavior, Ecology and Evolution.* Oxford University Press, New York.

568. Nunn, C.L., Altizer, S.M., Jones, K.E., and Sechrest, W. 2003. Comparative tests of parasite species richness in primates. *American Naturalist,* 162, 597–614.

569. Nunn, C.L., Altizer, S.M., Sechrest, W., and Cunningham, A.A. 2005. Latitudinal gradients of parasite species richness in primates. *Diversity and Distributions,* 11, 249–256.

570. O'Connell, J.F., and Allen, J. 1998. When did humans first arrive in Greater Australia and why is it important to know? *Evolutionary Anthropology,* 6, 132–146.

571. O'Meara, M. 1999. Reinventing cities for people and the planet. Worldwatch Paper #147. Worldwatch Institute, Washington, DC, 1–94.

572. Obendorf, P.J., Oxnard, C.E., and Kefford, B.J. 2008. Are the small human-like fossils found on Flores human endemic cretins? *Proceedings of the Royal Society B: Biological Sciences,* 275, 1287–1296.

573. Odling-Smee, F.J., Laland, K.N., and Feldman, M.W. 2003. *Niche Construction: The Neglected Process in Evolution.* Princeton University Press, Princeton, NJ.

574. Ognjanovic, S., Yamamoto, J., Maskarinec, G., and Le Marchand, L. 2006. NAT2, meat consumption and colorectal cancer incidence: an ecological study among 27 countries. *Cancer Causes and Control,* 17, 1175–1182.

575. Okie, J.G., and Brown, J.H. 2009. Niches, body sizes, and the disassembly of mammal communities on the Sunda Shelf islands. *Proceedings of the National Academy of Sciences* (USA), 106, 19679–19684.

576. Oppenheimer, S. 2003. *Out of Eden: The Peopling of the World.* Constable, London.

577. Oppenheimer, S. 2009. The great arc of dispersal of modern humans: Africa to Australia. *Quaternary International,* 202, 2–13.

578. Osborne, A.H., Vance, D., Rohling, E.J., Barton, N., Rogerson, M., and Fello, N. 2008. A humid corridor across the Sahara for the migration of early modern humans out of Africa 120,000 years ago. *Proceedings of the National Academy of Sciences* (USA), 105, 16444–16447.

579. Owens, I.P.F., and Bennett, P.M. 2000. Ecological basis of extinction risk in birds: habitat loss versus human persecution and introduced predators. *Proceedings of the National Academy of Sciences* (USA), 97, 12144–12148.

580. Parfitt, S.A., Ashton, N.M., Lewis, S.G., Abel, R.L., Coope, G.R., Field, M.H., et al. 2010. Early Pleistocene human occupation at the edge of the boreal zone in northwest Europe. *Nature,* 466, 229-233.

581. Parks, S.A., and Harcourt, A.H. 2002. Reserve size, local human density, and mammalian extinctions in U.S. protected areas. *Conservation Biology,* 16, 800–808.

582. Parmesan, C. 1996. Climate and species' range. *Nature,* 382, 765–766.

583. Parmesan, C., and Yohe, G. 2003. A globally coherent fingerprint of climate change impacts across natural systems. *Nature,* 421, 37–42.

584. Partridge, T., Wood, B.A., and deMenocal, P.B. 1995. The influence of global climatic change and regional uplift on large-mammalian evolution in East and southern Africa. In E.S. Vrba, et al., editors. *Paleoclimate and Evolution with Emphasis on Human Origins.* Yale University Press, New Haven, CT.

585. Passey, B.H., Levin, N.E., Cerling, T.E., Brown, F.H., and Eiler, J.M. 2010. High-temperature environments of human evolution in East Africa based on bond ordering in paleosol carbonates. *Proceedings of the National Academy of Sciences* (USA), 107, 11245–11249.

586. Patterson, W.P., Dietrich, K.A., Holmden, C., and Andrews, J.T. 2010. Two millennia of North Atlantic seasonality and implications for Norse colonies. *Proceedings of the National Academy of Sciences* (USA), 107, 5306–5310.

587. Paul, L.M., editor. 2009. *Ethnologue: Languages of the World,* 16th ed. SIL International, Dallas, TX.

588. Pearson, O.M. 2000. Postcranial remains and the origin of modern humans. *Evolutionary Anthropology,* 9, 229–247.

589. Pearson, O.M. 2004. Has the combination of genetic and fossil evidence solved the riddle of modern human origins? *Evolutionary Anthropology,* 13, 145–159.

590. Peng, Y., Shi, H., Qi, X., Xiao, C., Zhong, H., Ma, R.Z., et al. 2010. The ADH1B Arg47His polymorphism in East Asian populations and expansion of rice domestication in history. *BioMed Central Evolutionary Biology,* 10, 8 pp. Available from www.biomedcentral.com/1471-2148/10/15.

591. Peres, A.C., and Janson, C.H. 1999. Species coexistence, distribution, and environmental determinants of neotropical primate richness: a community-level zoogeographic analysis. In J.G. Fleagle, C.H. Janson, and K.E. Reed, editors. *Primate Communities.* Cambridge University Press, Cambridge.

592. Peres, C.A. 1990. Effects of hunting on western Amazonian primate communities. *Biological Conservation,* 54, 47–59.

593. Peres, C.A., and Dolman, P.M. 2000. Density compensation in Neotropical primate communities: evidence from 56 hunted and nonhunted Amazonian forests of varying productivity. *Oecologia*, 122, 175–189.

594. Perez, V.R., Godfrey, L.R., Nowak-Kemp, M., Burney, D.A., Ratsimbazafy, J., and Vasey, N. 2005. Evidence of early butchery of giant lemurs in Madagascar. *Journal of Human Evolution*, 49, 722–742.

595. Perry, G.H., and Dominy, N.J. 2009. Evolution of the human pygmy phenotype. *Trends in Ecology and Evolution*, 24, 218–225.

596. Perry, G.H., Dominy, N.J., Claw, K.G., Lee, A.S., Fiegler, H., Redon, R., et al. 2007. Diet and the evolution of human amylase gene copy number variation. *Nature Genetics*, 39, 1256–1260.

597. Petraglia, M., Koriettar, R., Boivin, N., Clarkson, C., Ditchfield, P., Jones, S., et al. 2007. Middle Paleolithic assemblages from the Indian subcontinent before and after the Toba super-eruption. *Science*, 317, 114–116.

598. Petraglia, M.D., Haslam, M., Fuller, D.Q., Boivin, N., and Clarkson, C. 2010. Out of Africa: new hypotheses and evidence for the dispersal of *Homo sapiens* along the Indian Ocean rim. *Annals of Human Biology*, 37, 288–311.

599. Pickford, M. 1987. The diversity, zoogeography and geochronology of monkeys. *Human Evolution*, 2, 71–89.

600. Pickford, M. 1990. Uplift of the roof of Africa and its bearing on the evolution of mankind. *Human Evolution*, 5, 1–20.

601. Pickford, M. 1992. Biogeography of hominoids and hominids. In T. Nishida et al., editors. *Topics in Primatology: 1. Human Origins*. University of Tokyo Press, Tokyo.

602. Pimentel, D., Lach, L., Zuniga, R., and Morrison, D. 2000. Environmental and economic costs of nonindigenous species in the United States. *BioScience*, 50, 53–65.

603. Pimm, S. 1995. Bird extinctions in the central Pacific. In J.H. Lawton and R.M. May, editors. *Extinction Rates*. Oxford University Press, Oxford.

604. Pitulko, V.V., Nikolsy, P.A., Girya, E.Y., Basilyan, A.E., Tumskoy, V.E., Koulakov, S.A., et al. 2004. The Yana RHA site: humans in the Arctic before the last glacial maximum. *Science*, 303, 52–56.

605. Plowright, W. 1982. The effects of rinderpest and rinderpest control on wildlife in Africa. In A. Edwards and U. McDonnell, editors. *Animal Disease in Relation to Animal Conservation*. Symposia of the Zoological Society of London 50. Academic Press, London.

606. Pollak, M.R., Genovese, G., Friedman, D.J., Ross, M.D., Lecordier, L., Uzureau, P., et al. 2010. Association of trypanolytic ApoL1 variants with kidney disease in African Americans. *Science*, 329, 841–845.

607. Pope, K.O. and Terrell, J.E. 2008. Environmental setting of human migrations in the circum-Pacific region. *Journal of Biogeography*, 35, 1–21.

608. Potts, R. 1998. Environmental hypotheses of hominin evolution. *Yearbook of Physical Anthropology*, 41, 93–136.

609. Potts, R. 1998. Variability selection in hominid evolution. *Evolutionary Anthropology*, 7, 81–96.

610. Poulakakis, N., A., P., Lymberakis, P., Mylonas, M., Zouros, E., Reese, D.S., et al. 2006. Ancient DNA forces reconsideration of evolutionary history of Mediterranean pygmy elephantids. *Biology Letters*, 2, 451–454.

611. Prebble, M., and Dowe, J.L. 2008. The late Quaternary decline and extinction of palms on oceanic Pacific islands. *Quaternary Science Reviews*, 27, 2546–2567.

612. Premo, L.S., and Hublin, J.-J. 2009. Culture, population structure, and low genetic diversity in Pleistocene hominins. *Proceedings of the National Academy of Sciences* (USA), 106, 33–37.

613. Preston, M.A., and Harcourt, A.H. 2009. Conservation implications of the prevalence and representation of locally extinct mammals in the folklore of Native Americans. *Conservation and Society*, 7, 59–69.

614. Prugh, L.R., Hodges, K.E., Sinclair, A.R.E., and Brashares, J.S. 2008. Effect of habitat area and isolation on fragmented animal populations. *Proceedings of the National Academy of Sciences* (USA), 105, 20770–20775.

615. Prugnolle, F., Manica, A., and Balloux, F. 2005. Geography predicts neutral genetic diversity of human populations. *Current Biology*, 15, R159–R160.

616. Pulquério, M.J.F., and Nichols, R.A. 2006. Dates from the molecular clock: how wrong can we be? *Trends in Ecology and Evolution*, 22, 180–184.

617. Purvis, A., Agapow, P.M., Gittleman, J.L., and Mace, G.M. 2000. Nonrandom extinction and the loss of evolutionary history. *Science*, 288, 328–330.

618. Purvis, A., Gittleman, J.L., Cowlishaw, G., and Mace, G.M. 2000. Predicting extinction risk in declining species. *Proceedings of the Royal Society B: Biological Sciences*, 267, 1947–1952.

619. Qin, Z., Yang, Y., Kang, L., Yan, S., Cho, K., Cai, X., et al. 2010. A mitochondrial revelation of early human migrations to the Tibetan Plateau before and after the last glacial maximum. *American Journal of Physical Anthropology*, 143, 555–569.

620. Quammen, D. 1996. *The Song of the Dodo: Island Biogeography in an Age of Extinctions*. Scribner, New York.

621. Quintana-Murci, L., Quach, H., Harmant, C., Luca, F., Massonnet, B., Patin, E., et al. 2008. Maternal traces of deep common ancestry and asymmetric gene flow between Pygmy hunter-gatherers and Bantu-speaking farmers. *Proceedings of the National Academy of Sciences* (USA), 105, 1596–1601.

622. Rabbie, J.M. 1992. The effects of intragroup cooperation and intergroup competition on in-group cohesion and out-group hostility. In A.H. Harcourt and F.B.M. de Waal, editors. *Coalitions and Alliances in Humans and Other Animals*. Oxford University Press, Oxford.

623. Rae, T.C., Koppe, T., and Stringer, C.B. 2011. The Neanderthal face is not cold adapted. *Journal of Human Evolution*, 60, 234–239.

624. Rahmstorf, S., and Ganopolski, A. 1999. Long-term global warming scenarios computed with an efficient coupled climate model. *Climatic Change*, 43, 353–367.

625. Raichlen, D.A., Armstrong, H., and Lieberman, D.E. 2011. Calcaneus length determines running economy: implications for endurance running performance in modern humans and Neandertals. *Journal of Human Evolution*, 60, 299–308.

626. Ralls, K., and Ballou, J. 1983. Extinction: lessons from zoos. In C.M. Schonewald-Cox et al., editors. *Genetics and Conservation*. Benjamin/Cummings Publishing Co., Menlo Park, CA.

627. Ramachandran, S., Deshpande, O., Roseman, C.C., Rosenberg, N.A., Feldman, M.W., and Cavalli-Sforza, L.L. 2005. Support from the relationship of genetic and geographic distance in human populations for a serial founder effect originating in Africa. *Proceedings of the National Academy of Sciences* (USA), 102, 15942–15947.

628. Rampino, M.R., and Self, S. 1992. Volcanic winter and accelerated glaciation following the Toba super-eruption. *Nature*, 359, 50–52.

629. Rapoport, E.H. 1982. *Areography: Geographical Strategies of Species*. Pergamon Press, New York.

630. Rasmussen, M., Y., L., Lindgreen, S., Pedersen, J.S., Albrechtsen, A., Moltke, I., et al. 2010. Ancient human genome sequence of an extinct Palaeo-Eskimo. *Nature*, 463, 757–762.

631. Raxworthy, C.J., et al. 2008. Extinction vulnerability of tropical montane endemism from warming and upslope displacement: a preliminary appraisal for the highest massif in Madagascar. *Global Change Biology*, 14, 1703–1720.

632. Reddy, S., and Dávalos, L.M. 2003. Geographical sampling bias and its implications for conservation priorities in Africa. *Journal of Biogeography*, 30, 1719–1727.

633. Reed, D.L., Light, J.E., Allen, J.M., and Kirchman, J.J. 2007. Pair of lice lost or parasites regained: the evolutionary history of anthropoid primate lice. *BMC Biology*, 5, 7.

634. Reed, F.A., and Tishkoff, S.A. 2006. African human diversity, origins and migrations. *Current Opinion in Genetics & Development*, 16, 597–605.

635. Reed, K.E. 1997. Early hominid evolution and ecological change through the African Plio-Pleistocene. *Journal of Human Evolution*, 32, 289–322.

636. Reed, K.E., and Fleagle, J.G. 1995. Geographic and climatic control of primate diversity. *Proceedings of the National Academy of Sciences* (USA), 92, 7874–7876.

637. Reich, D., Green, R.E., Kircher, M., Krause, J., Patterson, N., Durand, E.Y., et al. 2010. Genetic history of an archaic hominin group from Denisova Cave in Siberia. *Nature*, 468, 1053–1060.

638. Relethford, J.H. 2004. Boas and beyond: migration and craniometric variation. *American Journal of Human Biology*, 16, 379–386.

639. Relethford, J.H. 2004. Global patterns of isolation by distance based on genetic and morphological data. *Human Biology*, 76, 499–513.

640. Relethford, J.H. 2005. *The Human Species. An Introduction to Biological Anthropology*, 6th. ed. McGraw-Hill, New York.

641. Relethford, J.H. 2009. Race and global patterns of phenotypic variation. *American Journal of Physical Anthropology*, 139, 16–22.

642. Relethford, J.H. 2010. Population-specific deviations of global human craniometric variation from a neutral model. *American Journal of Physical Anthropology*, 142, 105–111.

643. Relethford, J.H., and Crawford, M.H. 1995. Anthropometric variation and the population history of Ireland. *American Journal of Physical Anthropology*, 96, 25–38.

644. Reséndez, A. 2007. *A Land So Strange: The Epic Journey of Cabeza de Vaca*. Basic Books, New York.

645. Richard, A.F., and Dewar, R.E. 1991. Lemur ecology. *Annual Review of Ecology and Systematics*, 22, 145–175.

646. Richards, C.L., Carstens, B.C., and Knowles, L.L. 2007. Distribution modelling and statistical phylogeography: an integrative framework for generating and testing alternative biogeographical hypotheses. *Journal of Biogeography*, 34, 1833–1845.

647. Richards, G.D. 2006. Genetic, physiologic and ecogeographic factors contributing to variation in *Homo sapiens*: *Homo floresiensis* reconsidered. *Journal of Evolutionary Biology*, 19, 1744–1767.

648. Richards, M., Oppenheimer, S., and Sykes, B. 1998. mtDNA suggests Polynesian origins in eastern Indonesia. *American Journal of Human Genetics*, 63, 1234–1236.

649. Richards, M.P., and Trinkaus, E. 2009. Isotopic evidence for the diets of European Neanderthals and early modern humans. *Proceedings of the National Academy of Sciences*, (USA), 106, 16034–16039.

650. Richerson, P.J., Bettinger, R.L., and Boyd, R. 2005. Evolution on a restless planet: were environmental variability and environmental change major drivers of human evolution? In F.M. Wuketits and F.J. Ayala, editors. *Handbook of Evolution: Evolution of Living Systems (including Hominids)*. Wiley-VCH, Weinheim, Germany.

651. Richerson, P.J., Boyd, R., and Bettinger, R.L. 2009. Cultural innovations and demographic change. *Human Biology*, 81, 211–235.

652. Richter, D., Moser, J., Nami, M., Eiwanger, J., and Mikdad, A. 2010. New chronometric data from Ifri n'Ammar (Morocco) and the chronostratigraphy of the Middle Palaeolithic in the Western Maghreb. *Journal of Human Evolution*, 59, 672–679.

653. Rightmire, G.P. 1998. Human evolution in the middle Pleistocene: the role of *Homo heidelbergensis*. *Evolutionary Anthropology*, 6, 218–227.

654. Rightmire, G.P. 2009. Middle and later Pleistocene hominins in Africa and Southwest Asia. *Proceedings of the National Academy of Sciences* (USA), 106, 16046–16050.

655. Roberts, D.F. 1978. *Climate and Human Variability*, 2nd ed. Cummings Publishing Co., Menlo Park, CA.

656. Roberts, N. 1984. Pleistocene environments in time and space. In R. Foley, editor. *Hominid Evolution and Community Ecology: Prehistoric Human Adaptation in Biological Perspective*. Academic Press, London.

657. Roberts, R.G., Flannery, T.F., Ayliffe, L.K., Yoshida, H., Olley, J.M., Prideaux, G.J., et al. 2001. New ages for the last Australian megafauna: continent-wide extinction about 46,000 years ago. *Science*, 292, 1888–1892.

658. Robins, A.H. 2009. The evolution of light skin color: role of vitamin D disputed. *American Journal of Physical Anthropology*, 139, 447–450.

659. Rogers, R.A., Rogers, L.A., Hoffmann, R.S., and Martin, L.D. 1991. Native American biological diversity and the biogeographic influence of Ice Age refugia. *Journal of Biogeography*, 18, 623–630.

660. Rohde, K. 1996. Rapoport's Rule is a local phenomenon and cannot explain latitudinal gradients in species diversity. *Biodiversity Letters*, 3, 10–13.

661. Rohde, K. 1997. The larger area of the tropics does not explain latitudinal gradients in species diversity. *Oikos*, 79, 169–172.

662. Rongo, T., Bush, M., and Van Woesik, R. 2009. Did ciguatera prompt the late Holocene Polynesian voyages of discovery? *Journal of Biogeography*, 36, 1423–1432.

663. Ronquist, F. 1997. Dispersal-vicariance analysis: a new approach to the quantification of historical biogeography. *Systematic Biology*, 46, 195–203.

664. Rosenberg, N.A., Mahajan, S., Ramachandran, S., Zhao, C., Pritchard, J.K., and Feldman, M.W. 2005. Clines, clusters, and the effect of study design on the inference of human population structure. *PLOS Genetics*, 1, 660–671.

665. Rosenberg, N.A., Pritchard, J.K., Weber, J.L., Cann, H.M., Kidd, K.K., Zhivotovsky, L., et al. 2002. Genetic structure of human populations. *Science*, 298, 2381–2385.

666. Rosenzweig, M.L. 1995. *Species Diversity in Space and Time.* Cambridge University Press, Cambridge.

667. Ross, C., and Regan, G. 2000. Allocare, predation risk, social structure and natal coat colour in anthropoid primates. *Folia Primatologica,* 71, 67–76.

668. Rossie, J.B., and Seiffert, E.R. 2006. Continental paleobiogeography as phylogenetic evidence. In S.M. Lehman and J.G. Fleagle, editors. *Primate Biogeography.* Springer, New York.

669. Rothman, J.M., P.J., V.S., and Pell, A.N. 2006. Decaying wood is a sodium source for mountain gorillas. *Biology Letters,* 2, 321–324.

670. Roulin, A., Wink, M., and Salamin, N. 2009. Selection on a eumelanic ornament is stronger in the tropics than in temperate zones in the worldwide-distributed barn owl. *Journal of Evolutionary Biology,* 22, 345–354.

671. Rowe, N. 1996. *The Pictorial Guide to the Living Primates.* Pogonias Press, East Hampton, NY.

672. Ruff, C. 2002. Variation in human body size and shape. *Annual Review of Anthropology,* 31, 211–232.

673. Ruff, C.B. 1994. Morphological adaptation to climate in modern and fossil hominids. *Yearbook of Physical Anthropology,* 37, 65–107.

674. Ruggiero, A. 1994. Latitudinal correlates of the sizes of mammalian geographical ranges in South America. *Journal of Biogeography,* 21, 545–559.

675. Ruggiero, A. 1999. Spatial patterns in the diversity of mammal species: a test of the geographic area hypothesis in South America. *Ecoscience,* 6, 338–354.

676. Ruggiero, A., Lawton, J.H., and Blackburn, T.M. 1998. The geographic ranges of mammalian species in South America: spatial patterns in environmental resistance and anisotropy. *Journal of Biogeography,* 25, 1093–1103.

677. Ruggiero, A., and Werenkraut, V. 2007. One-dimensional analyses of Rapoport's rule reviewed through meta-analysis. *Journal of Biogeography,* 16, 401–414.

678. Rylands, A.B., editor. 1993. *Marmosets and Tamarins: Systematics, Behaviour, and Ecology.* Oxford University Press, Oxford.

679. SAS Institute Inc. 2009. *JMP 8.0.* SAS Institute Inc., Cary, NC.

680. Savolainen, P., Leitner, T., Wilton, A.N., Matisoo-Smith, E., and Lundeberg, J. 2004. A detailed picture of the origin of the Australian dingo, obtained from the study of mitochondrial DNA. *Proceedings of the National Academy of Sciences* (USA), 101, 12387–12390.

681. Sax, D.F. 2001. Latitudinal gradients and geographic ranges of exotic species: implications for biogeography. *Journal of Biogeography,* 28, 139–150.

682. Sax, D.F., and Gaines, S.D. 2008. Species invasions and extinction: the future of native biodiversity on islands. *Proceedings of the National Academy of Sciences* (USA), 105, 11490–11497.

683. Scheinfeldt, L.B., Soi, S., and Tishkoff, S.A. 2010. Working toward a synthesis of archaeological, linguistic, and genetic data for inferring African population history. *Proceedings of the National Academy of Sciences* (USA), 107, 8931–8938.

684. Schmidt-Nielsen, K. 1997. *Animal Physiology: Adaptation and Environment,* 4th. ed. Cambridge University Press, Cambridge.

685. Schoenemann, P.T. 2004. Brain size scaling and body composition in mammals. *Brain, Behavior and Evolution,* 63, 47–60.

686. Scholander, P.F. 1955. Evolution of climatic adaptation in homeotherms. *Evolution,* 9, 15–26.

687. Schotterer, U., Fröhlich, K., Gäggeler, H.W., Sandjordi, S., and Stichler, W. 1997. Isotope record from Mongolian and Alpine ice cores as climatic indicators. *Climatic Change,* 36, 519–530.

688. Schroeder, K.B., Jakobsson, M., Crawford, M.H., Schurr, T.G., Boca, S.M., Conrad, D.F., et al. 2009. Haplotypic background of a private allele at high frequency in the Americas. *Molecular Biology and Evolution,* 26, 995–1016.

689. Schroeder, K.B., Schurr, T.G., Long, J.C., Rosenberg, N.A., Crawford, M.H., Tarskaia, L.A., et al. 2007. A private allele ubiquitous in the Americas. *Biology Letters,* 3, 218–223.

690. Schulte, P., Alegret, L., Arenillas, I., Arz, J.A., Barton, P.J., Bown, P.R., et al. 2010. The Chicxulub impact and mass extinction at the Cretaceous-Paleogene boundary. *Science,* 327, 1214–1218.

691. Schulte, P., Speiger, R.P., Brinkhuis, H., Kontny, A., Claeys, P., Galeotti, S., et al. 2008. Comment on the paper, "Chicxulub impact predates K-T boundary: New evidence from Brazos, Texas" by Keller et al. (2007). *Earth and Planetary Science Letters,* 269, 613–619.

692. Searle, J.B., Jones, C.S., Gündüz, I., Scascitelli, M., Jones, E.P., Herman, J.S., et al. 2009. Of mice and (Viking?) men: phylogeography of British and Irish house mice. *Proceedings of the Royal Society B: Biological Sciences,* 276, 201–207.

693. Segal, B., and Duffy, L.K. 1992. Ethanol elimination among different racial groups. *Alcohol,* 9, 213–217.

694. Segerstrale, U. 2000. *Defenders of the Truth: The Battle for Science in the Sociobiology Debate and Beyond.* Oxford University Press, Oxford.

695. Seiffert, E.R., Simons, E.L., and Attia, Y. 2003. Fossil evidence for an ancient divergence of lorises and galagos. *Nature,* 422, 421–424.

696. Semino, O., Passarino, G., Oefner, P.J., Lin, A.A., Arbuzova, S., and et al. 2000. The genetic legacy of Paleolithic *Homo sapiens sapiens* in extant Europeans: a Y chromosome perspective. *Science,* 290, 1155–1159.

697. Serrat, M.A., King, D., and Lovejoy, C.O. 2008. Temperature regulates limb length in homeotherms by directly modulating cartilage growth. *Proceedings of the National Academy of Sciences* (USA), 105, 19348–19353.

698. Shackleton, N.J. 1995. New data on the evolution of Pliocene climatic variability. In E.S. Vrba, et al., editors. *Paleoclimate and Evolution with Emphasis on Human Origins.* Yale University Press, New Haven, CT.

699. Shea, J.J. 2003. Neandertals, competition, and the origin of modern human behavior in the Levant. *Evolutionary Anthropology,* 12, 173–187.

700. Shea, J.J. 2008. Transitions or turnovers? Climatically-forced extinctions of *Homo sapiens* and Neanderthals in the east Mediterranean Levant. *Quaternary Science Reviews,* 27, 2253–2270.

701. Shen, G., Gao, X., Gao, B., and Granger, D.E. 2009. Age of Zhoukoudian *Homo erectus* determined with $^{26}Al/^{10}$ burial dating. *Nature,* 458, 198–200.

702. Shen, L., Chen, X.Y., Zhang, X., Li, Y.Y., Fu, C.X., and Qiu, Y.X. 2005. Genetic variation of *Ginkgo biloba* L. (Ginkgoaceae) based on cpDNA PCR-RFLPs: inference of glacial refugia. *Heredity,* 94, 396–401.

703. Shepher, J. 1971. Mate selection among second generation kibbutz adolescents and adults: incest avoidance and negative imprinting. *Archives of Sexual Behavior,* 1, 293–307.

704. Shkolnik, A., Taylor, C.R., Finch, V., and Borut, A. 1980. Why do Bedouins wear black robes in hot deserts? *Nature,* 283, 373–375.

705. Simmons, I.G. 1979. *Biogeography: Natural and Cultural.* Duxbury Press, London.

706. Simonson, T.S., Yang, Y., Huff, C.D., Yun, H., Qin, G., Witherspoon, D.J., et al. 2010. Genetic evidence for high-altitude adaptation in Tibet. *Science,* 329, 72–75.

707. Sinclair, A.R.E., Mduma, S., and Brashares, J.S. 2003. Patterns of predation in a diverse predator-prey system. *Nature,* 425, 288–290.

708. Sirugo, G., Hennig, B.J., Adeyemo, A.A., Matimba, A., Newport, M.J., Ibrahim, M., et al. 2008. Genetic studies of African populations: an overview on disease susceptibility and response to vaccines and therapeutics. *Human Genetics,* 123, 557–598.

709. Skutnabb-Kangas, T., and Harmon, D. 2002. Review of Daniel Nettle, *Linguistic Diversity,* Oxford: Oxford University Press, 1999. *Language Policy,* 1, 175–182.

710. Small, C., and Cohen, J.E. 2004. Continental physiography, climate, and the global distribution of human population. *Current Anthropology,* 45, 269–288.

711. Smith, R.J., and Cheverud, J.M. 2002. Scaling of sexual dimorphism in body mass: a phylogenetic analysis of Rensch's rule in primates. *International Journal of Primatology,* 23, 1095–1135.

712. Smith, R.J., and Jungers, W.L. 1997. Body mass in comparative primatology. *Journal of Human Evolution,* 32, 523–559.

713. Snodgrass, J.J., Leonard, W.R., Tarskaia, L.A., Alekseev, V.P., and Krivoshapkin, V.G. 2005. Basal metabolic rate in the Yakut (Sakha) of Siberia. *American Journal of Human Biology,* 17, 155–172.

714. Snow, R.W., Guerra, C.A., Noor, A.M., Myint, H.Y., and Hay, S.I. 2005. The global distribution of clinical episodes of *Plasmodium falciparum* malaria. *Nature,* 434, 214–217.

715. Soares, P., Ermini, L., Thomson, N., Mormina, M., Rito, T., Röhl, A., et al. 2009. Correcting for purifying selection: an improved human mitochondrial molecular clock. *American Journal of Human Genetics,* 84, 740–759.

716. Soares, P., Rito, T., Trejaut, J., Mormina, M., Hill, C., Tinkler-Hundal, E., et al. 2011. Ancient voyaging and Polynesian origins. *American Journal of Human Genetics,* 88, 239–247.

717. Spellerberg, I.F., and Sawyer, J.W.D. 1999. *An Introduction to Applied Biogeography.* Cambridge University Press, Cambridge.

718. Springer, M.S., Murphy, W.J., Eizirik, E., and O'Brien, S.J. 2003. Placental mammal diversification and the Cretaceous-Tertiary boundary. *Proceedings of the National Academy of Sciences* (USA), 100, 1056–1061.

719. Stanley, S.M. 1992. An ecological theory for the origin of *Homo. Paleobiology,* 18, 237–257.

720. Steadman, D.W. 1995. Prehistoric extinctions of Pacific island birds: biodiversity meets zooarchaeology. *Science,* 267, 1123–1131.

721. Steadman, D.W. 2006. *Extinction and Biogeography of Tropical Pacific Birds.* University of Chicago Press, Chicago.

722. Stearns, S.C., editor. 1999. *Evolution in Health & Disease.* Oxford University Press, Oxford.

723. Steele, T.E., and Klein, R.G. 2008. Intertidal shellfish use during the Middle and Later Stone Age of South Africa. *Archeofauna,* 17, 63–76.

724. Steiper, M.E., and Young, N.M. 2008. Timing primate evolution: lessons from the discordance between molecular and paleontological estimates. *Evolutionary Anthropology,* 17, 179–188.

725. Stevens, G.C. 1989. The latitudinal gradient in geographical range: how so many species coexist in the tropics. *American Naturalist,* 133, 240–256.

726. Stewart, C.-B., and Disotell, T.R. 1998. Primate evolution—in and out of Africa. *Current Biology,* 8, R582–R588.

727. Stewart, C.-B., Schilling, J.W., and Wilson, A.C. 1987. Adaptive evolution in the stomach lysozymes of foregut fermenters. *Nature,* 330, 401–401.

728. Stinson, S. 2000. Growth variation: biological and cultural factors. In S. Stinson et al., editors. *Human Biology. An Evolutionary and Biocultural Perspective.* Wiley-Liss, New York.

729. Stinson, S., and Frisancho, A.R. 1978. Body proportions of highland and lowland Peruvian Quechua children. *Human Biology,* 50, 57–68.

730. Stone, A.C., Griffiths, R.C., Zegura, S.L., and Hammer, M.F. 2002. High levels of Y-chromosome nucleotide diversity in the genus *Pan. Proceedings of the National Academy of Sciences* (USA), 99, 43–48.

731. Stoner, C.J., Caro, T.M., and Graham, C.M. 2003. Ecological and behavioral correlates of coloration in artiodactyls: systematic analyses of conventional hypotheses. *Behavioral Ecology,* 14, 823–840.

732. Storey, A.A., Ramirez, J.M., Quiroz, D., Beavan-Athfield, N., Addison, D.J., Walter, R., et al. 2007. Radiocarbon and DNA evidence for a pre-Columbian introduction of Polynesian chickens to Chile. *Proceedings of the National Academy of Sciences* (USA), 104, 10335–10339

733. Strait, D.S., and Wood, B.A. 1999. Early hominid biogeography. *Proceedings of the National Academy of Sciences* (USA), 96, 9196–9200.

734. Strassman, B.I., and Dunbar, R.I.M. 1999. Human evolution and disease: putting the Stone Age in perspective. In S.C. Stearns, editor. *Evolution in Health & Disease.* Oxford University Press, Oxford.

735. Strickland, R. 1986. Genocide-at-law: an historic and contemporary view of the Native American experience. *University of Kansas Law Review,* 34, 713–755.

736. Stringer, C.B. 1995. The evolution and distribution of later Pleistocene human populations. In E.S. Vrba, et al., editors. *Paleoclimate and Evolution with Emphasis on Human Origins.* Yale University Press, New Haven, CT.

737. Stringer, C.B., and Andrews, P. 1988. Genetic and fossil evidence for the origin of modern humans. *Science,* 239, 1263–1268.

738. Strong, W.L., and Hills, L.V. 2005. Late-glacial and Holocene palaeovegetation zonal reconstruction for central and north-central North America. *Journal of Biogeography,* 32, 1043–1062.

739. Stuart, A.J., Kosintsev, P.A., Higham, T.F.G., and Lister, A.M. 2004. Pleistocene to Holocene extinction dynamics in giant deer and woolly mammoth. *Nature,* 431, 684–689.

740. Su, B., Jin, L., Underhill, P., Martinson, J., Saha, N., McGarvey, S.T., et al. 2000. Polynesian origins: insights from the Y chromosome. *Proceedings of the National Academy of Sciences* (USA), 97, 8225–8228.

741. Su, B., Xiao, C., Deka, R., Seielstad, M.T., Kangwanpong, D., Xiao, J., et al. 2000. Y chromosome haplotypes reveal prehistorical migrations to the Himalayas. *Human Genetics,* 107, 582–590.

742. Suddendorf, R.F. 1989. Research on alcohol metabolism among Asians and its implications for understanding causes of alcoholism. *Public Health Reports,* 104, 615–620.

743. Summerhayes, G.R., Leavesley, M., Fairbairn, A., Mandui, H., Field, J., Ford, A., et al. 2010. Human adaptation and plant use in highland New Guinea 49,000 to 44,000 years ago. *Science,* 330, 78–81.

744. Surovell, T.A. 2000. Early Paleoindian women, children, mobility, and fertility. *American Antiquity, 65*, 493–508.

745. Surovell, T.A., Holliday, V.T., Gingerich, J.A.M., Ketron, C., Haynes, C.V., Hilman, I., et al. 2009. An independent evaluation of the Younger Dryas extraterrestrial impact hypothesis. *Proceedings of the National Academy of Sciences* (USA), 106, 18155–18158.

746. Sutherland, W.J. 2003. Parallel extinction risk and global distribution of languages and species. *Nature, 423*, 276–279.

747. Swisher, C.C., Curtis, G.H., Jacob, T., Getty, A.G., Suprijo, A., and Widiasmoro. 1994. Age of the earliest known hominids in Java, Indonesia. *Science,* 1118–1121.

748. Swisher, C.C., Rink, W.J., Antón, S.C., Schwarcz, H.P., Curtis, G.H., and Widiasmoro, A.S. 1996. Latest Homo erectus of Java: potential contemporaneity with Homo sapiens in southeast Asia. *Science, 274*, 1870–1874.

749. Tamm, E., Kivisild, T., Reidla, M., Metspalu, M., Smith, D.G., and et al. 2007. Beringian standstill and spread of Native American founders. *PLoS ONE, 2*, e829, 1–5.

750. Tattersall, I. 2009. Human origins: out of Africa. *Proceedings of the National Academy of Sciences* (USA), 106, 16018–16021.

751. Taylor, N.A.S. 2006. Ethnic differences in thermoregulation: genotypic versus phenotypic heat adaptation. *Journal of Thermal Biology, 31*, 90–104.

752. Templeton, A.R. 2002. Out of Africa again and again. *Nature, 416*, 45–51.

753. Terborgh, J., Lopez, L., Nunez, V.P., Rao, M., Shahabuddin, G., Orihuela, G., et al. 2001. Ecological meltdown in predator-free forest fragments. *Science, 294*, 1923–1926.

754. Terborgh, J., Lopez, L., and Tello, J.S. 1997. Bird communities in transition: the Lago Guri Islands. *Ecology, 78*, 1494–1501.

755. Terrell, J. 1986. *Prehistory in the Pacific Islands. A Study of Variation in Language, Customs, and Human Biology.* Cambridge University Press, Cambridge.

756. Terrell, J.E. Language and material culture on the Sepik coast of Papua New Guinea: using social network analysis to simulate, graph, identify, and analyze social and cultural boundaries between communities. *Journal of Island and Coastal Archaeology, 5*, 3–32.

757. Terrell, J.E. 1997. The postponed agenda: archaeology and human biogeography in the twenty-first century. *Human Ecology, 25*, 419–436.

758. Terrell, J.E. 2006. Human biogeography: evidence of our place in nature. *Journal of Biogeography, 33*, 2088–2098.

759. Terrell, J.E. 2010. Social network analysis of the genetic structure of Pacific Islanders. *Annals of Human Genetics, 74*, 211–232.

760. Terrell, J.E., Kelly, K.M., and Rainbird, P. 2001. Foregone conclusions? In search of "Papuans" and "Austronesians." *Current Anthropology*, 42, 97–124.

761. Tewksbury, J.T., Huey, R.B., and Deutsch, C.A. 2008. Putting the heat on tropical animals. *Science*, 320, 1296–1297.

762. The HUGO Pan-Asian SNP Consortium. 2009. Mapping human genetic diversity in Asia. *Science*, 326, 1541–1545.

763. Thompson, R.S., and Anderson, K.H. 2000. Biomes of western North America at 18,000, 6000 and 0 C-14 yr BP reconstructed from pollen and packrat midden data. *Journal of Biogeography*, 27, 555–584.

764. Thornhill, R., Fincher, C.L., and Aran, D. 2009. Parasites, democratization, and the liberalization of values across contemporary countries. *Biological Reviews*, 84, 113–131.

765. Tilkens, M.J., Wall-Scheffler, C., Weaver, T.D., and Steudel-Numbers, K. 2007. The effects of body proportions on thermoregulation: an experimental assessment of Allen's rule. *Journal of Human Evolution*, 53, 286–291.

766. Tilman, D., May, R.M., Lehman, C.L., and Nowak, M.A. 1994. Habitat destruction and the extinction debt. *Nature*, 371, 65–66.

767. Timmreck, C., Graf, H.-F., Lorenz, S.J., Niemeier, U., Zanchettin, D., Matei, D., et al. 2010. Aerosol size confines climate response to volcanic super-eruptions. *Geophysical Research Letters*, 37, L24705, 5 pp.

768. Tishkoff, S.A., Reed, F.A., Friedlaender, F.R., Ehret, C., Ranciaro, A., Froment, A., et al. 2009. The genetic structure and history of Africans and African Americans. *Science*, 324, 1035–1044.

769. Tishkoff, S.A., Reed, F.A., Ranciaro, A., Voight, B.F., Babbitt, C.C., Silverman, J.S., et al. 2007. Convergent adaptation of human lactase persistence in Africa and Europe. *Nature Genetics*, 39, 31–40.

770. Tishkoff, S.A., and Verrelli, B.C. 2003. Patterns of human genetic diversity: implications for human evolutionary history and disease. *Annual Review of Genomics and Human Genetics*, 4, 293–340.

771. Tishkoff, S.A., and Williams, S.M. 2002. Genetic analysis of African populations: human evolution and complex disease. *Nature Reviews Genetics*, 3, 611–621.

772. Toole, M.J. 1997. Displaced persons and war. In B.S. Levy and V.W. Sidel, editors. *War and Public Health*. Oxford University Press, New York.

773. Torrence, R. 1983. Time budgeting and hunter-gatherer technology. In G. Bailey, editor. *Hunter-Gatherer Economy in Prehistory*. Cambridge University Press, Cambridge.

774. Trejaut, J.A., Kivisild, T., Loo, J.H., Lee, C.L., He, C.L., Hsu, C.J., et al. 2005. Traces of archaic mitochondrial lineages persist in Austronesian-speaking Formosan populations. *PLOS Biology*, 3, e247, 1362–1372.

775. Trinkaus, E. 1981. Neanderthal limb proportions and cold adaptation. In C.B. Stringer, editor. *Aspects of Human Evolution*. Taylor & Francis, London.

776. Trinkaus, E. 2005. Early modern humans. *Annual Review of Anthropology*, 34, 207–230.

777. Turner, A. 1992. Large carnivores and earliest European hominids: changing determinants of resource availability during the Lower and Middle Pleistocene. *Journal of Human Evolution*, 22, 109–126.

778. Turner, A. 1999. Assessing earliest human settlement of Eurasia: Late Pliocene dispersions from Africa. *Antiquity*, 73, 563–570.

779. Turner, A., and Wood, B.A. 1993. Taxonomic and geographic diversity in robust australopithecines and other African Plio-Pleistocene larger mammals. *Journal of Human Evolution*, 24, 147–168.

780. Turney, C.S.M., Flannery, T.F., Roberts, R.G., Reid, C., Fifield, L.K., Higham, T.F.G., et al. 2008. Late-surviving megafauna in Tasmania, Australia, implicate human involvement in their extinction. *Proceedings of the National Academy of Sciences* (USA), 105, 12150–12153.

781. U.S. Census Bureau, G.D., 2000. American Indian and Alaska Native areas. In *Geographic Areas Reference Manual*. U.S. Census Bureau, Suitland, MD. 1–18.

782. Uinuk-Ool, T.S., Takezaki, N., and Klein, J. 2003. Ancestry and kinships of native Siberian populations: the HLA evidence. *Evolutionary Anthropology*, 12, 231–245.

783. Underhill, P.A., Passarino, G., Lin, A.A., Shen, P., Lahr, M.M., Foley, R.A., et al. 2001. The phylogeography of Y chromosome binary haplotypes and the origins of modern human populations. *Annals of Human Genetics*, 65, 43–62.

784. Upchurch, P. 2008. Gondwana break-up: legacies of a lost world. *Trends in Ecology and Evolution*, 23, 229–236.

785. Vajda, E. 2010. A Siberian Link with Na-Dene Languages. *Anthropological Papers of the University of Alaska, New Series*, 5, 33–99.

786. van Gemerden, B.S., Olff, H., Parren, M.P.E., and Bongers, F. 2003. The pristine rain forest? Remnants of historical human impacts on current tree species composition and diversity. *Journal of Biogeography*, 30, 1381–1390.

787. Van Vuren, D. 1998. Mammalian dispersal and reserve design. In T.M. Caro, editor. *Behavioral Ecology and Conservation Biology*. Oxford University Press, New York.

788. Vartanyan, S.L., Garutt, V.E., and Sher, A.V.N. 1993. Holocene Dwarf mammoths from Wrangel Island in the Siberian Arctic. *Nature*, 362, 337–340.

789. Verpoorte, A. 2009. Limiting factors on early modern human dispersals: The human biogeography of late Pleniglacial Europe. *Quaternary International*, 201, 77–85.

790. Verwimp, P. 2005. An economic profile of peasant perpetrators of genocide: micro-level evidence from Rwanda. *Journal of Development Economics*, 77, 297–323.

791. Victor, P.-E. 1964. *Man and the Conquest of the Poles*. Hamish Hamilton, London.

792. Vigilant, L. and Bradley, B.J. 2004. Genetic variation in gorillas. *American Journal of Primatology*, 64, 161–172.

793. Virah-Sawmy, V., Willis, K.J., and Gillson, L. 2010. Evidence for drought and forest declines during the recent megafaunal extinctions in Madagascar. *Journal of Biogeography*, 37, 506–519.

794. Voris, H.K. 2000. Maps of Pleistocene sea levels in Southeast Asia: shorelines, river systems and time durations. *Journal of Biogeography*, 27, 1153–1167.

795. Vrba, E.S. 1995. The fossil record of African antelopes (Mammalia, Bovidae) in relation to human evolution and paleoclimate. In E.S. Vrba, et al., editors. *Paleoclimate and Evolution, with Emphasis on Human Origins*. Yale University Press, New Haven, CT.

796. Vrba, E.S. 1996. Climate, heterochrony, and human evolution. *Journal of Anthropological Research*, 52, 1–28.

797. Vrba, E.S., Denton, G.H., Partridge, T.C., and Burckle, L.H., editors. 1995. *Paleoclimate and Evolution, with Emphasis on Human Origins*. Yale University Press, New Haven, CT.

798. Waguespack, N.M. 2007. Why we're still arguing about the Pleistocene occupation of the Americas. *Evolutionary Anthropology*, 16, 63–74.

799. Waide, R.B., Willig, M.R., Steiner, C.F., Mittelbach, G., Gough, L., Dodson, S.I., et al. 1999. The relationship between productivity and species richness. *Annual Review of Ecology and Systematics*, 30, 257–300.

800. Waldner, L.S. 2008. The kudzu connection: exploring the link between land use and invasive species. *Land Use Policy*, 25, 399–409.

801. Walker, J.D., and Geisman, J.W. 2009. Geological time scale. Geological Society of America, Boulder, CO. [Cited 2010].

802. Walker, R., Gurven, M., Hill, K., Migliano, A.B., Chagnon, N., De Souza, R., et al. 2006. Growth rate and life histories in twenty-two small-scale societies. *American Journal of Human Biology*, 18, 295–311.

803. Wallace, A.R. 1876. *The Geographical Distribution of Animals, Vol. I.* Harper and Bros, New York.

804. Wallace, A.R. 1876. *The Geographical Distribution of Animals, Vol. II.* Harper and Bros., New York.

805. Wallace, A.R. 1880. *Island Life*. Macmillan, London.

806. Wallace, B.L. 2009. L'Anse Aux Meadows, Leif Eriksson's Home in Vinland. *Journal of the North Atlantic*, 2(Special Issue 2), 114–125.

807. Walsh, R.P.D. 1996. The environment. In P.W. Richards, editor. *The Tropical Rain Forest*, 2nd ed. Cambridge University Press, Cambridge.

808. Walther, G., Post, E., Convey, P., Menzel, A., Parmesan, C., Beebee, T.J.C., et al. 2002. Ecological responses to recent climate change. *Nature*, 416, 389–395.

809. Watts, D.P., Muller, M.N., Amsler, S.J., Mbabazi, G., and Mitani, J.C. 2006. Lethal intergroup aggression by chimpanzees in Kibale National Park, Uganda. *American Journal of Primatology*, 68, 161–180.

810. Weaver, T.D. 2003. The shape of the Neanderthal femur is primarily the consequence of a hyperpolar body form. *Proceedings of the National Academy of Sciences* (USA), 100, 6926–6929.

811. Weaver, T.D. 2009. The meaning of Neanderthal skeletal morphology. *Proceedings of the National Academy of Sciences* (USA), 106, 16028–16033.

812. Weaver, T.D., and Roseman, C.C. 2008. New developments in the genetic evidence for modern human origins. *Evolutionary Anthropology*, 17, 69–80.

813. Weaver, T.D., Roseman, C.C., and Stringer, C.B. 2007. Were Neanderthal and modern human cranial differences produced by natural selection or genetic drift? *Journal of Human Evolution*, 53, 135–145.

814. Weaver, T.D., and Steudel-Numbers, K. 2005. Does climate or mobility explain the differences in body proportions between Neanderthals and their Upper Paleolithic successors? *Evolutionary Anthropology*, 14, 218–223.

815. Webb, S. 2008. Megafauna demography and late Quaternary climatic change in Australia: a predisposition to extinction. *Boreas*, 37, 329–345.

816. Weber, J., Czarnetzki, A.E., and Pusch, C.M. 2005. Comment on "The brain of LB1, Homo floresiensis." *Science*, 310, 236b.

817. Weber, W.W. 1999. Populations and genetic polymorphisms. *Molecular Diagnosis*, 4, 299–307.

818. Wells, J.C.K., and Stock, J.T. 2007. The biology of the colonizing ape. *Yearbook of Physical Anthropology*, 50, 191–222.

819. Welsch, R.L., and Terrell, J. 1992. Language and culture on the north coast of New-Guinea. *American Anthropologist*, 94, 568–600.

820. West, G.B., Brown, J.H., and Enquist, B.J. 1997. A general model for the origin of allometric scaling laws in biology. *Science*, 276, 122–126.

821. Weston, E.M., and Lister, A.M. 2009. Insular dwarfism in hippos and a model for brain size reduction in *Homo floresiensis*. *Nature*, 459, 85–88.

822. White, A.W., Worthy, T.H., Hawkins, S., Bedford, S., and Spriggs, M. 2010. Megafaunal meiolaniid horned turtles survived until early human settlement in Vanuatu, Southwest Pacific. *Proceedings of the National Academy of Sciences* (USA), 107, 15512–15516.

823. White, T.D. 1995. African omnivores: global climatic changes and Plio-Pleistocene hominids and suids. In E.S. Vrba, et al., editors. *Paleoclimate and Evolution with Emphasis on Human Origins.* Yale University Press, New Haven, CT.

824. White, T.D., Ambrose, S.H., Suwa, G., Su, D.F., DeGusta, D., Bernore, R.L., et al. 2009. Macrovertebrate paleontology and the Pliocene habitat of *Ardipithecus ramidus. Science,* 326, 87–93.

825. White, T.D., Ambrose, S.H., Suwa, G., and WoldeGabriel, G. 2010. Response to comment on the paleoenvironment of *Ardipithecus ramidus. Science,* 328, 1105-e.

826. Whittaker, R.J. 2003. Climatic-energetic explanations of diversity: a macroscopic perspective. In T.M. Blackburn and K.J. Gaston, editors. *Macroecology: Concepts and Consequences.* Blackwell Publishing, Oxford.

827. Whittaker, R.J. 2007. *Island Biogeography: Ecology, Evolution and Conservation,* 2nd. ed. Oxford University Press, Oxford.

828. Whittaker, R.J., Araújo, M.B., Jepson, P., Ladle, R.J., Watson, J.E.M., and Willis, K.J. 2005. Conservation biogeography: assessment and prospect. *Diversity and Distributions,* 11, 3–23.

829. Wickler, S., and Spriggs, M. 1988. Pleistocene human occupation of the Solomon-Islands, Melanesia. *Antiquity,* 62, 703–706.

830. Wiggins, D.A., Møller, A.P., Sorensen, M.F.L., and Brand, L.A. 1998. Island biogeography and the reproductive ecology of great tits *Parus major. Oecologia,* 115, 478–482.

831. Wiktionary. 2006. Wiktionary:Frequency lists/TV/2006/1-1000. Wikimedia. Available from http://en.wiktionary.org/wiki/Wiktionary:Frequency_lists/TV/2006/1-1000. [Cited 2010.]

832. Wilkins, J.F. 2006. Unraveling male and female histories from human genetic data. *Current Opinion in Genetics & Development,* 16, 611–617.

833. Wilmshurst, J.M., Anderson, A.J., Higham, T.F.G., and Worthy, T.H. 2008. Dating the late prehistoric dispersal of Polynesians to New Zealand using the commensal Pacific rat. *Proceedings of the National Academy of Sciences* (USA), 105, 7676–7680.

834. Wilmshurst, J.M., Hunt, T.L., Lipo, C.P., and Anderson, A.J. 2011. High-precision radiocarbon dating shows recent and rapid initial human colonization of East Polynesia. *Proceedings of the National Academy of Sciences* (USA), 108, 1815–1820.

835. Wilson, E.O. 1993. Is humanity suicidal? *New York Times Magazine,* May 30, 1993, 24–29.

836. Wilson, M.L., Hauser, M.D., and Wrangham, R.W. 2001. Does participation in intergroup conflict depend on numerical assessment, range location, or rank for wild chimpanzees? *Animal Behaviour,* 61, 1203–1216.

837. Winterhalder, B., and Lu, F. 1997. A forager-resource population ecology model and implications for indigenous conservation. *Conservation Biology,* 11, 1354–1364.

838. Wolfheim, J.H. 1983. *Primates of the World: Distribution, Abundance and Conservation.* University of Washington Press, Seattle.

839. Wollstein, A., Lao, O., Becker, C., Brauer, S., Trent, R.J., Nürnberg, P., et al. 2010. Demographic history of Oceania inferred from genome-wide data. *Current Biology,* 20, 1983–1992.

840. Wood, B. 2010. Reconstructing human evolution: Achievements, challenges, and opportunities. *Proceedings of the National Academy of Sciences* (USA), 107, 8902–8909.

841. Wood, B., and Harrison, T. 2011. The evolutionary context of the first hominins. *Nature,* 470, 347–352.

842. Wood, B., and Strait, D. 2004. Patterns of resource use in early *Homo* and *Paranthropus. Journal of Human Evolution,* 46, 119–162.

843. Woodward, F.I., and Kelly, C.K. 2003. Why are species not more widely distributed? Physiological and environmental limits. In T.M. Blackburn and K.J. Gaston, editors. *Macroecology: Concepts and Consequences.* Blackwell Scientific Publishers, Oxford.

844. World Resources Institute. 2009. *EarthTrends.* The Environmental Information Portal. World Resources Institute, Washington DC. Available from http://earthtrends.wri.org/. [Cited 2009.]

845. World Resources Institute. 2010. *EarthTrends.* Agriculture and Food. World Resources Institute, Washington DC. Available from http://earthtrends.wri.org/. [Cited 2010.]

846. Wrangham, R. 2009. *Catching Fire: How Cooking Made Us Human.* Basic Books, Perseus Books Group, New York.

847. Wrangham, R., and Carmody, R. 2010. Human adaptation to the control of fire. *Evolutionary Anthropology,* 19, 187–199.

848. Wright, J.S., Muller-Landau, H.C., and Schipper, J. 2009. The future of tropical species on a warmer planet. *Conservation Biology,* 23, 1418–1426.

849. Wright, P.C., and Jernvall, J. 1999. The future of primate communities: a reflection of the present? In J.G. Fleagle, C.H. Janson, and K.E. Reed, editors. *Primate Communities.* Cambridge University Press, Cambridge.

850. Wroe, S., and Field, J. 2006. A review of the evidence for a human role in the extinction of Australian megafauna and an alternative interpretation. *Quaternary Science Reviews,* 25, 2692–2703.

851. Wroe, S., Field, J., and Grayson, D.K. 2006. Megafaunal extinction: climate, humans and assumptions. *Trends in Ecology and Evolution*, 21, 61–62.

852. Wurster, C.M., Bird, M.I., Bull, I.D., Creed, F., Bryant, C., Dungait, J.A.J., et al. 2010. Forest contraction in north equatorial Southeast Asia during the Last Glacial Period. *Proceedings of the National Academy of Sciences (USA)*, 107, 15508–15511.

853. Yi, X., Liang, Y., Huerta-Sanchez, E., Jin, X., Cuo, Z.X.P., Pool, J.E., et al. 2010. Sequencing of 50 exomes reveals adaptation to high altitude. *Science*, 329, 75–78.

854. Yoder, A.D., and Nowak, M.D. 2006. Has vicariance or dispersal been the predominant biogeographic force in Madagascar? Only time will tell. *Annual Review of Ecology, Evolution and Systematics*, 37, 405–431.

855. Yokley, T.R. 2009. Ecogeographic variation in human nasal passages. *American Journal of Anthropology*, 138, 11–22.

856. Young, J.H. 2007. Evolution of blood pressure regulation in humans. *Current Hypertension Reports*, 9, 13–18.

857. Young, J.H., Chang, Y.-P.C., Kim, J.D.-O., Chretien, J.-P., Klag, M.J., Levine, M.A., et al. 2005. Differential susceptibility to hypertension is due to selection during the out-of-Africa expansion. *PLOS Genetics*, 1, 0730–0738.

858. Yu, J., and Dobson, F.S. 2000. Seven forms of rarity in mammals. *Journal of Biogeography*, 27, 131–139.

859. Yule, J.V., Jensen, C.X.J., Joseph, A., and Goode, J. 2009. The puzzle of North America's Late Pleistocene megafaunal extinction patterns: test of new explanation yields unexpected results. *Ecological Modelling*, 220, 533–544.

860. Zhivotovsky, L.A., Rosenberg, N.A., and Feldman, M.W. 2003. Features of evolution and expansion of modern humans, inferred from genome-wide microsatellite markers. *American Journal of Human Genetics*, 72, 1171–1186.

861. Zhu, R.X., Potts, R., Xie, F., Hoffman, K.A., Deng, C.L., Shi, C.D., et al. 2004. New evidence on the earliest human presence at high northern latitudes in northeast Asia. *Nature*, 431, 559–562.

General Index

If no species is stated, the topic is humans. "Primate" is nonhuman primate.

adaptation, 89–127, 156–159, 193–204
  Allen effect, 96–99, 103–113,
    126–127
  altitude, elevation, 89, 91, 120–123,
    126
  anatomy, morphology, 89, 91–113,
    123–127
  animals, 92–93, 98–100, 119, 194
  athleticism, 106, 112–113
  Bergmann effect, 96–103, 107–113,
    126–127
  body size, 96–103, 107–113, 126–127,
    155–162. *See also* Island effect
  climate, latitude, 89–119, 126–127,
    227–228
  diet, 193–204
  drugs, 193, 201–202, 204
  evidence for origins, 111
  genetics, 15–17, 19, 96, 115–117, 122
  Gloger effect, 92
  geographic range size, 165–166
  mammals, 98–100, 120 165, 167, 194
  Neanderthals, 21–22, 111–113
  nutrition, 95–97, 107–110, 118,
    122–123, 127
  primates, 98–101, 140, 165, 167, 194
  physiology, 89, 113–127, 193–204
  sex differences, 89, 123–127
  skin color, 89, 90, 92–96, 126–127
  temperature, 89, 90–93, 96–120,
    123–127
  thrifty genotype, 119–120, 123–126
Africa
  and adaptation, 95, 96, 119. *See also*
    Allen effect; Bergmann effect
  barriers and dispersal, 81
  barriers to primate distributions, 78
  biodiversity, 130, 186, 233, 235
  blood groups, 17–19
  climate, 57–63, 150
  competition, 208, 221–222, 226
  continent of origin, 15, 23–30, 111
  cultural diversity, 133, 212
  diet, 39, 194–198, 203
  disease, parasites, pathogens, 19,
    215–219, 221, 234, 235
  dispersal within, 30–31, 210
  genetic distance to other regions, 24–29
  geographic range size, 191
  hunter-gatherers, 39, 66, 150, 168, 191
  Madagascar, 43
  morphology, 20, 21
  nutrition, 109
  out of Africa. *See* Out of Africa
  population density, 191
  potato consumption, 194
  primates, 22, 23, 166, 186, 236
  pygmy peoples, 107–108, 116, 160

agriculture, 30, 37–38, 47, 60, 66, 74, 75,
140, 149, 171
Allen effect, 96–99, 103–113, 126–127. *See
also* adaptation
altitude, elevation
adaptation to, 89, 91, 120–123, 126
barrier to dispersal, 77, 79, 80, 139, 171,
172
Americas
adaptation, 121–123, 124–125
barriers and dispersal, 80, 81
biodiversity, 130, 186, 240–241
blood groups, 17–20
climate, 53, 74–75, 227, 230–233
cultural diversity, 130–134, 143–144,
145, 174, 185, 187
diet, nutrition, 83, 108, 201
disease, parasites, pathogens, 19, 188,
217, 219–221
dispersal, 43–46, 47, 49, 53, 74–75, 80
extermination, extinction, 208, 211–212,
228, 230–233, 235
genetic distance to other regions, 24–26
geographic range size, 169, 170, 174,
191, 209
hunter-gatherers, 44, 47, 83, 130, 132,
133, 134, 150, 152, 167–170, 185,
191
latitude and geographic range size, 167,
169
travel distances, 170
morphology, 20, 21
population density, 150, 185, 191
primates, 78, 166, 188, 235, 238
sex differences, 124–125
anatomy. *See* morphology
Andes, and adaptation to high altitudes, 89,
91, 121–123
animals. *See also* biodiversity, birds,
mammals; disease, parasites, pathogens;
primates (nonhuman)
Allen effect, 96, 99, 101
Bergmann effect, 96, 98–100
coat colors, 91–93
and dietary adaptation, 194
diversity and latitude (Forster effect), 90,
128–130, 134–146, 184–186
as evidence for dispersal, 48–49
extinction, 228–239
and geographic range size, 189
population size, 147
Rapoport effect, 165, 185
species-area relationship, 155, 178–180,
183

anthropology's value to biogeography, 2, 3
archeology
dispersal dates, routes, population sizes,
etc., 28–46
Arctic. *See* latitude
Arctic peoples (Eskimo, Inuit)
adaptation, 95–96, 227
animals, 117
cultural diversity
diet, 62, 64, 202–204
dispersal, distribution, 75
morphology, 101–104
physiology, 95–96, 117–118
technology, 62, 65
Ardipithecus, 58, 59, 63
area effects, 155–192
areography, 2
Asia
and adaptation, 121–122
barriers and dispersal, 81
biodiversity, 130, 179–180, 186, 231,
232, 238, 240, 241
blood groups, 18
climate, 37, 68, 69, 72–74, 150, 175
cultural diversity, 133, 134, 212
diet, 200
disease, parasites, pathogens, 189
dispersal, 31, 32, 33, 35–37, 41–43, 45,
49, 68, 72–74, 145
extermination, extinction, 212, 228, 231,
235–236
genetic distance to other regions, 24–26,
29, 34, 37
geographic range size, 191, 241
other *Homo* species, 31, 32, 33, 36, 51,
159–162
hunter-gatherers, 150, 168, 191
morphology, 20, 21, 96
physiology, 121–122
population density, 150, 151, 152, 191
primates, 22, 23, 78–79, 120, 145, 165,
166, 179, 186, 187, 238
pygmy peoples, 160
Australia. *See also* Tasmania
adaptation, 114, 117–118
biodiversity, 228, 229, 232, 233
blood groups, 18
climate, 150, 175, 230, 232, 233
dispersal, 34, 35, 38–40, 49, 50, 79, 80
extermination, extinction, 211, 228, 229,
230–236
genetics, 26, 34, 38, 40
geographic range size, 169, 185, 186,
191, 235, 236

hunter-gatherer, 150, 168, 191
language, 40
morphology, 40
population density, 191
technology, 181
Australopithecus, 4, 54, 59, 63

bacteria, distribution, 2, 49, 240, 241–242
barriers to dispersal and distribution,
    77–85
  and diversity, 80–84, 130, 176–177, 181,
    183–184
  geographic barriers, 33, 68, 77–81, 139,
    171, 172, 179, 184
  pathogens, 81
  and primates, 78–79
  and technology, 181, 183
  xenophobia, 81–85, 138
Bergmann effect, 96–103, 107–113,
    126–127. *See also* adaptation
Beringia. *See* Siberia
biodiversity, 128–130, 134–146, 172–178,
    184–188, 228–244
  area available, 141, 155, 178–180,
    183
  area used (Darwin-Rapoport effect),
    184–186
  barriers, 176, 177, 183–184
  climate, 90, 128–130, 134–146,
    172–178, 211, 230–233, 236–237
  community structure, 239, 228–230,
    234
  comparison with cultural diversity,
    129–130, 135–144, 172–178, 211,
    238
  competition, 143–144, 145, 173–175
  and disease, parasites, pathogens, 84, 92,
    136–138, 173–174, 189, 240
  extinction, 225, 228–239, 242–244
  geographic range size, 136, 137, 139,
    145, 172–178, 184–186, 225,
    228–242
  hotspots, 90, 139, 211, 237–238
  islands, 155, 178–180, 187–188
  isolation, 183–184
  latitude (Darwin-Rapoport effect),
    184–186
  latitude (Forster effect), 84, 90, 128–130,
    134–146, 155, 172–178, 184–188,
    211, 238
  latitude (Rapoport effect), 136, 139, 155,
    172–178
  population density (humans), 237–238
  and predation, 239

primates, 84, 91, 129–130, 139,
    140–141, 145, 157–158, 179,
    185–186
  productivity, 136, 139, 142, 146, 173,
    238
  regional-local, 145
  sampling error, 185, 238
  species-area relationship, 140–141, 155,
    178–180
biogeography
  field of enquiry, 1, 2
  medical relevance, 5
  value to anthropology, 3
birds, 72, 79, 92, 96, 98, 99, 117, 128, 156,
    185, 188, 210, 228–229, 240
blood groups, 17–19
body size
  Bergmann effect, 96–103, 107–113,
    126–127
  extinction, 225, 228–230, 232, 239
  island effect, 155–162
  primates, 98–100, 140, 155, 157–159,
    161
  pygmy peoples, 102–103, 107–108,
    160–162

climate. *See also* adaptation; dispersal;
    diversity; geographic range size; latitude;
    productivity
  blood groups, 19
  and dispersal, distribution, 59, 60, 63–75,
    79
  doubts about effects on origins and
    dispersal, 53, 55–56, 60–61, 70, 75
  and extinctions, 227–228, 230–233,
    236–237
  and hominin, *Homo*, human origins,
    57–66
  measures, 11
  and population density, size, 90,
    149–152, 185–186
  primate origins, 57–61
  and range change, 235–237
  sea levels, 33–34, 40, 69, 70, 75, 78,
    179
  volcanoes, 67, 69, 72
  warming, 57, 223
Clovis, 44
color and climate, sun, 5–7, 92–96
communication, barriers and diversity,
    80–84
competition, 203–212, 218–222, 225–239
  between human populations, 136, 143,
    145, 163, 207–213

competition *(continued)*
  between species, 136, 137, 145, 156–157,
    215, 221–222, 225–228
  and disease, 218–220
  and dispersal, out of Africa, 210–211,
    221–222
  and diversity, 84, 136–137, 139, 141,
    143–145, 173–174
  and diversity by latitude (Forster effect),
    136–137, 173–174
  and extermination, 208–209, 211–212,
    225–228, 234–236
  and geographic range size, 84, 137, 143,
    163–165, 170, 173–174, 207–213,
    215
  on islands, 156–157, 187–188
  and population density, size, 187–188
  xenophobia, 83
cooperation, 44, 70–71, 84
  and diversity, 140, 144
  and geographic range size, 84, 171, 207,
    212–213, 239–242
creationism, 1, 23
cultural diversity, 128–146, 172–178,
    180–185. *See also* hunter-gatherer
    society; individual continents
  area available (area effect), 130, 141,
    155, 180–183
  area used (Darwin-Rapoport effect), 185,
    187
  and barriers, 81–85, 130, 177, 184
  and biodiversity, 129–130, 135–144,
    172–178, 211, 238
  climate, 90, 128–146, 150–151,
    172–178
  competition, 84, 141–144, 173–175
  cooperation, 84, 140, 144
  and disease, parasites, pathogens, 83–84,
    136–138, 173–174, 189–190,
    218–219
  evolution, 145
  extinction, 207–209, 211–212
  geographic range size, 136, 137, 139,
    141, 143–145, 155, 172–178,
    184–187, 207–210, 211–212
  hotspots, 90, 130, 134, 139, 211
  hunter-gatherers, 130, 132, 134, 139,
    142–144
  impoverishment, 181, 183, 207–209
  islands, 80, 130, 155, 180–184,
    188–190
  isolation, 80, 184
  latitude (Darwin-Rapoport effect), 185,
    187

latitude (Forster effect), 90, 128–146,
    172–178
latitude (Rapoport effect), 136, 139,
    143–144, 172–178
population size (Tasmania effect), 181,
    183
and productivity, 90, 136, 139, 140,
    142–144, 146, 149–151, 173,
    175–176, 185–186, 238
regional-local influence, 145
sampling error, 144
of technology, tools, 71, 181–183
xenophobia, 83–84
culture. *See also* cultural diversity;
    hunter-gatherer society
  as barrier to mating, movement,
    81–85
  definitions, 9, 10
  and diet, 193–204
  and disease, 223
  diversity. *See* cultural diversity
  as evidence for dispersal, 25, 31, 40, 42,
    45, 46
  evolution, 145
  number members. *See* population size
  tools and dispersal, 62–65, 70–71

Darwin-Rapoport effect, 176, 184–187
Denisova, 32, 33, 34, 226
diet, 193–204
  adaptation to, 193–204
  alcohol, 199–201
  Arctic peoples, 202–204
  Asia, 199–201
  as barrier to movement, 83, 138
  fat, 202–204
  health, 193, 200–204
  hunter-gatherers, 39, 62–65
  milk (lactose, lactase), 194–198
  Neanderthal, 70, 227
  physiology, 193–204
  primates, 115, 194
  seaweed, 199
  skin color, 95–96
  starch (amylase), 198–199
disease, parasites, pathogens, 215–221
  barrier to dispersal, 81, 83, 215, 218–220
  blood groups, 19
  and coat color (Gloger effect), 92
  distribution, 5, 216–218, 241–242
  and diversity, 83–84, 92, 136–138, 173,
    174, 215–218
  as evidence for dispersal, 49
  and extermination, 221, 234

and geographic range size, 137, 173, 174,
   189, 215, 218–220, 240
HLA (human leucocyte antigen),
   218–219
on islands, 156, 188–190, 192
latitude, 84, 92
malarial diseases, 5, 19, 189, 215–217,
   219–220, 223
medical practice, 5
and mortality, 107
and physiology, 215–219
and population density, size 156,
   188–190, 221
primates, 84, 189
and territoriality, xenophobia, 83–84
trypanosomiasis (sleeping sickness),
   219–220
yellow fever, 219–220
dispersal, 15–51. *See also* geographic range
   size; out of Africa
   into Africa, 35
   within Africa, 30–31, 72
   into Americas, 43–46
   into Asia, 36–37
   barriers, 33, 68, 77–85
   climate, 66–76
   culture, language with genes, 31, 38
   and diet, habitat, 39, 63–66, 68
   and disease, parasites, pathogens,
      218–220, 240
   into Europe, 37–38, 111
   global, 33–50, 212–213
   introduced species, invasives, 49–50,
      239–241
   into Oceania (Pacific), 40–43, 46
   other species as evidence, 46, 48–50, 240
   into Sahul, 38–40
   sex differences, 43, 47–48
   speeds, 39, 44
   technology, tools, 62–65, 70–71
diversity. *See* biodiveristy; cultural diversity
drugs, 193, 201–202

environmental influence, 7, 19, 29. *See also*
   adaptation; climate
Eskimo. *See* Arctic peoples
Eurasia. *See* Asia, Europe
Europe
   adaptation, 95, 111, 114–115, 117–119,
      195–199
   barriers, 79–81, 83
   blood groups, 18
   climate, 28, 67, 72–74, 175
   cultural diversity, 133

diet, 83, 195–198
disease, parasites, pathogens, 217–220
dispersal, 28, 31, 37–38, 50, 72–74, 111,
   219
extermination, extinction, 227–228,
   230–232
genetic distance to other regions, 24–26,
   29
hunter-gatherers, 38
Neanderthal, 32, 67, 68, 111, 227–228
other *Homo* species (excluding
   Neanderthal), 31, 32, 67, 227
physiology, 219
technology, 171
extermination, extinction
   biodiversity, 187–188, 225–239,
      242–244
   body size, 228–230, 232, 239
   climate, 227–228, 230–233, 236–237
   cultures, 207–209, 211–212
   and disease, 234
   Homo (nonhuman), 226–228,
      242–244
   of human species, 242–244
   Neanderthal, 70, 161, 226–228

Floresiensis, 155, 159–162, 192
   pathology, 161–162
Forster effect, 90, 128–146, 155, 176–178.
   See also animals; biodiversity; cultural
   diversity; mammals; plants; primates
   (nonhuman)

genes, genetics, genotype
   and adaptation, 96, 115–117, 122
   and African origins, 23–30
   ancestry, history, origins, 15–30, 33–43,
      45–50, 69, 71–73
   bottlenecks, 27, 28
   and dispersal dates and routes, 15–17,
      33–43, 45–50, 69, 71–73
   drift, 16–17, 21–22, 95, 96, 110, 201
   founder effects, 16, 17, 21
   molecular clock, 50–51
   population homogeneity, variability, size,
      6, 24, 28, 36–37, 69, 71, 72
   thrifty genotype, 119–120
geographical ecology, 2
geographic range size, 31–46, 162–187,
   189, 190–192
   and biodiversity, 136, 137, 139, 145,
      172–178, 184–186
   climate, 155, 167, 169–170, 171, 175,
      177, 185–186

geographic range size *(continued)*
competition, 84, 137, 143, 163–165, 170,
173–174, 207–213
contraction, 70, 207–212, 235–237
cooperation, 84, 171, 207, 212–213
and cultural diversity, 136, 137, 139,
141, 143–145, 155, 172–178,
184–187, 207–210, 211–212
and disease, parasites, pathogens, 189,
218–221, 240
distances moved, 150–151, 169–170
expansion, 31–46, 70–71, 207, 210,
212–213, 239–241. *See also* out of
Africa
extinction. *See* extermination, extinction
hunter-gatherers, 143, 151, 152,
163–168, 170, 173–176, 190–191
introduced species, invasives, 49–50,
239–241
languages, 164, 166–169, 171, 173–176,
184–187
latitude (Darwin-Rapoport effect), 176,
184–187
latitude (Forster effect), 136, 155,
176–178
latitude (Rapoport effect), 136, 151, 155,
165–173, 176–178
Neanderthal, 32, 33, 70, 226–227
population density, 190–192
primates, 155, 157–158, 162–163,
165–166, 179, 185–186
productivity, 143, 149–151, 167, 169,
173, 175–176, 185–186
rarity, 163–164, 211–212
skew, 163–165, 211–212
territoriality, xenophobia, 84, 170, 208
and variability, 165–167
glacials, inter-glacials. *See* climate
Gloger effect, 92

health, medicine, 5, 194–204, 215–221
hominid, hominin, human taxonomy, 4, 32
Homo species (nonhumans), 30–33, 34, 36,
38, 51, 54, 58–63, 66–70, 97, 111–113,
155, 159–162, 226–228, 229, 236,
242–244
hunter-gatherer society, 9, 38, 47, 83. *See
also* cultural diversity; cultures
Bergmann effect, morphology, 101
climate, 101, 130, 134, 150, 151
density, 66, 90, 143, 150–152
diet, 62, 64–65, 83
distances moved, 39, 44, 150–151,
diversity, 130, 132, 134, 142–144

geographic range size, 143, 151, 152,
163–165, 167–168, 170, 173–176,
190–191
latitude, 101, 130, 134, 150–152
population density, size, 147–148,
150–152, 185, 188, 191
technology, tools, 62, 64–65, 71

immigration, 145
imperialism, 141, 145, 163–165
inbreeding, incest, 82, 84
India, 26, 33, 35
Inuit. *See* Arctic peoples
island effect (body size), 155–162
animals, birds, mammals, primates,
156–159
and competition, 156–157
Flores "hobbit," Floresiensis, 155,
159–162
and predation, 157
primates, 157–159
islands,
body size (island effect), 155–162
diversity, 155, 178–184
disease, parasites, pathogens, 188–1190
isolation, 181–1184
physiology, 119–1120
population density, 187–1190
thrifty genotype, 119–1120

Janzen mountain pass, 79, 80–181, 171,
176, 177

language. *See also* cultural diversity;
cultures; hunter-gatherer society
and area available, 180–1182
barrier, 77, 80, 82–184
as culture, 9, 10
Darwin-Rapoport, 185, 187
diversity, 25, 129–134, 137, 138, 140, 141,
144, 145, 147–149, 152, 173–176,
180–182, 184, 185, 207–210
evidence for dispersal, 40, 42, 43, 45–46,
48
extinction, 31, 211–212, 277
and genes, 31
geographic range size, 164, 166–169,
171, 173–177, 184–187, 211–212
islands, 80, 130, 180–184, 188
and latitude (Forster effect), 129–134,
152, 173–177
and latitude (Rapoport effect), 166–169,
173, 176–177
rarity, 211–212

latitude, 95–120, 128–146, 149–152, 172–178. *See also* adaptation; climate; productivity
   Allen effect, 96–99, 103–113, 126–127
   anatomy, morphology, 89, 91–113, 123–127
   Bergmann effect, 96–103, 107–113, 126–127
   Darwin-Rapoport effect, 176, 184–187
   dispersal out of Africa, 53, 72
   diversity cultures, 89–90, 128–129, 130–146, 155, 172–178
   diversity species, 128–130, 134–142, 145, 172–178
   geographic range size (Rapoport effect), 136, 139, 155, 165–173, 176–178, 184–188
   Gloger effect, 92
   physiology, 89, 113–119
   and population density, size, 149–152
   productivity, 90, 136, 139, 142–144, 146, 149–151
   as a valid correlate, 129
Levant. *See* Middle East

macroecology, 2
Madagascar
   barriers, 78, 80
   body size, 230, 234
   climate, 232, 233
   dispersal, 43, 80
   diversity, 185, 186, 187
   extermination, extinction, 229, 230, 233, 234
   geographic range size, 166
   primates, 22, 23, 166, 185–187, 229, 230, 233, 234
malaria, 5, 19, 189, 215–217, 219–220, 223
mammals. *See also* primates (nonhuman)
   adaptation, Allen effects, Bergmann effects, 98–101, 107, 120
   adaptation, coat colors, 91–93
   bovids as evidence for evolution, 54, 61–62, 66–67
   disease, parasites, pathogens, 84, 137, 189
   diversity and latitude, 84, 91, 128–130, 137, 140–141
   as evidence for dispersal, 48–50
   extinction, 228–239
   geographic range size, 137, 162–163, 165–166, 240
   Gloger effect, 92
   influence on dispersal, 221–222

measles, 189–190, 221
medicine, health, 5, 194–204, 215–221
Melanesia. *See* Oceania
methods, 9–11. *See also* sampling error
Micronesia. *See* Oceania
Middle East
   climate and dispersal, 66–71
   genetic distance to other regions, 26
   route out of Africa, 33–34
   morphology, 95–113, 155–162. *See also* Allen effect; Bergmann effect; body size; Gloger effect; sex differences
   Australian, 40
   ear wax, 20
   as evidence for African origins, 24
   finger prints, 20, 21
   head shape, 21
multi-regional hypothesis, 24, 29

Neanderthal
   activity, strength, 112–113
   adaptations to cold, 22, 111–113
   climate, 59, 67, 68, 70, 73, 111–113
   competition with humans, 208, 226–228, 234
   diet, 70, 227
   distribution, 32, 33, 67, 68, 70, 73, 113, 226–227
   duration, 243
   extinction, 70, 161, 225–228, 234
   morphology, 21, 22, 68, 111–113
   pathology, 161
   origin, 32, 59, 67
   technology, tools, 68, 70, 73, 113
New Guinea. *See also* Oceania
   barriers to dispersal, 82
   blood groups, 17–18
   cultural diversity, 130, 139
   disease, parasites, pathogens, 217
   dispersal to, 33–34, 38–42, 49, 175
   genetics, 26, 34, 38, 40–41
   geography, 41
   language, 42, 130
   morphology, 124
New Zealand. *See also* Oceania
   dispersal to, 46, 49, 50
   extermination, extinction, 229, 233
   introduced species, 240, 241
nutrition. *See* adaptation, diet

Oceania (Melanesia, Micronesia, Polynesia, Pacific islands)
   archeology, 41
   barriers, 82–83

Oceania *(continued)*
  blood groups, 18
  cultural diversity, 133, 180, 182
  diet, 203
  disease, parasites, pathogens, 189, 221
  dispersal, 40–43, 46, 49–50, 80, 82–83
  extermination, extinction, 212, 221
  genetics, 24–26, 29, 40–42
  geography, 41
  language, 42, 168, 180, 212
  morphology, 20
  physiology, thrifty genotype, 119–120
oceans, seas, as barriers, 77–80
origins, 17–51, 54–62. *See also* out of
    Africa
  Africa as continent of, 23–30
  of Asians, 36–37
  of Australians, New Guineans, Tasma-
      nians, 38–40
  of Austronesian speakers, 40–43, 49–50
  and climate, 53–66
  dates. *See* regions
  of Europeans, 37–38
  and grassland, 63–66
  of Malagasay, 43
  multi-regional hypothesis, 24, 29
  of Native Americans, 43–46
  and population size (bottleneck), 27–28
  primates, 22, 23, 57–61
Orrorin, 4, 58
out of Africa, 15, 23–35. *See also* dispersal;
    geographic range size
  barriers, 33, 68
  climate, 53, 66–72, 175
  competition, 210–211
  dates, 15, 31–35, 54, 67–71
  disease, parasites, pathogens, 49
  evidence, 15–21, 23–30, 31–35, 48–49,
      54–56
  mammals, 48–49, 66, 221–222
  other *Homo* species, 31–32, 36, 53–54,
      66–67
  population size, 27–28, 72, 210
  routes, 15, 31–35

Pacific islands. *See* Oceania
Paranthropus, 4, 54, 59, 62, 63
parasites, pathogens. *See* disease, parasites,
    pathogens
phylogeography, 1, 21
physiology, 113–127, 193–204
  adaptation to climate, temperature, 89,
      113–127
  and diet, 193–204

  and disease, parasites, pathogens, 5, 19,
      201–202, 215–219
  as evidence for African origins, 24
plague, 221
plants, 46, 62, 65, 74, 128, 129, 138, 142,
    165, 175, 178, 185, 194, 199, 229,
    240–241
Polynesia. *See* Oceania
population density, size, 147–153,
    185–192
  and biodiversity, 237–238
  bottlenecks, 27, 28
  climate, 69, 72, 90, 149–152
  competition, 187–188, 209–210
  contraction, extinction, 207–212, 221,
      237–239
  cultural loss, 183
  and disease, parasites, pathogens, 156,
      188–190, 215, 221
  founding, original, 27–28, 72, 82, 210
  geographical influences (not climate,
      latitude), 66, 90, 149
  and geographic range size, 190–192
  on islands, 156, 187–190
  latitude, 90, 149–152, 176
  and out of Africa, 27–28, 72, 210
  primates, 147, 187–191
  pygmy peoples, 66
  rarity, 147–149
  skew, 90, 147–149
predators,
  and dispersal, out of Africa, 221–222
  extinction, 239
  on islands, 156–157, 159, 187–188
prey, 221–222
primates (nonhuman)
  adaptation, 120
  Allen, Bergmann effects, 99–101
  barriers, 78–79, 82
  body size, 97–100, 140, 155, 157–159,
      161
  climate and origins, 57–61
  coat color, 91, 92
  competition between species, 145
  diet, 194
  disease, parasites, pathogens, 84, 189
  dispersal, 63, 78–79
  diversity and area, 179
  diversity and geographic range size
      (Darwin-Rapoport effect), 185–186
  diversity and latitude (Forster effect), 84,
      91, 129–130, 140–141
  diversity hotspot, 139
  extinction, 229–230, 237

genetic variability, 6
geographic range size, 162–163,
    165–166, 240
on islands, 155, 157–158, 179, 187–188
latitude and geographic range size,
    165–166
latitude and morphology, 98–101, 140
latitude and parasites, 84
latitude and variability, 165–167
origins, 22–23
    population density, size, 147, 149,
        151, 187–188, 190, 191
    Rapoport effect, 165–166
    taxonomy, 4
productivity. *See also* climate; latitude
and diversity, 90, 128, 136 139, 142–144,
    146, 149–151, 173, 175–176,
    185–186, 237–238
and geographic range size, 143, 149–151,
    167, 169, 173, 175–176, 185–186
measures, 142
and population density, size, 149–151,
    238
pygmy peoples, 31, 66, 102–103, 107–108,
    116, 160–162

race, racism, 5–7, 85, 119
Rapoport effect, 136, 139, 155, 165–173,
    176–178
rarity, 147–149, 153, 155, 162–164,
    211–212

Sahelanthropus, 4, 58
Sahul. *See* Australia; New Guinea; Tasmania
sampling error, lack, 50, 53, 54–55, 144,
    185, 190, 238, 242. *See also* methods
sea levels, 33–34, 40, 69, 78, 161, 181
sex differences
adaptations to energy balance, 89,
    123–127
chivalry, 126
dispersal, 43, 47–48
Shakespeare, 147

Siberia
    climate, 73, 74
    dispersal into, 37, 45, 53, 73–74
    dispersal into Americas, 43–46, 53,
        74–75
    extermination, extinction, 211, 236
    genetics, 26, 73
    Homo (nonhuman), 34, 68, 73, 226
    language, 46
    Neanderthal, 68, 73
    origin of Amerindians, Native Americans,
        27, 43–46, 74
sickle cell anemia, thalassemia, 19, 216–217
skin color. *See* adaptation
society. *See* cultural diversity; culture;
    hunter-gatherer society; language
speciation, 128, 135–139, 173, 177
species. *See* animal; biodiversity; bird,
    disease, parasite, pathogen; diversity;
    Homo species (nonhuman); mammal;
    plant; primate (nonhuman); shellfish
species-area relationship, 155, 178–180

Tasmania. *See also* Australia
cultural poverty (Tasmania effect), 181,
    183
dispersal to, 38–40
extermination, extinction, 208, 209, 236
technology, tools, 62–65, 69, 70–71, 101,
    181–183
temperature. *See* adaptation; Allen effect;
    Bergmann effect; climate; geographic
    range size; latitude; productivity
territoriality, 170, 176, 208. *See also*
    competition; xenophobia
tropics. *See* latitude
Tibet, and adaptation to high altitudes, 89,
    91, 121–122

xenophobia, 5–7, 77, 81–85, 119, 170. *See
    also* competition; territoriality

yellow fever, 219–220

Author Index

Agetsuma, N., 99
Agustí, J., 55, 73
Aiello, L. C., 160, 162
Aldenderfer, M. S., 121
Allen, J. A., 96, 111
Alroy, J., 232, 233
Altizer, S. H., 189
Ambrose, S. H., 69, 71, 72
Amos, W., 27, 72
Anderson, R. M., 221
Anikovich, M. V., 37
Archibald, J. D., 50
Archibold, O. W., 62
Argue, D., 160, 161
Arita, H. T., 190
Armitage, S. J., 32, 55, 67, 68
Arseniev, V. K., 211
Atkinson, Q. D., 25, 27
Austin, C. C., 50
Avise, J. C., 1
Ayres, J. M., 78

Baab, K. L., 160
Bailey, G. N., 62
Bailey, R. C., 66, 107
Baker, P. T., 114, 117, 118, 123, 124, 189,
    190, 203, 209, 221
Balmford, A., 139, 238
Banning, C., 125, 126
Barbujani, G., 38, 80, 83
Barlow, N. D., 71

Barnosky, A. D., 231, 232, 233
Barraclough, G., 208
Barthell, J. F., 241
Barton, R. N. E., 30
Basell, L.S ., 55, 63, 67, 72
Baz, A., 183
Beall, C. M., 119, 121, 122
Beals, K. L., 110
Beard, K. C., 22
Beehner, J. C., 120
Bell, G., 135, 147
Bennett, A., 122, 123
Bennett, P. M., 185, 232
Benson, L., 71
Benton, M. J., 22
Berger, L. R., 54, 61
Bergmann, C., 96, 98, 111
Bindon, J. R., 5, 98, 119, 120
Binford, L. R., 9, 10, 11, 39, 44, 62, 64, 66,
    101, 130, 132, 134, 142, 143, 144, 147,
    148, 150, 151, 152, 164, 166, 167, 168,
    169, 170, 174, 186, 188, 190, 191, 203,
    204, 210, 239
Bininda-Emonds, O. R. P., 122
Bird, D. W., 39
Bischof, N., 82
Blackburn, T. M., 97, 112, 128, 129, 241
Bliege Bird, R. L., 229, 232
Boeskorov, G. G., 237
Bogin, B., 108
Böhm, M., 2, 129, 141

Bomford, M., 240
Bonnefille, R., 55, 60, 63
Borregaard, M. K., 190, 191
Boulton, A. M., 241
Bouzouggar, A., 71
Bowler, J. M., 35, 38
Bradley, B. J., 91, 92
Bradley, S. B., 97, 98
Bramanti, B., 38, 74
Bramble, D. M., 63
Brandon-Jones, D., 63, 78
Brès, P. L. J., 81
Brierley, C. M., 60
Brillat-Savarin, J. A., 193, 194
Broadhurst, C. L., 39
Bromage, T. B., 54, 60, 63
Bromham, L., 22, 157, 158, 161
Brook, B. W., 229, 230, 233, 234
Brown, J. H., 2, 90, 137, 190, 232, 237
Brown, P., 31, 54, 159, 160, 161
Browne, J., 1, 2
Brumm, A., 160, 161
Brunet, M., 58
Burney, D. A., 43, 80, 228, 229, 230, 232, 233, 234
Burnham, K. P., 11

Cahill, M., 213
Campbell, D. G., 240
Campbell, M. C., 5, 28, 30, 34, 201, 204
Cann, R. L., 23, 25, 26, 35
Carbone, C., 157
Cardillo, M., 229, 239
Carey, J. W., 110
Carto, S. L., 55, 67
Cashdan, E., 9, 10, 81, 129, 137, 138, 140, 142–143
Caughley, G., 228, 229, 233, 236
Cavalli-Sforza, L. L., 2, 3, 5, 17, 18, 19, 23, 24, 25, 26, 28, 30, 31, 35, 36, 37, 38, 40, 41, 42, 43, 45, 46, 74, 80, 83, 102, 103, 210, 218
Chakraborty, D., 99, 101
Channell, R., 209, 235, 236
Chaplin, G., 94, 95, 96
Chase, J. M., 142
Chatters, J. C., 208
Chen, F. C., 58
Cherry-Garrard, A. G. B., 119
Chiaroni, J., 16, 17
Chivers, D. J., 194
Churchill, W., 208
CIESIN, 147
Cincotta, R. P., 238

Ciochon, R. L., 36, 51, 63
Colinvaux, P. A., 63
Collard, I. F., 129, 133, 134, 139, 142–143, 144, 149, 150, 151, 166, 168, 176
Coller, M., 34, 78
Colyn, M., 233
Committee on the Earth System Context for Hominin Evolution, 55
Cooper, A., 229
Coppeto, S. A., 185
Cordain, L., 202, 203
Corvinus, G., 243
Costello, E. K., 2, 242
Cotgreave, P., 185
Cowlishaw, G., 129, 143, 162, 165, 179, 190, 236
Cox, C. B., 128
Cox, M. P., 42
Crawford, M. H., 43
Currie, T. E., 141, 142, 144, 152, 166, 169, 171, 175, 176, 178, 186, 209
Curtin, P. D., 221

Dalby, A., 211
Darwin, C., 5, 6, 16, 23, 43, 49, 77, 79, 81, 82, 89, 90, 91, 96, 184, 187, 207
de Filippo, C., 47
de Lumley, M-A., 31, 36
de Vos, J., 161
deMenocal, P. B., 54, 55, 60, 61, 62, 63
Dennell, R., 31, 49
Destexhe, A., 208
Dewar, R. E., 43, 80, 229
Diamond, J. M., 130, 175, 177, 208, 209, 221, 229, 241
Dillehay, T. D., 44
Disotell, T. R., 23, 27, 30, 33, 34, 35, 36
Ditlevsen, P. D., 55, 68, 69
Dodo, Y., 113
Doherty, D. A., 185
Donoghue, P. C. J., 50
Dow, M. M., 82
Draper, H. H., 202, 203
Driscoll, C. A., 241
Dunbar, R. I. M., 99, 120, 237
Duncan, J. R., 238
Durham, W. H., 5, 96, 194, 195, 197

Eckhardt, R. B., 162
Eeley, H. A. C., 162, 165, 190
Efron, B., 147
Egan, T., 71
Elton, C. S., 240
Emmons, L. H., 187

Enattah, N. S., 197
Endicott, P., 50, 51
Enserink, M., 223
Erlandson, J. M., 46
Eshleman, J. A., 43, 45, 46
Estrada-Mena, B., 20

Fagan, B., 71
Fahrig, L., 178, 184
Faith, J. T., 228, 231
Falk, D., 160, 162
Falush, D., 30, 49, 240
FAO, 194
Fernandez, M. H., 97
Field, J. S., 33, 34, 69
Finarelli, J. A., 22, 69
Finch, V. A., 92, 93
Fincher, C. L., 83, 137, 138, 218
Finlayson, C., 23, 30, 31, 32, 33, 34, 36, 37,
    55, 56, 60, 62, 63, 68, 70, 71, 72, 73,
    74, 111, 112, 113, 161, 171, 226, 227
Fitzhugh, W. W., 3, 46, 75
Flannery, T., 158
Fleagle, J. G., 23, 30, 31, 57, 97, 229
Foley, R. A., 3, 30, 35, 53, 54, 55, 57, 58,
    61, 62, 67, 72, 142–143, 144, 151, 221,
    222, 243
Fooden, J., 99, 101, 105
Ford, J., 219
Forster, J. R., 128, 178
Forster, P., 28, 29, 30, 33, 34, 35, 37, 60, 69,
    72, 73
French, A. R., 157, 239
Frey, B. S., 126
Frisancho, A. R., 5, 97, 108, 109, 114, 115,
    117, 118, 121, 123
Froehle, A. W., 117
Frumkin, A., 67, 69, 70

Gage, K. L., 240
Gagneux, P., 6, 24
Garcia-Martin, E., 201, 204
Garn, S. M., 123
Garrard, A. N., 222, 226, 234
Gaston, K. J., 2, 128, 139, 165, 176, 177,
    190, 191, 243
George, D., 44
Ghalambor, C. K., 79, 114, 177
Gilbert, M. T. P., 44
Gill, J. L., 231, 235
Gillespie, R., 38, 41, 50, 228, 229
Gillooly, J. F., 50
GlobalSecurity.org, 208, 210
Gloger, C. W. L., 92

God, 207
Goddard, I., 130, 133, 134, 144, 167, 169,
    174, 185, 187, 211
Godfrey, L. R., 188, 229, 230, 231, 232,
    233
Goebel, T., 37, 43, 44, 45, 46, 68, 73, 74, 75
Goedde, H. W., 200, 201
Goodman, S. M., 78, 229
Gordon, R. G., 9, 80, 129, 130, 131, 132,
    133, 145, 147, 148, 149, 152, 164, 166,
    167, 168, 174, 180, 181, 182, 184, 188,
    211, 212
Gravlee, C. C., 21
Gray, R. D., 41, 42, 43, 80
Grayson, D. K., 124, 125, 228, 229, 230,
    232, 233, 234, 242
Green, R. E., 226
Greenberg, J. H., 45
Greksa, L. P., 121
Groube, L., 39
Groves, C. P., 4, 78
Grubb, P., 48, 63, 78
Guernier, V., 137
Gunz, P., 30

Haak, W., 38, 74
Hage, P., 43, 48
Hall, R., 46, 75, 110
Hammer, M. F., 25, 28, 34, 35
Hancock, A. M., 5, 19, 96, 115, 117, 194,
    198, 199, 204
Harcourt, A. H., 63, 79, 84, 99, 100, 112,
    128, 129, 130, 139, 140, 143, 149, 150,
    151, 158, 159, 162, 163, 165, 166, 167,
    172, 176, 179, 184, 185, 186, 187, 190,
    191, 194, 208, 229, 230, 236, 238, 239
Harlan, J. R., 241
Harmon, D., 9, 129, 142, 211
Harpending, H., 27
Harris, E. E., 16, 195, 197, 199
Harris, L. D., 178
Harrison, G. A., 5, 18, 21, 216, 217, 218, 219
Harrison, K. D., 211
Harrison, T., 23, 58, 179
Harvey, P. H., 108, 178, 190
Haug, G. H., 71
Hawkes, K., 44, 198
Hawkins, B. A., 129
Heads, M., 22
Heaney, L. R., 63, 78, 79, 158, 179, 180
Heesy, C. P., 1, 22
Hehemann, J. E., 199
Henderson, D. A., 240
Henn, B. M., 30

Henrich, J., 181, 183
Hershkovitz, I., 161
Hibbert, C., 208
Hibbs, D. A., 238
Hiernaux, J., 159
High Level Committee on the Indian
  Diaspora, 145
Hill, K. R., 47
Hillebrand, H., 128
Hinde, R. A., 244
Hintjens, H. M., 208
Hoberg, E. P., 240
Hockett, B., 227
Hodgkinson, A., 50
Hoffecker, J. F., 37, 46, 68, 70, 71, 73, 74,
  75, 111, 171
Hofmeyr, M. D., 92, 93
Holden, C., 195, 197
Holliday, T. W., 68, 103, 104, 105, 111, 113
Holton, N. E., 113
Holy Bible, 207
Hooper, E., 240
Hrdy, S. B., 44, 227
Hubbell, S. P., 135, 147
Hublin, J. J., 31
Hudjashov, G., 38, 40, 80
Huffman, O. F., 36
HUGO Pan-Asian SNP Consortium, 2, 36
Hunter, J. P., 22
Huntford, R., 119
Huxley, T. H., 1, 3, 23, 51

Ingram, C. J. E., 195, 196, 197, 199
Institute of Historical Research, 145
Isaac, N. J. B., 229, 232, 239
IUCN, 228, 235

Jablonski, D., 144, 172
Jablonski, N. G., 57, 94, 95
Jacob, T., 161
Jacobs, Z., 71
James, F. C., 98, 99
Janson, C. H., 170
Janzen, D. H., 79, 114, 117, 140, 171, 176,
  177, 237
Jernvall, J., 239
Jobling, M. A., 27, 30, 35, 41, 42
Johnson, C. N., 229, 230, 234, 235, 236
Johnson, R., 241
Johnson, S., 83
Jolly, A., 229, 230
Jolly, D., 233
Jones, C. B., 162
Joordens, J. C. A., 39

Jordan, F. M., 48
Jungers, W. L., 162

Kamilar, J. M., 92
Kaneko, A., 189
Kari, J., 46
Karim, A. K. M. B., 201, 202
Kayser, M., 43, 48
Kirch, P. V., 40, 41, 43, 228
Kirkpatrick, R. C., 99
Klein, R. G., 4, 23, 28, 30, 31, 32, 36, 39,
  40, 43, 58, 60, 63, 71, 72, 73, 111, 227,
  230
Kline, M. A., 183
Knowles, L. L., 29
Koch, P. L., 228, 230, 231, 233
Köhler, M., 161
Korkea-Aho, T. L., 188
Krause, J., 32, 33, 34, 54, 226
Krebs, J. R., 44
Kroeber, A. L., 170
Kruess, A., 188
Kuklick, H., 3
Kumar, S., 58
Kunin, W. E., 163
Kupfer, J. A., 179
Kurlansky, M., 115
Kuzmina, E. E., 37

Ladle, R. J., 3
Lahr, M. M., 23, 27, 28, 30, 31, 32, 33, 35,
  36, 48, 55, 60, 68, 72
Lai, Y-C., 92
Lande, R., 16, 17
Laurance, W. F., 178, 180, 184
Lavely, W., 81
Lawes, M. J., 145
Lawton, J. H., 228
LeJeune, K. D., 241
Lell, J. T., 45
Lentz, C., 201
Leonard, W. R., 117, 123, 124
Lewis, M. P., 9, 211, 212
Li, J. Z., 23, 25, 27
Lindstedt, A. L., 97
Liu, H., 25, 27
Liu, W., 32, 216
Loh, J., 130
Lomolino, M. V., 1, 2, 3, 11, 34, 92, 96,
  128, 135, 137, 138, 145, 156, 157, 158,
  161, 174, 175, 178, 184, 187, 232, 235,
  236, 237, 241, 242
Lowen, C., 170
Luck, G. W., 238

Lum, C. L., 20
Lyell, C., 183, 184

MacArthur, R. H., 2, 62, 92, 137, 178, 184
Macaulay, V., 33, 34, 36, 37, 39
Mace, G. M., 235, 236
Mace, R., 129, 134, 138, 163, 166
MacPhee, R. D. E., 228, 233, 234
Maffi, L., 9, 129, 138, 211
Magalon, H., 202
Malhi, R. S., 47
Malthus, T. R., 238
Mandryk, C. A. S., 75
Manne, L. L., 129, 141, 145, 172, 176
Marcello, A., 39
Marler, P., 83
Marlowe, F. W., 47, 62, 106, 202, 203, 222
Marquet, P. A., 147
Marr, K. L., 74
Marsh, L. K., 179
Martin, P. S., 228, 230, 232, 236
Martin, R. D., 22, 57, 161, 162
Martínez-Navarro, B., 32
Martinón-Torres, M., 31
Masters, J. C., 22, 43
Matisoo-Smith, E., 49
Matthysen, E., 210
Maudlin, I., 215, 219
McCain, C. M., 79, 114, 171, 177
McComb, K., 209
McCurdy, S. A., 124
McDougall, I., 30, 51
McHenry, H. M., 97, 98, 159, 160, 161
McMahon, T. A., 106
McNab, B. K., 90, 97, 99, 107, 117, 119,
    120, 156, 157
McNeill, J. R., 81, 209, 219, 220
Meijaard, E., 78, 79
Meijer, H. J. M., 160
Meiri, S., 96, 98, 99, 107, 156, 157,
    159, 161
Mellars, P., 31, 32, 37, 69, 71, 72, 183, 227
Mercador, J., 66
Messier, W., 194
Meyer, D., 2, 200, 201, 202
Mielke, J. H., 5, 18, 19, 20, 21, 24, 25, 28,
    41, 80, 95, 97, 98, 110, 216, 218
Migliano, A. B., 107, 108
Mijares, A. S., 38
Milanich, J. T., 170
Miller, G. H., 228, 229, 232, 233, 235
Milton, K., 83, 84, 138
Mitani, J. C., 170
Mithen, S., 55, 56

Mittermeier, R. A., 78
Molnar, S., 5, 19, 21, 80, 95, 101, 110, 120,
    121, 217
Moodley, Y., 49
Moore, J. L., 129, 139, 141, 176
Moore, L. G., 121, 122
Moran, E. E., 114, 115, 117
Morgan, G. S., 229
Moritz, C., 57, 232, 237
Morwood, M. J., 31, 159, 160, 161, 162, 192
Mourre, V., 71
Munguía, M., 137
Munoz, S. E., 56
Murphy, W. A., 208

Nash, L. T., 3
Nee, S., 135, 147, 148
Nelson, B. W., 185, 238
Nettle, D., 10, 84, 129, 140, 143, 144, 171,
    208, 211, 212
Nijman, V., 79, 179
Nitecki, M. H., 228
Niven, J. E., 160
Nogués-Bravo, D., 231, 237
Norton, H. L., 96
Nudds, R. L., 99
Nunn, C. L., 84, 92, 137, 189

Obendorf, P. J., 161
O'Connell, J. F., 38
Odling-Smee, F. J., 198
Ognjanovic, S., 201, 202
Okie, J. G., 156
O'Meara, M., 149
Oppenheimer, S., 23, 31, 33, 34, 36, 37, 38,
    40, 41, 43, 44, 45, 46, 67, 68, 69
Osborne, A. H., 67
Owens, I. P. F., 229, 232

Parfitt, S. A., 31
Parks, S. A., 184, 238
Parmesan, C., 57, 232, 237
Partridge, T., 62, 63
Passey, B. H., 114
Patterson, W. P., 71
Pearson, O. M., 23, 29, 30, 40, 111
Peng, Y., 200
Peres, C. A., 162, 188, 229, 232, 239
Perez, V. R., 229
Perry, G. H., 102, 103, 107, 108, 159,
    160, 198
Petraglia, M. D., 32, 33, 69
Pickford, M., 62, 63, 94, 128, 129, 162
Pimentel, D., 240

Pimm, S., 233
Pitulko, V. V., 37, 45, 73
Plowright, W., 234
Pollack, M. R., 217
Pope, K. O., 33, 36, 37, 38, 67, 68, 69, 73
Potts, R., 62
Poulakakis, N. A. P., 156
Prebble, M., 229
Premo, L. S., 27, 82
Preston, M. A., 222, 228
Prugh, L. R., 184
Prugnolle, F., 25
Pulquério, M. J. F., 50
Purvis, A., 143, 229

Qin, Z., 121
Quammen, D., 229
Quintana-Murci, L., 47

Rabbie, J. M., 5, 85
Rae, T. C., 113
Rahmstorf, S., 237
Raichlen, D. A., 112
Ralls, K., 82
Ramachandran, S., 2, 24, 30
Rapoport, E. H., 2, 155, 156, 165, 166, 171, 174, 176, 185, 187
Rasmussen, M., 43, 46
Raxworthy, C. J., 237
Reddy, S., 185, 238
Reed, D. L., 240
Reed, F. A., 30
Reed, K. E., 63, 179
Reich, D., 32, 33, 34, 226
Relethford, J. H., 5, 17, 19, 21, 24, 29, 38, 110, 216, 223
Reséndez, A., 170, 208
Richard, A. F., 229, 230
Richards, C. L., 1
Richards, G. D., 159, 160, 161
Richards, M., 42, 43
Richards, M. P., 227
Richerson, P. J., 62, 73, 183
Richter, D., 30
Rightmire, G. P., 30, 31
Roberts, D. F., 96, 97, 100, 103, 105, 110, 114, 117, 123
Roberts, N., 55
Roberts, R. G., 233
Robins, A. H., 95
Rogers, R. A., 46
Rohde, K., 141, 165
Rongo, T., 220
Ronquist, F., 1

Rosenberg, N. A., 6, 29
Rosenzweig, M. L., 130, 140–141, 142, 172, 180
Ross, C., 91
Rossie, J. B., 1, 22
Rothman, J. M., 115
Roulin, A., 92
Rowe, N., 78, 91
Ruff, C., 97, 100, 101, 102, 103, 104, 105, 106, 108, 111, 123
Ruggiero, A., 139, 165
Rylands, A. B., 78

Savolainen, P., 40, 50
Sax, D. F., 73, 241, 243
Scheinfeldt, L. B., 30, 31, 35
Schmidt-Nielsen, K., 90, 120
Schoenemann, P. T., 107
Scholander, P. F., 97
Schotterer, U., 37
Schroeder, K. B., 43
Schulte, P., 54
Searle, J. B., 50
Segal, B., 200
Segerstrale, U., 4
Seiffert, E. R., 51
Semino, O., 37
Serrat, M. A., 127
Shackleton, N. J., 55, 60
Shea, J. J., 68, 69, 70, 71, 113, 226, 227
Shen, G., 51
Shen, L., 242
Shepher, J., 82
Shkolnik, A., 92
Simmons, I. G., 3
Simonson, T. S., 122
Sinclair, A. R. E., 159
Sirugo, G., 5
Skutnabb-Kangas, T., 144
Small, C., 148, 149, 150
Smith, R. J., 97, 98, 100
Snodgrass, J. J., 117
Snow, R. W., 216
Soares, P., 32, 41, 42, 44, 45
Spellerberg, I. F., 241
Springer, M. S., 50
Stanley, S. M., 63
Steadman, D. W., 228, 236
Stearns, S. C., 5
Steele, T. E., 39
Steiper, M. E., 50
Stevens, G. C., 139, 165, 171, 176
Stewart, C-B., 22, 194
Stinson, S., 108, 109, 123, 124

Stone, A. C., 6, 24
Stoner, C. J., 92
Storey, A. A., 41, 46
Strait, D. S., 30, 48, 62
Strassman, B. I., 221
Strickland, R., 145
Stringer, C. B., 23, 68
Strong, W. L., 232
Stuart, A. J., 231, 232
Su, B., 42, 121
Suddendorf, R. F., 200
Summerhayes, G. R., 39
Surovell, T. A., 44, 230
Sutherland, W. J., 129
Swisher, C. C., 36, 161, 226, 236

Tamm, E., 43, 44, 45
Tattersall, I., 31
Taylor, N. A. S., 113, 114
Templeton, A. R., 29
Terborgh, J., 157, 187, 188, 239
Terrell, J. E., 2, 3, 40, 41, 43, 80, 180, 182
Tewksbury, J. T., 79, 114, 171, 177, 237
Thompson, R. S., 3
Thornhill, R., 83
Tilkens, M. J., 98, 104
Tilman, D., 225
Timmreck, C., 69
Tishkoff, S. A., 5, 25, 28, 30, 31, 33, 72, 195, 197, 199
Toole, M. J., 210
Torrence, R., 62, 64, 71, 171, 202, 222
Trejaut, J. A., 42
Trinkaus, E., 23, 29, 33, 37, 38, 68, 103, 104, 105, 111
Turner, A., 48, 66, 68, 222
Turney, C. S. M., 236

Uinuk-Ool, T. S., 24, 37, 43, 45
Underhill, P. A., 33
Upchurch, P., 1, 22

Vajda, E., 46
van Gemerden, B. S., 240
Van Vuren, D., 211
Vartanyan, S. L., 236
Verpoorte, A., 73, 74
Verwimp, P., 208
Victor, P-E., 228
Vigilant, L., 58
Virah-Sawmy, V., 230
Voris, H. K., 78
Vrba, E. S., 31, 54, 60, 61, 62, 63, 67

Waguespack, N. M., 44, 46
Waide, R. B., 142
Waldner, L. S., 241
Walker, J. D., 57
Walker, R., 107, 108
Wallace, A. R., 1, 31, 53, 77, 78, 79, 128, 129, 145, 225, 228, 243
Wallace, B. L., 80
Walsh, R. P. D., 62
Walther, G., 57
Watts, D. P., 209
Weaver, T. D., 17, 21, 22, 23, 25, 28, 110, 111, 112, 113, 227
Webb, S., 230
Weber, W. W., 5, 161, 201, 202
Wells, J. C. K., 44, 62
Welsch, R. L., 9
West, G. B., 90
Weston, E. M., 160
White, A. W., 228
White, T. D., 55, 63
Whittaker, R. J., 139, 142, 145, 178, 184, 226
Wickler, S., 41
Wiggins, D. A., 188
Wiktionary, 147
Wilkins, J. F., 47
Wilmshurst, J. M., 41, 50
Wilson, E. O., 242
Wilson, M. L., 209
Winterhalder, B., 233
Wolfheim, J. H., 78
Wollstein, A., 41, 42
Wolpoff, M., 24
Wood, B., 4, 32, 54, 58, 62, 243
Woodward, F. I., 185
World Resources Institute, 81, 109, 151
Wrangham, R. W., 198, 221
Wright, J. S., 237
Wright, P. C., 239
Wroe, S., 230, 232, 233
Wurster, C. M., 63

Yi, X., 122
Yoder, A. D., 43
Yokley, T. R., 110
Young, J. H., 5, 115, 116, 117, 204
Yu, J., 162
Yule, J. V., 234

Zhivotovsky, L. A., 27, 28
Zhu, R. X., 36, 51, 73

COMPOSITION
Westchester Book Group

TEXT
10/13 Sabon

DISPLAY
Sabon